Bruce King

# An Interesting Life, So Far

## Memoirs of Literary and Musical Peregrinations

# To Adele

Bruce King

# AN INTERESTING LIFE, SO FAR

Memoirs of Literary and Musical Peregrinations

*ibidem*-Verlag
Stuttgart

**Bibliografische Information der Deutschen Nationalbibliothek**
Die Deutsche Nationalbibliothek verzeichnet diese Publikation in der Deutschen Nationalbibliografie; detaillierte bibliografische Daten sind im Internet über http://dnb.d-nb.de abrufbar.

**Bibliographic information published by the Deutsche Nationalbibliothek**
Die Deutsche Nationalbibliothek lists this publication in the Deutsche Nationalbibliografie; detailed bibliographic data are available in the Internet at http://dnb.d-nb.de.

∞

Gedruckt auf alterungsbeständigem, säurefreien Papier
Printed on acid-free paper

ISBN-13: 978-3-8382-0956-2

© *ibidem*-Verlag
Stuttgart 2017

Alle Rechte vorbehalten

Das Werk einschließlich aller seiner Teile ist urheberrechtlich geschützt. Jede Verwertung außerhalb der engen Grenzen des Urheberrechtsgesetzes ist ohne Zustimmung des Verlages unzulässig und strafbar. Dies gilt insbesondere für Vervielfältigungen, Übersetzungen, Mikroverfilmungen und elektronische Speicherformen sowie die Einspeicherung und Verarbeitung in elektronischen Systemen.

All rights reserved. No part of this publication may be reproduced, stored in or introduced into a retrieval system, or transmitted, in any form, or by any means (electronical, mechanical, photocopying, recording or otherwise) without the prior written permission of the publisher. Any person who does any unauthorized act in relation to this publication may be liable to criminal prosecution and civil claims for damages.

Printed in the EU

# Contents

Preface: An Interesting Life; an autobiography .............. 7

Chapter 1: Family Stories .................................................. 13

Chapter 2: Fleeing ............................................................ 37
*Gloucester, Philadelphia and Camden 1933–50*

Chapter 3: The Big Apple ................................................. 59
*New York and Columbia 1950–54*

Chapter 4: Beginner's England ......................................... 85
*Leeds, England 1954–56*

Chapter 5: New England ................................................... 99
*Leeds, England 1957–60*

Chapter 6: The Golden Years ........................................... 119
*New York Again 1960–61 Calgary 1961–62*
*Ibadan 1962–65*

Chapter 7: Lecturing and Professing ............................... 145
*Bristol, England 1966–68 Lagos, Nigeria 1968–70*

Chapter 8: Swinging Sixties, Drugs and Graves ............ 165
*Deia, Mallorca 1964–86*

Chapter 9: Wandering Scholars ...................................... 191
*Windsor, Canada 1970–73*
*Zaria, Nigeria 1973–76 Columbia, Missouri 1976–77*

Chapter 10: Paris Once More and After ......................... 217
*Paris, France 1977–78*
*Sterling and Dollar, Scotland 1978–79*

Chapter 11: Trying to Begin Anew .................................. 233
*Christchurch, New Zealand 1979–83*
*Florence, Alabama 1983–86*

Chapter 12: Poetical India .................................................. 257
*India 1984*

Chapter 13: In Adele's Footsteps to Middle America ... 293
*Muncie, Indiana 1986–2003*

Chapter 14: Unholy and Holy Lands ............................... 323
*Calgary Again, 1987*
*Beersheva, Israel 1988–89*
*India II 1989*

Chapter 15: An influential Conference ........................... 347
*Paris, 1990–91 Walcott 1987–91*

Chapter 16: My Derek Walcott Decade .......................... 371
*Boston, Trinidad, St. Lucia 1992–2003*

Chapter 17: India III
Two More Visits 2005–2006 .......................................... 397

Chapter 18: Home ........................................................... 415
*Paris, France 2001–15*

Chapter 19: Second Home ............................................. 439
*New Orleans 1988–2015*

Chapter 20: Summer Home ............................................ 467
*Hvar, Croatia 2005–2015*

Chapter 21: Reprise and an
Unconclusive Conclusion ............................................... 493
*Paris, France 2015–?*

Index .................................................................................. 507

Bruce King: List of Publications ..................................... 517

# Preface
# An Interesting Life; an autobiography

My title comes from a time when Adele, my wife, was diagnosed with advanced lung cancer and was told by the surgeon that she had only a few months left to live. She said something to the effect that it had been an interesting life. She need not have been so resigned. After telephoning friends and other doctors around the United States we found in the same city and hospital a surgical team that decided it was best to do an emergency operation and removed one of her lungs. That was over two decades ago and she is still living, swimming, dancing, having sex and I think enjoying life.

I had not much thought about our lives before that, but I began to see that it could be regarded as interesting, especially as over the past decade friends have suggested that I write an autobiography. They included a Nobel prize in literature laureate, and one of the most famous black professors of literature in America. A well-known professor of English and black literatures in France said I was the best scholar of my generation (which seemed ironic as I was then as often unemployed, without any university job) and knew famous writers from around the world (which was true). I had taught at universities in England, Scotland, Canada, Nigeria, France, Israel, New Zealand and the United States and written or edited some early books about Indian, Nigerian and Caribbean literatures. I sometimes even heard from people who remembered my earlier books on English seventeenth-century literature, and there were a smaller group of scholars who claimed to have been influenced by an article I had written about West African High-Life music (supposedly the first essay on the topic).

I could not see how to write about my life, as I did not want it to consist of bragging about people I had met or the self-justifications and complaints that fill many autobiographies. I was not a celebrity, a rock, TV or movie star, I avoided limelights. I was known in several small areas of research, but not too many people. I was a literary critic and who would want to read about that and my seemingly endless futile attempts to fit into the academic life.

Besides, I wanted to write about other sides of myself that I considered more important, such as my life-long involvement with music or that somehow my social dancing was highly regarded wherever I lived. African women would tell my wife that I was a great dancer; on the streets of Indianapolis, New Orleans, and Paris there were people who stopped us to mention that they admired our dancing recently at some club. Even in our eighties when slow and arthritic we would be asked by young couples to change partners and dance with them. I thought of ourselves as the Old Smoothies. We had been asked were we retired Broadway dancers. How could I write about that or even about my cooking of which I was proud. In Paris we knew the best new inexpensive restaurants and some of the best chefs. Young Parisians said that they would never know Paris as we did. That was my new accomplishment, but it was not the sort of thing for an autobiography.

I tried several times. Once I began with the years when I was a student at Columbia College and would regularly sit in as a drummer at Felton's Lounge, Harlem, with the great blues musicians Brownie McGhee and Sonny Terry. That was a decade before other young whites began playing with black blues musicians, a period soon followed by many famous white blues bands. I thought I would tell about when Brownie's brother Sticks thought I was trying to steal one of his women and threatened to cut off my balls. "She's bread, man, bread".

The story would continue with me a student in England going to a party in London where Brownie was being feted and my annoyance that the drummer would not let me play. Didn't he know I was Brownie's drummer? I did not know that I was annoyed with Charlie Watts, soon to be famous as the drummer for the Rolling Stones. Decades later Brownie was performing at a white folk club in Detroit (I was then teaching at Windsor University across the river). He said to blind Sonny something like do you remember Bruce that "white boy" who used to play drums in New York. Then the story would move on to Auckland New Zealand when we had to return from holiday to

Christchurch where I was a professor of English and my daughter Nicole was in school. Brownie and Sonny would arrive for a festival a few days after we left and my daughter stunned me by asking why I thought I should have stayed on to greet them at the airport. She was not born when I knew them in New York.

Still later I was working on a biography of Derek Walcott and interviewing a black patron of the arts in Harlem who mentioned that she was born, I think it was, above Felton's Lounge. I told her I regularly sat-in there during the early 1950s and was asked what a "white boy" like me was doing there. I thought it would make a good way to start a "white boy's" autobiography, but when I looked up on Google some of those whom I thought of as very minor musicians I had played with during that period I discovered that they were not so unknown and had later opened for the Rolling Stones in New York and were often written up in the *New York Times*. Had all the seeming minor characters in my life later become famous? I felt deflated and gave up.

Some years later I tried once more to write my autobiography. An important friend in my youth, who had introduced me to Sonny, Brownie, and other musicians, was a black trumpet player with whom I had gone to school and who had tried to blow up the Statue of Liberty and been sent to Sing-Sing. This seemed a good way to begin an autobiography. I remembered that I had known two other revolutionary "bombers", but as I tried to recall the other two I once more felt this a dead end to pursue.

One, almost comically, was a Scots who placed explosives in postal boxes in sympathy for Irish Celtic nationalism and spent time in prison for it. As I had met him so briefly in Mallorca, where we were spending the summer as usual in Deia (yes the Deia of Robert Graves and other writers as well as such groups as The Soft Machine and Gong), he really could not be included. The other, a South African who fled to Israel, where you guessed it, I was a visiting professor, was now married to an American, had children and was a respected university professor of African literature, an expertise we shared as I had taught at three different Nigerian universities for a decade and edited the first book about Nigerian literature. My South African friend no longer seemed revolutionary, although the American government continued to regard him as a threat and would only grant him visiting visas to speak at conferences. I invited him to speak on my Modern Language Association special panel about Nadine Gordimer, which formed the core of a book I edited, and dedicated to him, by which time he had died.

To explain his story would have involved the stories of others including a famous labor historian who had been his revolutionary leader, whom I had met when we were both giving lectures in Amherst, and the complicated relationship he had with his family. I now knew the son, a poet and translator, in Paris where I lived. As for my trumpet player threat to the Statue of Liberty I became so befuddled trying to trace his life after his having served prison time, and the ways in which he had become an unlikely hero of American Black revolutionaries, that I felt this was not the way to go. Either I did not really know my famous people or I knew them for a short time and was puzzled trying to see them in more depth.

Then not long ago when I was finishing my third book about Indian literature (which for me means literature written in English) and wondering whether to move on to Pakistani literature, several younger people I knew said that I should write about my life. I kept saying that I had tried and failed. While retelling a story about Allen Ginsberg and the Beats reading in Paris during 1958, which I attended, in which I had a different, minor, part than that portrayed by literary historians, I could see that my listener's eyes were widening; this was a fabulous time before she was born. Aha, how to start that autobiography. So I began with that and intended to move on to another amusing story about asking Ginsberg in Tel-Aviv about his Calcutta contacts as I was returning to India to do research for my second book about Indian poetry. As Ginsberg was surrounded by admirers I, streetwise, got behind him and whispered what I wanted. He appeared to ignore me but at the end of the evening he nodded to me to follow him and his friends to a café where they were eating. I quietly sat at the end of table until he told the person across from him to make room for me. After telling me who to see in India he said that space is needed for others and I was dismissed.

This would move on to stories I was told in Calcutta about him drinking tea while his well brought up Indian literary friends learned the ways of prostitutes. There also could be related stories about a minor but culturally significant group of Indian Bengali writers, the Hungryalists with whom Ginsberg became involved as had I. Later while my wife was teaching at a university in Muncie Indiana I had unintentionally assembled the largest collection of Hungryalist materials that writers kept sending to me. The main significance of this movement was that it represented the provincial lower classes outside of Calcutta. They wrote badly in English, a language they did not know, as a protest against the literary and cultural refinement of the Bengali elite. I had become the one who knew their history, had interviewed them, and had copies of their works. I had even attended a seminar in Calcutta concerning their social significance.

One day I had a letter from a student at Bowling Green University who to my amazement was writing an undergraduate dissertation about the Hungryalists—his teacher Howard McCord edited some of the obscure American literary magazines that had published Indian literature during the 1960s. I told him I could not take time or afford to photocopy everything. Later he turned up at our door with a taxi waiting to take him to a photocopy shop and then back to the private airplane that (I forget) he rented or owned, which I thought would be a proper American conclusion to my story of the impoverished Hungryalists protest movement's afterlife in the American mid-west.

But, always but, as I began to write about the Ginsberg and the Beats poetry reading in Paris I felt I had to explain why I was in Paris, that I was then studying at Leeds University, England, why I was at Leeds after leaving Columbia, and soon I was writing about my high school teachers, my home town, my family, origins, a regular Victorian biography of a kind that I had often mocked, but what I could I do? This was the story that would become my autobiography whether I wanted it or not. I knew enough about writing to know that when material starts coming you do not block it, do not even question it; later you can edit and try to structure it, but once a dam burst open you would be flooded and had little choice of what was given unless you lied. Even someone like me can imagine he has been seduced by a muse, perhaps rather the muse of foolishness than poetry.

What follows is the result. It has much more about origins, social distinctions, pride, fears, and about being a Jew than I would have chosen, but that is of interest even though it lacks the texture of fiction. Writers I know claim that the borders between fact and fiction have become indistinct, but not for me.

Bruce King
Hvar-Paris

## Chapter 1: Family Stories

I twice heard Mary McCarthy tell of an Italian family she knew that progressed in three generations from immigrants, through Mafia, to proud parents of professionals including a university professor in the Humanities. Although the stages were not identical (we belonged to no Mafia, but a distant relative was said to have worked for them as an accountant and ran off with a million to France; he was a wanted man), and it took two rather than three generations, our family story is similar, a passage from impoverished Jewish immigrants to university professor, with the earning of money and the rise in social position on the way. I would not have seen it that way until I started to write this chapter. It is the kind of story that Saul Bellow and Philip Roth wrote novels about earlier when I was too engrossed in my own life to read them.

I hardly knew my father who died of a blood clot after a long illness when I was twelve. Before then he and my mother worked day and evening in the pharmacy and after closing we would sometimes drive to a White Castle for oniony hamburgers which we would sit in the car and eat. Other meals often were prepared by the cleaning lady; my mother worked in the pharmacy downstairs. My parents were very different from each other. My father was a short unattractive man who wore his dark brown hair in an Adolph Hitler lick in front. His old fashioned neckties, although patterned appeared brown to me as did his shoes and eyes. He followed the horse races (which he listened to on the radio and on which he bet and early taught me to bet), liked to play poker with fellow Jewish pharmacists, and I am told was a good dancer when young. He wanted to be a medical doctor but chose pharmacy since the course at Temple University in Philadelphia was shorter and less expensive. He also wanted to be a writer and sent some stories to journals that specialized in detective and crime tales. He mailed copies to himself to prove authorship, but no one ever wanted them.

One of my memories of him was his making up stories that he told me when I was very young. I also learned from him to get a second medical opinion and not to bring food into a men's room. As an old fashioned pharmacist he concocted medicines from varying ingredients when doctors wrote a prescription. He had cures for such problems as the phlegm brought on by winter colds and sinus infections that felt as if they were going to choke me, concoctions I wish I could get from pharmacists today as I suffer from chronic sinusitis and spend parts of every year complaining about my doctors who futility prescribe antibiotics in an attempt, I suppose, to create the ultimate superbug. I did find one of the medicines that my father would give me and was told that it was now banned as dangerous.

People say my father was a gentle and good man, but my lasting memory is of him kicking me repeatedly after I had fallen and tried to protect myself on the floor after some argument. The quarrel had something to do with my waiting on customers at our soda fountain in the front of the pharmacy. I did not make up the being kicked which remains in my mind, but I may have made up the cause as I was always under pressure by older teenagers to give them larger servings of ice-cream, not charge them for cokes or cigarettes. If you are a ten-year-old, one of the few Jews in a small town, and feel an outsider you are easily intimated by those in their late teens threatening to beat you up and seeming to jeer at you. I was not toughened by the experience, I was becoming afraid. I hated being told to serve at the soda fountain and many decades later I remain uncomfortable when asking people to pay their debts.

My father was born off Staten Island on a ship of Russian Jewish refugees. I do not know his original family name and assume that an immigration official renamed the family "King". I would sometimes overhear him telling my mother about pogroms in Russia or strange burial customs in Eastern Europe, but he would have heard about them from his parents. He was born in America. Similar to many Americans I have little information about my father's past except that he sometimes spoke of violent fights between various ethnic groups when he was growing up. There was a synagogue in Gloucester where Jews from outlying towns sometimes came together. I remember nothing about it except that one year my father read in Hebrew from sacred scripture, a privilege for which he had to pay. I seem to remember there was a competition, even bidding, for the right, although I may have imagined it. My mother was then proud of him. He must have learned Hebrew as a child; I had no Hebrew or other religious instruction in my youth and did not wish for any. I would be puzzled by people saying that they had gone to this or that church and only felt I had missed out when I saw that they had learned to read music to sing in the choir, a skill I lacked and never really learned, despite my increasing interest in music as I grew older.

One year my father bought our first car, a small, undistinguished Plymouth. That summer we took a trip to New York and Rhode Island. We stopped outside a building on the lower east side of Manhattan that had some association with my father's youth. (Perhaps a decade later I sometimes saw a Barnard girl who lived in the same Jewish immigrant area and whose brother graduated from Harvard and Harvard Law School at an impressively young age. They seemed an exotic foreign branch of what could have been my family tree. ) We also visited Coney Island where I burnt a hole in my trousers on a dangerous slide. During one summer we sought a motel in northern New Jersey that did not have any sign saying "No Niggers or Jews Allowed". Instead we stayed at a hotel in the New York borsch belt. There were kayaks, which I had not heard of previously, games such as shuffleboard, and I got lost, wandered around the paths for seeming hours and was afraid of being eaten by a bear; I sat on a bee and was warned that I could die if I sat on another. I only over-came that fear after having been stung by a wasp; I survived easily although with pain.

Most of those years exist as free-floating fragments in my memory. I am puzzled by how we had gasoline for our car to take vacations during the war; I can recall a family trip to Atlantic City when the train was packed with soldiers. I cannot understand why my father allowed notorious anti-Semitic literature to be sold on the shelves of the magazine rack in the drugstore, or what I thought I was doing collecting (was it?) aluminum foil or searching with a flash light the sky at night for enemy airplanes. Those years make no sense to me now. Did I really belong to a group of school children trying to do something for the war effort? I remember my parents objecting to paying its dues; they felt that those who worked in the local shipyard were earning more than them and often expressed resentment about this. We might not have been very middle class but we were middle class professionals and they were working class, although they had more liquid in more than one sense. They were heavy drinkers, often drunk at nights in the local bars, while my father occasionally had a quiet whiskey among friends. My mother did not drink. I was stunned when Irish neighbors, whose children I sometimes joined in games, told us that anti-British they were pro-German. Were they really Irish or did we just term them that because of their family origin? I also cannot make any sense of some national radio program in which I was honored for supporting the armed forces and was quoted as saying when I grew up I wanted to attend West Point. What could this have been about, my Victory Garden, some attempt to sell war bonds by school children? That must have been the height of my patriotism, perhaps explainable by Army and Navy then having the best national football teams, even better than Notre Dame. When the war in Europe ended I took a bus to Philadelphia and joined the celebrating crowds and watched women kissing solders. A few years later, perhaps when I was 14 rather than 12 years old, I was already an alienated intellectual who felt that real life was among the prostitutes, alcoholics, and artists of Paris. I had somehow got my hands on books by Henry Miller.

My father's family consisted of bright intellectuals who at the social events I was made to attend discussed Zionism, the Bund, Israel, and varieties of Marxism while eating what seemed to me strange smelly foreign foods, and drinking tea from samovars. I would now consider such a family a privilege, but I was an American boy, living in a small provincial town, and I wanted nothing to do with such alien exoticism, which in my mind somehow seemed dirty, Jewish. I could not help, however, but be impressed by an uncle—I think I was partly named after him—who taught at one of the better schools in Philadelphia and who was said to have written an MA thesis of such length that he was told he would be granted a doctorate if he enrolled at Temple University (where my father's family went as it was public and far less expensive than the University of Pennsylvania) a further year, which he refused as having no need for a doctorate to teach History in school and as it, a common theme, would cost too much.

There appeared even more exotic creatures I was to learn about on that side of the family. When my father was alive we visited Cape Cod one summer and I recall going into a large netted enclosure on the ocean in which fish had been trapped after the tide retreated. I was afraid of a small squid or baby ray that I thought was a poisonous stingray. My aunt's husband was a card carrying Communist, who after she died, moved to California to raise chickens, one of many Zionist socialists who thought that God would redeem Jews who returned to the land in California. He next thought Guatemala could be a socialist paradise but his new wife refused to go. He later moved to Israel where he once more failed as a chicken farmer. It is easy to mock him but he had progressed during his life from a non-English speaking immigrant from Poland who began during the 1930s as a peddler in Cape Cod carrying what he could on his back.

Over the years members of that side of the family would turn up and for a time be part of my life. There was Dorothy who worked as an administrator at the WMCA and as a translator of foreign movies in New York. Was she the one who much younger played saxophone boringly at one of those to me seeming interminable family gatherings? She had a rebellious brother who rejected his mother's married name, returned to being a King, left for California, then as exotic for me as Timbuktu, to become a radio disc Jockey and who I was told became later big in the recording industry.

Another cousin would have an even more erratic academic career than my own. Unable to afford studying medical school at Tufts, she had gone to France hoping to enter a Swiss medical school, but returned to USA and while teaching in the New York school system was taking graduate courses in English literature at New York University when we met. We remained occasionally in contact as she moved through various university teaching, administrative, and even presidential positions from New York, to New Jersey, Indiana, Ohio, and back to New York. Each step up the academic and administrative ladder seemed to make her more dissatisfied with life and those with whom she worked. She was the one who eventually inherited the property next to the Kennedy's at Hyannis Port and which she found was mortgaged and had to sell to pay her father's debts. About once every two or three years we still exchange an email message to tell each other that we are still alive. She complains, I brag, a regular hee-haw. She helped me through some of the intricacies of family genealogy although like me she remains uncertain who begat whom and did what.

My father had a brother who also became a pharmacist, and owned a small drug store in Philadelphia He, his wife, and two daughters lived in an apartment behind it. Sometime in my later teens, when I had a car, I began visiting when I discovered that the elder of the daughters, Mimi, had a strong interest in literature and knew about the so-called New Critics who were then the champions and explicators of modern literature and its cultural bye ways including the attractions of Roman Catholicism in contrast to the Marxism that had attracted the previous generation.

Mimi had attended Michigan University where she studied with Austin Warren famous for collaborating with Rene Wellek on *Theory of Literature* and as author of a book on the poet Richard Crashaw and the Baroque sensibility. She also studied at the University of North Carolina in Chapel Hill, one of centers at that time of the New Critics. She and her friends, who taught at such places as Swarthmore and Beaver colleges, had evenings with rosé wine, gouda cheese and crackers, and talked about books, politics and foreign films. Finishing their doctoral dissertations seemed a high hurdle, and they spoke admiringly of the then famous poet critic Peter Viereck who when told that he needed a Ph.d to gain tenure at Mount Holyoke College added footnotes to one of his already published books. To me this was a glimpse of civilization, although perhaps a bit effete.

Such impressions were misleading. Mimi had told me about stealing books at college, about having friends who were communists, and first mentioned to me Kenneth Burke and Stanley Edgar Hyman's *The Armed Vision*. I think she had had lovers although she was not physically attractive. I doubt that she ever thought of herself as a literary critic in contrast to someone for whom the critical debates of the period were part of culture and she married someone who she knew would never make a name for himself as a writer, intellectual or as a teacher at a major university. It was a marriage that puzzled me but she knew what she was choosing. She was rational and self-limiting and I think my innocence and energy amused her. I also seemed to have money and had such material goods as my own new automobile, which was beyond her means. She once said I was given financial freedom so my mother could run the pharmacy and her life without looking after me.

Mimi became well-known almost by accident. Having moved to Trenton, New Jersey, where her husband was teaching at a local teacher's college, she was casually employed by The Educational Testing Service, which thought it should have some females on its staff, an up and coming idea, and before long was in charge of its English literature tests, and then moved further up in the hierarchy of what became an influential international testing organization.

On my occasional returns to the USA after teaching abroad I would visit her, be impressed by her largish collection of pre-Columbian art, follow the progress of the daughter that she and her husband adopted and raised. The direction my life was taking also amused her. I was one of those writing about the new national literatures of the British Commonwealth; Mimi remarked that she thought I was still a scholar of Seventeenth Century British Commonwealth literature. That was better than those who asked if "Commonwealth Literature" referred to the Pennsylvania or Massachusetts Commonwealth.

I also learned that her younger sister, Claire, whom I had wrongly dismissed after her marriage to a high school physical education instructor as someone who telephoned mother about cake recipes, was actually a mathematical genius who gave advanced courses sponsored by the government to other young scientific geniuses who needed mathematical stimulation between high school and college.

Al, the father of Mimi and Claire, seemed to me quietly unambitious, but his wife, Bella, was sharp, clever, and looked down on the family that my father had married into as socially beneath her. Jews from Vienna were highly assimilated German speakers unlike the Yiddish speakers of most of Eastern Europe. My mother's cooking, when she cooked, was to me uneatable; I would sneak out for a hamburger at a nearby diner. I enjoyed the few meals I had at Mimi's made by her mother. Something as simple as a Greek salad, the dressing made with mint, oregano, olive oil and lemon juice was a discovery. I once used Mimi's influence when I was a professor in New Zealand and my marriage was troubled: my daughter decided to apply to American universities for the coming year but it was late for her to apply for the tests that she needed. I telephoned Mimi and an exception was made for Nicole who soon escaped to Bryn Mawr College. I know little of Mimi's later years. I several times telephoned and was told by her husband Milton that she was suffering from Lyme disease.

There were some very successful genes on my father's side of the family, although my father seemed not to have inherited them. He started a pharmacy in Trenton after marrying my mother, and supposedly was cheated out of everything by his partner. He had a small pharmacy in Gloucester, New Jersey, to me the end of the world. I think that he had no competition at first; then a large chain opened a bigger drug store. I would be embarrassed when my father and mother, when going out after work, stopped to compare prices of items in the competitor's windows.

My mother and her family were different from my father. She and her sisters and brothers were physically attractive, energetic, unbookish, not well educated. The men were athletic. I have the impression that they were impoverished and at some point they anglicized their name Goldberg to Gilbert. My mother's father was from Manchester, England. Grey eyed, cheerful, amusing, he would walk daily across the Delaware River Bridge from Camden to Philadelphia where he sold fruit and vegetables from a pushcart. I hardly ever saw him and heard that my grandmother never forgave him for leaving her to travel around the world as a sailor for a decade. Typical of the confused family history inherited by many Americans I have heard two conflicting accounts for his leaving his wife for a decade. In one version he was one of the Wobblies, heroic, militant, Socialist radicals, and was threatened by strikebreakers that if they saw him again he would be killed. In the other version he was one of the strikebreakers whom the Wobblies threatened to kill. From what I remember of my grandmother I would have been happy for either excuse to get out of town.

She was from someplace on the German Polish border and spoke only Yiddish. She found America disagreeable; as she grew older she seldom left her old fashioned wooden house in a poorer section of Camden. When she wanted something she would telephone her daughters and threaten to throw herself down the stairs if they did not come immediately to help her. My mother was a sucker for such intimidation and took care of her mother during the final decades of my grandmother's life, an expectation she had about herself and other mothers left on their own in old age. Her two sisters were more hardhearted. I have no idea of what my grandmother was like when younger and I so disliked her that until now the question never came up. On the basis of no actual information I assume my grandmother and grandfather, both immigrants, were brought together by some rabbi, family friend, or marriage broker. My mother spoke with her mother in broken Yiddish and would sometimes use Yiddish expressions when speaking with my father. This was part of their Jewish background of which I wanted no part. Years later when I was having difficulties learning German my mother would insist it was an easy language.

My mother was an attractive grey-eyed woman who had about six years of schooling in Camden followed by a short course in nursing. I understand that she had a crush on some other man but was made to marry my father by her mother who saw him as potentially more successful. I do not know how much of this is true or for that matter how much of anything about my parents is true as they seldom spoke in front of me about their history.

My mother must have married in her late teens and had me about six years later. I know that she felt anxiety around non-Jews and thought that she should have observed Jewish traditions. Her few friends were mostly Jews, and she seemed uncomfortable when I started dating non-Jews. She told me that at the first argument I would be called a "damn Jew". (It has never happened.) If she ever met someone Jewish whom I was seeing she would praise her as a potential daughter-in-law, for which she had a curiously old fashioned term, someone you would "be proud to bring home". Yet when I did marry Adele, from a very religious Methodist family, Lillian took to her immediately, unlike Adele's family who did everything it could to break up the marriage including disinheriting Adele. Lillian justified my marriage by saying that everyone born in America was a "Christian". Try to unpack that.

She once surprisingly claimed my father had said gentiles were for bedding not marrying. Surprising because my mother was extremely puritanical about sex, spoke of it as a disease, and the main source of my information about her came from a family friend who sarcastically said that poor Joe, my father, had to give her a fur coat to take her to bed. She had no fur coats, desired them badly, (often saying that when I grew up and became wealthy I would buy her a white mink coat).

She had a boyfriend for many years, after my father's death, whom she hoped to marry until he married someone else. He was tall, well built, played tennis, had a trim old fashioned moustache, and a legal practice which was housed in the Camden offices of Myers Baker, the husband of my aunt Florence. He had an instant tour of Europe and returned saying Oxford and Cambridge are over-rated as the buildings were so old. I cannot believe he was satisfied romancing my mother without results although I have the impression that that is what she would have preferred. There were other men romancing her who also decided to marry elsewhere and she would indicate that despite the money spent on expensive restaurants and trips, these were just friends. She was always well dressed, lady-like, and appeared to make friends easily, but kept turning down offers of employment or marriage that would have taken her away from the world she knew. I also suspect that while she wanted to be desired by men and liked their company she felt one marriage was enough. She was probably afraid of sexual demands, but possibly also afraid of losing her freedom.

She was sensitive about her limitations and tried to hide them. Her lack of educational qualifications troubled her. During the 1960s when there was much talk about university degrees based on life experience, my mother found this perhaps the only part of that period's liberating ideas with which she agreed. She had to employ trained pharmacists to work in the drugstore after my father's death, and I think resented that. In a pinch she would illegally fill prescriptions on her own and was fined by the state pharmaceutical body. It was not something she ever told me about and I did not want to know, as her insecurities were troubling, embarrassing, almost insane. If she received a general mailing from the Red Cross or some other body seeking money she would fantasize that General Eisenhower or whoever signed the appeal was writing directly to her personally. Such declarations were depressing and when my mother visited Europe my friends would say "poor you", but while I was unhappy seeing her I knew I was badly indebted to her. I had inherited her sense of order and cleanliness. I do not like chairs out of place, I cannot leave dishes in the sink overnight, I have seldom been drunk. Although I enjoy sex and feel driven by it, I have been told by women that I am puritanical. Offered a drunken fling I will back out. I am more likely to need seducing than to seduce. Even that I had become a good dancer was because in my teens she was willing to pay for me to take Arthur Murray dancing lessons and also wanted me, and later Adele, to be well dressed. I would have become a less presentable and less sociable creature without her influence.

She wanted me to study pharmacy like my father at Temple University and manage the drug store, for which after his death she had taken a bank loan and remodeled. I had no intention of taking on her or his life and wanted to get to Paris or New York as soon as I could. She paid for my four years at Columbia College, paid for my years as a doctoral student at Leeds University in England and for a time supported my marriage. Wanting to be free from her I broke the financial connection after my first university appointment. How can one speak of disliking someone who gave me an Oldsmobile when I was still in high school, who arranged a wedding party when I married Adele in St Louis (a marriage which her parents tried to prevent and then hoped to break up), and gave us financial support until I earned my doctorate and had my first job, someone who gave us money to furnish our first apartment?

Yet, in truth, I disliked her, at times hated her, as I felt she was always ignoring, or more likely was ignorant of, what I was accomplishing and was continually trying to make me come back into her small world of her Jewish friends, her neighbors, people I had gone to high school with and long forgotten, and of course family relations. Whereas others had moved on, bought apartments in Florida, divorced, remarried, travelled, Lillian stayed in the same area all her life, saw the same or similar people, and depended on a narrowing circle of family. There were her two sisters, a brother, maybe two pharmacists who were friends of my father. She would end with her younger brother's daughter taking care of her, a destructive relationship for both.

My mother unchallenged came to believe her own fantasies. Fainting in a supermarket and becoming wheel chaired she blamed her fall on a banana skin in one version and in another version on a soapy floor. The supermarket had offered to pay for her hospital and other expenses but when sued in court had no trouble showing that she had a history of falls and conditions likely to provoke falls. She never accepted that she was wrong.

My cousin, Uncle Henry's daughter, who took care for her never worked a day in her life, came increasing to live in her own imagined world, and rather than being saved from sex and drugs became a bitter aging adolescent who having run through her parents and my mother's money, and having a series of automobile accidents, kept telephoning relatives asking them to fund her life. My wife, Adele, who is unusually sympathetic to the troubles of others said after a series of such telephone calls that once my cousin got through our door I would never get her out.

About that Oldsmobile. The size, make, and vintage of your car were significant for several decades and for all I know might still be in some social circles. I have since usually had friends where an unassertive smaller car is a sign of good taste, just as at present the educated and sensitive ride bicycles, a skill I never picked up when young. Not only was my mother conscious of the social status of flashy expensive cars, she had moved up to a Cadillac, and assumed that young women would be impressed if I had my own Oldsmobile, She was part of her time in her views. If you telephoned a woman you did not or hardly knew for a date, you would be asked whether you had a car, the brand, the age. It was no doubt a substitute for inquiring about other less public matters. I know that a taste for fancy automobiles continues. I have friends in New Orleans who originally came from small towns in Mississippi and who wear gold jewelry, gamble at casinos, do not read books, and are proud of their cars that they often change. We remain friends and enjoy each other's company although outside New Orleans with its jazz clubs, restaurants, and amusing bohemian characters we would find little in common. As I sometimes have difficulty explaining to people what I do for a living I mumbled to one that I knew and had written about two Nobel laureates in literature. She looked delighted and said I was the first person she knew who had won two Nobel prizes. An Oldsmobile would not have led to such confusion.

Among the ecologically minded do you demand more or less fancy bicycles of your partners?

I can see that my problems with my mother started shortly after my father died and she began to renew herself. I mentioned the bank loan that allowed her to remodel the pharmacy and placed emphasis on selling cosmetics rather than filling prescriptions. She bought a new, younger, fashionable wardrobe, dyed her long graying hair a bright blondish red that she continued to have over the decades until her death and which she insisted was her natural hair color. I immediately smelt a rat when told that I was now her young man expected to escort her to gatherings. I did not need to read Freud to feel sexually threatened. One of my confused memories of childhood was of her kissing my penis (for which she had an affectionate Yiddish term). (Another memory is of being with my mother and her two sisters in a changing room on the beach in Atlantic City where I was conscious of their uncovered breasts, no doubt the start of my wanting to be surrounded by attractive women, a characteristic that others have noticed.) There was often sexual tension. If I closed the door of my bedroom at night she wanted it left open, no doubt suspecting me of masturbation. More likely I was trying to listen to a late night radio show from New York that I followed and which allowed me to feel I was part of a larger world of singers, actors, and other stars.

I did not want to be my mother's young man; I wanted to be left alone to my own interests including girls my own age. I was at that time discovering literature and jazz music and learning about intellectual topics and I wanted to keep as far away from my mother and her world as possible. Our tastes were opposites. If I wanted a jazz record she would claim it was "sad" and instead buy me something I hated. We rapidly grew apart except for the financial connection.

Yet it was not like this earlier. When I was learning to read and count it was she who taught me rather than my father. I remember her helping me read an illustrated children's version of *Gulliver's Travels*. How was this possible? Where did she find the book? She could not have any notion of the author Jonathan Swift. She also taught me to count. When I went to kindergarten I was upset when the teacher told the class that 900 was followed by ten hundred. I knew she was lying. It should be a thousand. This was the start of my distrust of the academy, and its attempt to reach the less able, and I was, what? Six years old? I already felt a rebel. I had told my mother that I did not believe in God when she said that God would not like something I had done. And I remember in first grade staring out the window and thinking how many more years am I expected to sit through this? Not a good start for someone who would become a university professor.

Lillian had two attractive sisters and three brothers. The sisters were achievers, the men were not. Her elder sister, Florence, was a tall blonde who somehow, I have not the least idea how, became part of the Republican Party establishment in New Jersey. She had lived the life of the jazz age during the depression, knew famous people, attended polo matches, and travelled to Paris and bought clothes there while others were selling apples on the street corners of America. Could her husband Myers somehow have opened the world to her? I doubt it as he always seemed a small figure with an obscure legal practice; I think he was in Real Estate. I see on Google that by 1936 Florence was already an alternate delegate to the Republican National Convention. She was several other times a delegate or alternate. I never discussed politics with her but thought that she was on the liberal Eastern Rockefeller wing of the Party. I cannot imagine her approving of the new moneyed trash that came into, and took over, the Party from the West and South. That was not her taste.

I admired her and her selfish self-centeredness, although she could be absurd, as when at a family party she tragically lost the label of a Parisian dress in the toilet. She lived in Pennsauken, New Jersey, just outside Merchantville, an area now a poor ghetto, then ritzy. I was impressed by the size of her house (we lived above our pharmacy), by the fishpond in the garden and by the two stone lions on the steps. She was childless and I became her favorite nephew. When I was small she gave me a Pekingese dog, Mickey, which I loved. When it was run over by a car my parents bought me another but it was no replacement and

I lost interest in dogs. Although I insisted that my parents get rid of the third dog, my mother claimed throughout her life that I loved dogs.

I do not know what she did in the Republican Party but Florence would offer me tickets to the Republican party convention in Philadelphia and their meetings in Camden with people like Adolphe Menjou, an older movie star who appeared absurd in his formal tails, stiff shirt, and dead white make up. Florence claimed Menjou was a model for how to dress. For a funeral? My mother complained that her sister would enter Christian churches but then my mother would love ham sandwiches although avoiding pork chops. I gather my Aunt Florence had enemies who said she kept her good looks through face-lifts. I doubt that they were right as there was no evidence of it. She had style. She always had a new Lincoln Continental that in appearance seemed to me to belong to the 1930s, with a raised trunk in back and a spare tire on the side. When she stopped driving she kept her last one garaged for many years as a collector's piece.

She had influence. When I graduated Columbia and was waiting to go to England I received a puzzled telephone call from Princeton University Admissions saying that a Senator had recommended me for graduate work but they could find no record of my application. In Paris my wife and myself were invited to a July 4th party at the Embassy where I remember slyly letting someone's baby sip our champagne. Florence had a stroke, was moved into my mother's apartment and cared for by her until a long time later she died without recovering her speech or ability to walk, which a physical therapist claimed was the result of laziness. The money she willed to me was used to pay for her expenses which were later paid for by Rose, my mother's younger sister. The property that I admired was illegally sold for a nominal cost by my mother, who assumed power of estate, to her brother Henry's perpetually hopeless daughter who after waiting for years for some movie or musical star to discover her unexplored talents would be company for my mother in her old age. The classic Continental, now over 25 years old and in mint condition, my mother assumed was junk and sold as such.

If Florence had seemingly effortlessly risen above her origins and joined the political WASP establishment, the youngest sister, Rose, seemed to glorify in her working class roots. She was not petite bourgeois. While the owner of a million dollar meat packing plant she would face strikes by her workers by joining them and discussing their mutual problems. Her personality was stronger than any union. A short pretty dynamic woman with reddish blonde hair and a loud crude voice she dominated her surroundings and those who surrounded her. She had left school in 9th grade, married at 16 Harry Cook, a butcher. He was said to demand sex when on the job, and if she was not immediately available there were female employees who were. He made a fortune during the Second World War by selling meat on the black market.

Despite his crudeness I liked Harry who was easy to like. There were parties at his large house in West Philadelphia, then fashionable before the Jews and most other ethnic groups fled the city, and many other cities, especially Camden, before that period's black invasion. (The Italians in South Philadelphia were the only whites who stayed.) He gave me my first tastes of various alcoholic drinks (my mother disapproved) and I remember us cooking steaks with onions, mustard and much black pepper. He had some notion that after my father's death I should stay with them while going to a synagogue and recite Hebrew prayers, a foolish idea that was tried and soon dropped.

Rose and Harry had two sons, a few years apart. Buddy (Herbert), the younger, was two years older than me, and although he had inherited his father's crudeness was attached to his and my mother. Albert, the older son, seemed more relaxed, knowledgeable, sociable. He would go on to study at the University of Pennsylvania, and I could discuss history and literature with him, unlike Buddy who acted as if culture was pretentious and to be jeered at. After his father's death in 1950 Buddy worked with Rose to expand their meat business into one of the world's largest pork processors. Buddy was head and Rose vice-president. They became millionaires by processing pork into Ham that they sold under the Bluebird Food label. Bluebird kept acquiring other processors and became the largest of its kind in the USA. When the company went public my mother was given several thousand shares (I wonder what became of them?). After Bluebird Food was sold to others, Rose, Buddy and friends seemed stymied and tried to start a similar company, the fate of which I know little. My mother would not eat pork. Rose loved pigs and loved to be given figurines of them for her collection. Rose also enjoyed an active social life and men. She bought an apartment in Fort Lauderdale and commuted to Philadelphia where she had another apartment on Rittenhouse Square. Her friends remained Jewish—I suppose many immigrant families stayed close to their ethnic communities, my generation may have been the first to move beyond—and she remained close to my mother who spoke of Rose as her younger sister.

Rose remarried, changed her name to Rose Cook Small, and after her second husband died had other men friends, but felt it would create too many legal problems if she remarried once more. My mother would be upset by Rose living with her lover, but Rose, then one of the higher paid woman executives in the USA, winner of a Horatio Alger award in 1977 and retired in 1980, was beyond such pettiness. She enjoyed dancing and when we stayed with her in Florida she gave us the keys to the apartment and said that she and her boyfriend would be out all night at a party. There was food in the fridge but in case we wanted to go out she gave us the name of a good place to dance.

Her men were the kind who gave head waiters big tips for the best tables and seemed to display their wealth, but Rose would save coupons from newspapers for discounted foods and was always alert for bargains. She had much commonsense. After meeting and listening to Henry's articulate but flighty daughter, Rose observed that the young woman would talk herself into and out of a job at the same interview. I enjoyed Rose's company the few times I saw her. We would talk about the raising of pork in Poland, why Italian fennel sausages would not be commercially successful in the USA, and even detective novels that she had started to read. I had the feeling that she thought

Florence, her elder sister, was a snob, and regarded herself as an example how you could rise in America from an impoverished immigrant background to success without losing yourself to other social models.

She was Democratic in her politics and life. Like many Jewish families of that time her hired help at home were black—although as time passed they were more likely Haitian than African-American—and she would take an interest in their lives and sit down and eat with them. She was loyal. Friends of her sons often became central to managing Bluebird Food; they would remain so in her son Buddy's businesses. She listed my mother as Bluebird staff so my mother would have medical coverage. Rose paid for the apartment my mother took to care for Florence after her stroke and she paid for Florence's medical care. Later, I have the impression that she or Buddy paid for my mother's care when ill. Although I admired her I seldom saw Rose as she aged but she had become the head of a large family of grandchildren and spent her final years suffering from Alzheimer's disease. I realized how little I had known of her early life when to check my facts I looked her up on a Google search and learned that she had pawned her engagement and wedding rings to help Harry's meat businesses to expand.

Buddy and I always seemed to be at war. He admired my mother, I admired his. In my early teens, after my father died, I would sometimes stay during the summer at an apartment Rose rented in Atlantic City for the use of her sons, their friends, and family. Buddy would mock the books I was reading or trying to read, and I felt he wanted to get at me. He had a reputation as a tough who could bang three heads together and whom it would be foolish to cross. His friends spoke of him as almost crazy when angered. One day, however, when a group of us were playing basketball Buddy and I got into a punch up. I don't remember the cause but he kept shouting that I mistreated my mother. I survived bruised but wonderfully surrounded by the women who were part of the group, and became for a time the hero who had fought Buddy as an equal.

His friends were older and more experienced than me, Albert's even older, and knew one another from Philadelphia. They shared cultural and social assumptions of which I was unaware. I felt an outsider until the fight on the basketball court, an incident lost in the fog of memory until about a year or two ago someone spoke to Buddy and he complained that he still had a scar from when, he claimed, I bit him on a finger during that fight. Did I really do that? Good for me.

That might have been the same summer when I once more found myself promoted from ugly duckling to swan as some of the group had heard me on a late night jazz record radio program. By my mid-teens I had become a serious jazz fan, had become a friend of the announcer on the one major Philadelphia jazz show, and as I was underage to go to jazz clubs I would accompany him unquestioned to late night places. I would first go to the station and talk with him on air about records before going to a night club with its visiting band followed by a jam session. I would not get home until morning. It did not happen often but to be heard on the radio by high school and college students gave me status.

While Buddy went on to head Bluebird Food and other companies and was, I think, on the board of several banks, and his children increased and multiplied and became parents themselves; his older brother Albert had become a gynecologist, had an affair with his nurse, was divorced by his wife whom had dated since school days and whose family were friends of the Cooks, a divorce which cost Rose much money and the scandal risked Albert being debarred from practice. He moved from Philadelphia to New Jersey, near Trenton, married an orthodox Jewish woman who insisted that she be called "Doctor" because of her Ph.d. and had a brilliant child who was not encouraged to read children's books. When I visited him he had become a practicing orthodox Jew, or at least a well off American version of one. He wore a kippah and he and his wife had two ovens, two dish washers, two of everything, because it was forbidden to mix this with that. His life as an orthodox American-Jew gynecologist was supported by his mother's and brother's decades of processing pork into ham. He had many forbidden swine to atone for.

Albert still remained relaxed, good natured, full of advice and I was glad to be given a book by his wife about mid-life crisis as this was such a period in my life and I had never heard the term before and did not understand why I was unhappy and quarrelsome. Albert was over-weight and had been to a Duke University program for those who had heart problems. I was sorry to learn some years later that he died of a heart attack.

Lillian, Rose and Florence had three brothers. Al, was an athlete who married, moved to Allentown Pennsylvania, and died young. I think he played baseball professionally. I do not remember the story and I doubt that there is anyone still living who can fill in the picture. Harry married a non-Jew, became a butcher, travelled in a butcher's truck and lived in a poor section of Camden. I think my mother looked down on him for his marriage. I disliked him the few times I saw him because of his crude speech and his expressions of dislike for "niggers". As I was a jazz fan we quarreled when we met, which thankfully was only when he dropped by our pharmacy and that was seldom. He had a son and daughter whom I occasionally met when I transferred from Gloucester to Camden High School and whom I would sometimes run into at a restaurant when I returned to New Jersey to see my mother. I think they became accountants. I knew little about them but we were friendly towards each other and my mother regarded them as model children who remained near home close to their parents.

Henry, the other brother, the youngest of the six, grey eyed, good looking, played basketball when younger, always appeared relaxed, had a quiet sense of humor, and a narrow moustache, was sought by women, and was evasive. I have the impression he lived with his mother when he was in the area. In Atlantic City when he sometimes stayed in the apartment Rose rented for the summer, we knew to say that he was not there when women telephoned for him, which they often did. He worked in the betting stalls at horse racing tracks which he followed around the country, in his car, as the season shifted to

Florida, New York, California, wherever. He would work in New Jersey at Cherry Hill, Atlantic City and Monmouth. He was always in contact with his sisters Lillian and Rose, part of the immediate family. In my teens I thought him cool and enjoyed listening to him until one time he began telling me about a young woman whom he was interested in and with whom he got no place. I never told him that she was someone whom I had gone to school with in Camden and who I thought was interested in me but whom I regarded as not having enough good looks to make up for her lack of brains and cultural knowledge. One day I had driven her home and felt offended by the vulgar mirrored halls and walls of the family's apartment. She was impressed that I was going to Columbia, but I ignored suggestions that we meet more often. Could Henry really have been attracted to her?

He seemed the perpetual bachelor, someone without a home who drifted here and there. When he did marry it was to a hairdresser who already had children by her first marriage. I at first thought she had a nice hard experienced cynical tone, but soon discovered that she was a middle class wannabe without taste of her own. If her neighbors had tat she felt she also must have tat. She seemed to have been destined for suburban life. Poor Henry, how could he have made such a mistake? Especially as after he was injured she mocked his manhood. He became house-ridden and depressed after a fire broke out at a race track and he was badly affected by smoke inhalation. He spent the remainder of his years a semi-cripple stuck in a stuffed soft chair listening to a snappish wife and a pampered daughter who blamed him for all her own failures.

As my mother felt she owed Henry—there may have been a loan of money at some time—the daughter became my replacement in her life. The loan might have been one of those stories that my mother invented to justify what she wanted. Besides thinking she could save Henry's daughter from the vices of the world, she also wanted someone to care for and comfort her as she had her mother and as she had taken care of Florence after her stroke. It was the fate I was always fleeing.

So unlike McCarthy's Italians we had only taken two generations to progress from impoverished immigrants to assimilated Americans and being spoiled, rebellious and resentful children. Progress or early decadence?

# Chapter 2: Fleeing

*Gloucester, Philadelphia and Camden 1933–50*

I imagine that there is some place, person, incident, or situation from which every person flees throughout his or her life. For me it is Gloucester, New Jersey where I lived during my youth until I left for New York and Columbia College. Gloucester, outside Camden, across the Delaware River from Philadelphia, was a small working class town where no one went on to university or sought further education after graduation unless it were a teacher's training college, and where my class mates, when not part of warring gangs or beating up homosexuals, would graduate to the local ship yard, work as mechanics in garages, join the police, and get drunk every Friday and Saturday night before going to church on Sunday.

Gloucester was my equivalent of the dry dusty deserted southwestern tumbleweed exhausted inbred otherwise forgotten village from which writers, movie stars, and serial killers flee. When I lived there I had a recurring fear that I would be slipping away one morning when I would be noticed and told I had to throw a party to celebrate my departure. The drink and food would cost everything I had saved and I would need start again and the next time I would be caught once more and still need to start preparing for my departure once more. Every dream of leaving seemed to come to the same anti-climax. I travelled around the world for decades fearing I would be forced to return to Gloucester. When I did return on a visit half a century later I was stunned that it appeared just an uninteresting, seeming harmless small town. There was now even a black bank guard.

I was born in Philadelphia's Hahnemann Hospital on New Year's Day, January 2nd. That is correct. Philadelphia was notorious for its "blue laws" and one was that no drinks could be sold and that there could be no public celebrations on a Sunday, and New Year's Day fell on a Sunday in 1933. So on New Year's Day, Monday, January 2nd, I was born to the sounds of the Mummer's Parade, a parade famous for attracting cross dressers, transsexuals, and other gays. Although I am not sexually transgressive, it seems an appropriate start for my life in which things have been often a bit odd. I was also born as Alvin G King, the Alvin compounded of Al and Irving, two of my father's brothers, and the G for my mother's family name, Americanized from Goldberg to Gilbert. Before many years I also wanted to change my name to something more American or British and I chose and have stuck with Bruce (sometimes Bruce A.) on the model of a comic strip character such as Clark Kent who was really, like me I wished, Superman. I would spend years seeking my Lois Lane.

    I have a few confused early memories beginning with that awful tiny room in Trenton, New Jersey, with the iron pipe bed held up on one corner by a rope or tied together clothes. My mother told me it had existed but I could not have remembered it and must have heard about it. I remember being at her mother's house in Camden, New Jersey where there was a large brown chow dog in the garden, a memory that merges with one in my early teens of being taken by a family friend to visit a farm and being told to open the front door. A massive dog knocked me over and began licking my face, while people laughed. On my back facing upward all I could see were the dog's two immense balls and the green fields beyond. Perhaps there were cows or goats but of little importance. That is still my image of farmland, a perspective of greenery with animals seen beyond dog balls. I am an urban dweller.

    I had a pet duck which I must have been given soon after its birth and which was attracted towards me as its parent. I understand the correct term is imprinted. It followed me to school. Years later when I felt that I could no longer eat four legged animals I decided I did not want to be a pure vegetarian and should still consume fish and fowl. Ducks, however, presented an emotional problem. I had to harden myself. Even today, living by the Canal St Martin in Paris, I often become excited and exclaim about migrating ducks that are a feature of the canal. I also had a pet turtle that would mysteriously disappear for months and then when it rained reappear. I still regard turtles as creatures of mystery, ritualistic. (A chef I know in California also had a pet duck. Hers was named "Lash Larue". It must have been when she was older.)

Other kindergarten memories include awareness that the other children had large erector sets and other building materials, often passed down from brothers and sisters. We were middle class and I could read and count, but they already had mechanical skills that I would never master. The boys would mistreat girls in the schoolyard, hitting them, pulling their hair, and demanding the names of their boyfriend. I rescued the two prettiest, and for years thought of them as somehow mine. Then in Junior High School both disappeared. The working class brunette ran off to be a dancer and when she reappeared several years later her older brother, who was a friend of mine, remarked that a woman would always love the man who took her virginity. That left me out. The other one, a blue eyed blond from a middle class family, whom I thought resembled the Norwegian figure skater and movie star Sonja Heine, moved to another town, developed a disease, lost her looks, and married young, by which time I had other women to idealize.

School seemed a waste of time. The teachers were only concerned with getting the others through, who often left as soon as they legally could, indeed sometimes before. I was supposed to educate myself. If the class was learning from one volume of world history during the year, I was told to read on my own the dozen or so volumes in the set. If the others found algebra difficult I went on to advanced mathematics, calculus, and imaginary numbers. Left to my own superiority I learned little and would be stunned a few years later when I met those who had genuinely mastered areas of which I had only a superficial knowledge. It seems to me that I had very little schooling. Latin was the worst. Although we were the elite-stream only one other person would go to on to further education (at a small college in Pennsylvania). There was no pretense of teaching or learning any Latin.

How could we? We had never been taught even English grammar, which would trouble me later when I tried to learn French.

Some of my ignorance was my own fault. For Biology I lost the dried frog we were supposed to dissect and assembled instead on a board for display a miscellany of chicken and other bones. I labeled the parts with seemingly appropriate names that I found in my biology book. Faced by this absurdity the biology teacher started laughing and dismissed the class. What else do I remember of Biology except the teacher regularly selling pickled cucumbers from a barrel and a required dissertation which we mindlessly copied from one written decades ago, each year increasing the misspellings, datedness and obscurity of the text.

I indiscriminately read E. E. Cummings, William Carlos Williams, Albert Camus, Jean-Paul Sartre, comic books, Major Douglas' books on Social Credit economics, New Directions annual anthologies, Lucretius, Homer, any books I could get my hands on provided that my high school teachers had not mentioned them. That was easy as they seemed to know nothing. For a time while in high school I subscribed to books published by one of the Classics Editions that had proliferated on the model of the Harvard five-foot shelf of great works that everyone needed to be educated. I wrote a book review (our project) of William Bradford's history *Of Plymouth Colony*, which was a Classic Club selection, supposedly the first book written in America, and was told by my English teacher that the book did not exist and I had made it up. She was one of the more lettered teachers in Gloucester; I disliked her intensely. At one point she was trying to teach poetry and did indeed explain the common forms of a sonnet. We were told to write a poem and I produced a not so subtle satire about her that she told me to read before the class; she said some people now write such stuff and consider it poetry. I do not remember whether it rhymed (probably did in odd ways), or its rhythm but it was certainly lineated like verse, and would have shown such influences as e e cummings,

William Carlos Williams, and whoever was in that year's New Directions Anthology. My attempt was no doubt horrible, but yes some people did consider my models poetry.

Most of my teachers were even worse. We were required to take state examinations in some subjects but were never told the results. When I hear complaints of terrible American public schools I feel I was already there, although now they are even more violent.

Gloucester was once part of history, a history recalled by a few old cannons at the city park facing the Delaware River. I cannot remember those Revolutionary War events that I looked up beyond a general who crossed the Delaware into New Jersey in search of clothing and horses. A hotheaded disciplinarian who led his troops into battles he was called Mad Anthony Wayne. Betsy Ross had lived in Gloucester and I made up a story that she was George Washington's mistress and had used his underwear for material to sew the first American flag. I told the story so many times over the years that I forgot I had made it up as a joke until my daughter said that her classmates claimed it was nonsense. Nonsense? Then I remembered, of course I had invented it. And in my version Mad Anthony Wayne kept crossing and re-crossing the Delaware, a parody of Washington's famous crossing.

Gloucester was a Jim Crow town with no blacks allowed. Black taxi drivers from over the river in Philadelphia would object when asked to take passengers to Gloucester, and any black with sense would make certain to be out of the town before night fall if not sooner. Teaching *The Merchant of Venice*, with its emblematic choice between three caskets, the vice- principal, appealing to racial horror of miscegenation, told my class what it must feel like if the Prince of Morocco had picked the one that would have made Portia wed the Negro. A girl with kinky hair who we thought of as Greek was not taken on the school trip to Washington DC because it was feared she might be part-black.

In the late 1930s a KKK cross had burned in front of a Jewish store in nearby Brooklawn. We were one of the few Jewish families in

Gloucester and throughout junior and senior high school I was called a Jew whenever someone was angry with or wanted to embarrass or intimidate me. A school friend's father once asked me whether the world's troubles would end if "all the Jews were drowned". The English teacher with whom I had quarreled told my mother that she would not accept such conduct from "one of her own". During a school argument over a presidential election I was told that all Jews were Republicans (actually our family was divided with my father being extreme left Democrat and my mother loyal to the Republicans because of her sister); a fist fight developed and I was chased home by a rapidly formed mob shouting "get the Jew" or a similar chant. My mother found me hiding and told me to return to school and face the mob. A few years later when I had a close black friend from another city and brought him home, my mother treated him as if this were not unusual although I knew that some of the customers in our drug store were making comments.

We were not practicing Jews and being called one felt strange, as it would often seem when it occurred later in life. The Irish and German Catholics who made up most of the town appeared like inhabitants of another planet with their punch ups, warring gangs, drunken evenings, worship of sports, and lack of study. A short period of memorizing what needed studying was bound to get me an A, while they hung out in gangs on street corners or were seeing available girls. I felt that I would have more and better girls later after I graduated school. There were often bitter territorial battles between the Catholics who went to the Catholic school and those who attended what Americans term public (state) school. Some battles were with chains and knives in the local park at night. To go to a dance at the wrong school was risky, punishment included being forced head down into a filthy toilet bowl. I tried to walk the thin line between the two groups but there were times I had to defend myself against louts, usually including at least one of my classmates.

In Junior High School new students arrived by bus from Brooklawn and other suburbs. They were less violent and I became friends with a small group that had a sense of humor. We mocked the platitudes of popular songs. "I love you for seventy mental reasons". We sang to the tune of the "Blue Danube"

> I like Rival Dog Food,
> ARF ARF,
> ARF ARF

Such parodies were common to an era of Spike Jones and his City Slickers. During the 1990s my wife taught a student from Montreal who could also remember singing to the melody of "Jealousy"

> Leprosy, my god I've got leprosy,
> There goes my eye ball into your highball
> Leprosy, my god I've got leprosy

I still cannot keep distinct the original lyrics of some songs from their parodies. "Laura, the face on the bar room floor" has long ago become the original in my mind.

Our sexual jokes were signs of anxieties and inexperience. A group of male students are late for class and say that they were outside "blowing bubbles". A female student arrives and announces that she is "Bubbles". A girl is driven further and further outside of town in her boyfriend's car and when told to "put out or get out" says that's her "legs are her best friend". Eventually she is driven so far away that she decides "even the best of friends must part."

I was good at sports, although disappointing at track and field meets where, instead of living up to my practice records, I performed terribly. I still shy off from competitions. In high school I even had a doctor's excuse to avoid Physical Education although I played as a guard on an amateur football team that included the best players from the school's team. I hated the culture that sports involved. There was always something homoerotic going on in the shower room after practices. Team members thought it amusing to step on your running shoes with their own spiked shoes. I remember once being selected to box with the class bully, someone larger and several years older. I hated boxing and expected to be murdered. Instead he whispered to me that he had sinus problems and "please do not hit my nose. Could we just pretend to fight?" It is sometimes true that freedom is to be free of fear itself.

My alternate life was in literature. I found such journals as *Atlantic* and *Harper's* in the magazine shelves of my father's pharmacy. When I was ten or eleven I asked for a two volume edition of Edgar Allan Poe's works for a Christmas present. I progressed to borrowing books from the public library. My tastes were at first limited, mostly middlebrow, but I was impressed by any sophistication, wit, or clever malice. Later I would discover Saki and Ronald Firbanks, but in my early teens I thought the novels of Thorne Smith highly amusing and loved the Mr and Mrs North crime stories of Frances and Richard Lockridge. I tried to write my own but after hearing a gun shot while drinking many cocktails at a country club I did not know what my characters were to do next.

I discovered a book shop in Philadelphia—was it called the Ben Franklin?-- where someone kindly explained that when I was older I would prefer Stendhal to Kenneth Patchen's *The Memoirs of a Shy Pornographer* and *The Journal of* Albion *Moonlight*, and Henry Miller's *The Air Conditioned Nightmare*. Other books I bought and read included Thomas Merton's *The Seven Storey Mountain*, translations of Albert Camus's *The Stranger* and *The Plague*, and such works of Jean-Paul Sartre in translation as *The Respectful Prostitute*, *No Exit*, *The Flies*, *Nausea*, and *Being and Nothingness* (of which I could make little sense)

Meanwhile I hung around with my friends at a garage at night and discussed how members of the local police force were related to the criminals who held up the gas station recently, how the police told us that the way to handle the situation was to get the criminals in an alley and dump them in the river afterwards, and how someone had died in an automobile accident and was revenged by his friends through trick driving that forced the guilty one off the road and left him a physical wreck near dead. I sometimes carried during those nights a pocketknife as a form of self-protection. I also sometimes carried a copy of a book of works by Friedrich Nietzsche that I had found among my father's four or five books. This was another kind of self-protection. I knew little of the world outside Gloucester except what I read; I assumed everyone lived by trying to avoid being beaten up, knifed, insulted. I was surprised when I met the son of one of the other three Jewish families who was visiting after they had moved to New York. In Gloucester he played center in the high school football team; he told me of his school in New York where there was a chess club (no one in Gloucester played chess) and classmates read books and discussed ideas. I was stunned. There was life elsewhere without constant assertions of manhood, threats, anger, rudeness.

I did have another life in Gloucester besides reading. Being an only child with parents usually working in their pharmacy I drifted to other homes like a stray dog. Wanda Morrison had been at school with my mother in Camden and also moved to Gloucester. She was from a Polish immigrant family and married a Scots. Cheerful, amusing, outgoing, gossipy, she was connected to local civic life and such celebrations as Easter and July 4th; he was a caricature Scots, withdrawn, quiet, serious. He worked in Philadelphia as a printer, she usually had a professional link with the schools in Gloucester as would her daughter Pat. From Wanda I learned bits of my family's history, of the relationship between my mother and father, and gossip about others. She spoke to me as if I were an adult and told me that as X's wife was frequenting bars he made her pregnant to stay at home. Y had her tubes tied which shocked the other teachers, but why should she not enjoy herself? It was at Wanda's I learned to garden (an interest that was useful in Nigeria where I grew European vegetables for my family as well as teaching; I was the mad white man out in the afternoon sun), had my wartime Victory Garden, and learned how to color and decorate Easter eggs, for which I won a few prizes. Around her life seemed normal and after I married I several times took Adele to meet Wanda and then Pat and her husband as if I wanted to show that I had a real family.

By then the Schroeders had left. I did not know where they had gone. I was told that he had had a stroke and Betty gave up her New

York trips to care for him. My mother, who had protectively bought the property after the Schroeders claimed they could not anymore afford the rent, felt that she had lost the opportunity to expand the pharmacy to make it more appealing to buyers before she retired.

They were our next-door neighbors and where I first wandered seeking family. The husband worked as a mechanic in Philadelphia where the two daughters, Betty and Jean, worked in offices. The wife stayed home, cleaning house, baking, preparing evening meals. Every New Year she made a delicious orange cake, moist and intensely flavored, for my birthday. People remarked that she used natural ingredients including fresh oranges. The husband had a dartboard in the basement where I brought my friends. He built the large folding table with legs on which I mounted annually a prize winning train and lighting display in our apartment for the Christmas season. Although I stopped having model trains before I left school, I still feel attracted when I hear of a display. I understand the story about the famous jazz trombonist Jack Teagarden who carried a train set in the trunk of his car when touring. A tall handsome man of American Indian origin he attracted women who when invited back to his motel room found they were expected to watch model trains.

Betty Schroeder, the elder daughter, introduced me to jazz music. She had been a Gene Krupa swing band fan and graduated to following jazz in New York where she would visit for weekends once or twice a month to hear musicians on 52$^{nd}$ Street and later, after she had become part of the scene, in Harlem. An attractive, dark haired and dark eyed slender woman, Betty was about ten years older than me and had an erotic charge. Whenever she returned from work she would be greeted by approving wolf whistles from the war vets who seemed perpetually to mill in front of our drug store at night. Although she had no interest in them I could see her appreciative smile. She knew that she was a desirable woman. She had, however, no one in Gloucester with whom she could share and brag about her prized photographs taken in New York with such famous black musicians as Buck Clayton, Teddy Wilson, Roy Eldridge, Don Byas, Big Sid Catlett, and Jo Jones. No one else, besides me, listened with her to her increasing collection of 78 rpm jazz records, discussed the difference between Dixieland and Swing and the new Be Bop. There was no one else to whom she would recount her musicians' gossip about this drummer being heavy footed, that drummer appearing always relaxed and unperturbed on stage. Once she returned from New York with stitches above an eyebrow from an automobile accident and I assumed her relationships had progressed beyond being a listener and fan of black jazz music.

I used to listen with her family to Superman, The Shadow and other radio series. I spent Christmas nights with them listening to radio announcements of Santa's coming arrival. Betty quizzed me on my readings for school tests. She had taught me such mnemonics as "A rat in Tommy's house may eat Tommy's ice cream" as a way of spelling "arithmetic ". For a history test I learned "Columbus sailed across the ocean blue in fourteen hundred and ninety two", which rapidly was parodied as "Columbus sailed the deep blue sea in fourteen hundred and ninety three". I progressed with her from Gene Krupa to a knowing jazz fan. Her skin darkened as she spent more time in New York. Was she taking some pill, was it her make up? I assumed that even in New York a pretty young white woman in the company of older black musicians might bring police attention. I was by then reading *Down Beat* and knew of the arrests of many musicians for using drugs and the brutal pistol-whipping of musicians, such as Miles Davis, with little cause. Police were enemies to be avoided.

Although jazz was a minority taste there was a continuing audience for big bands and renewed interest in small combos. In school we danced to Artie Shaw, Stan Kenton and Woody Herman. There were interesting arrangements by such bands as Claude Thornhill on the jukeboxes. The jazz world was dividing into traditionalists and Be Boppers. The rediscovery of older jazz was as much a revelation as the strange chords and choppy rhythms of the boppers. A weekly radio show began called "This is Jazz", which broadcast live jam sessions that I recorded on a machine I acquired for just that purpose. Because of 'This is Jazz" I travelled to New York one weekend in April 1948 to hear Kid Ory's Creole Jazz Band at Carnegie Hall during its short visit from California where it broadcast regularly on a radio program introduced by Orson Welles. I went to Stuyvesant Casino, then a jazz venue, and I went on a "river boat" jazz concert where I spoke with Pops Foster about his slapping the strings of the bass. I returned to Philadelphia and to Gloucester with a suitcase just in time to go to school. I was then 15 years old and in tenth grade. I had been out of my depth in the hotel one night when a crying woman began banging on the door mistakenly pleading to be let in. She had left her man over an argument and was trying to return. If she had wanted to discuss jazz I would have known what to do.

I can see now that jazz significantly shaped my life. It led to an interest in black American and then African art, music and culture that took me to Africa and the new African writing and eventually into Commonwealth and postcolonial literatures. If I became a literary critic and scholar of Indian, African and Caribbean literatures, it was partly an extension of my interest in American jazz. Jazz also led to dancing to live music, the source of many acquaintances over the decades. My first publications were reviews and articles in jazz magazines.

I was discovering that there was plenty of good jazz to be heard in Philadelphia. (Even around Camden there were good jazz clubs although I did not know them.) Besides touring groups like Louis Armstrong, and a few bars for jazz fans (one of which was owned by Red Rodney, a well-known white bebop trumpeter), there were occasional big modern jazz concerts at the Academy of Music where I heard Charlie Parker, Dizzy Gillespie, Art Tatum, Billie Holiday, Tiny Grimes, Slam Stewart and other living giants when I was 15–16 years old. More often on Sunday afternoons there were Dixieland concerts in the Academy's Foyer with such musicians as Sidney Bechet, Danny Barker, Wild Bill Davidson, Muggsy Spanier, Willie the Lion Smith, Sammy Price, Jimmy Archie, Baby Dodds, Pops Foster, James P Johnson, and Danny Alvin. I even sat in one time with the Wilbur and Sidney de Paris brothers' band when their drummer was late getting on stage. I had no idea I was listening to and meeting musicians who when I was older would be legends.

Through the Foyer concerts I met Walter Bowe, a black trumpet player my age who lived in Camden, and through Walter I came to know other musicians in the Philadelphia area, several of whom would figure in my life during the next decade or more. There was the clarinetist Dick Hadlock already modeling his playing on the eccentric great Pee Wee Russell, and a trombonist Norman Finkelstein, both studying in the business school at Penn. Dick's father, a jazz fan, worked for an American company in Brazil where Dick was brought up and studied saxophone. Hadlock drove an antique chain-driven Rolls Royce in which we once made a trip to Bridgeport, Connecticut where he had lived until he was 12. He still had friends there in, was it?, the White Eagle Jazz Band.

He would move to New York, live in a Greenwich Village loft owned by the black painter Beauford Delaney, made famous by Henry Miller, would buy and edit *The Record Changer*, a then obscure jazz publication to which I contributed from Europe. Like much that he did, *The Record Changer* became part of American cultural history. The complete run was republished some years back by the University of California Press.

Dick studied clarinet or saxophone at various times with Sidney Bechet, Garven Bushell and Lee Konitz, three greats. He wrote the often republished *Jazz Masters of the Twenties* and for decades had radio jazz programs in California while he continued to play clarinet in or lead different groups. Although as a critic and broadcaster his range was wide his personal tastes were limited mostly to the white Dixieland of the so-called Chicago school of the twenties. He admired Bix Beiderbecke and Frank Trumbauer. He intensely disliked the New Orleans jazz revival that started in the late 1930s and which had a strong influence on many, including Walter Bowe and myself.

He travelled to Paris one year and was given a demonstration of over-blowing and other "free jazz" post-modern reed techniques by Steve Lacy, whom he knew in New York during the 1950s and who had also taken lessons with Sidney Bechet. Lacy had famously been asked by pianist Cecil Taylor around 1955 "why such a young man as you should be playing such old man's music" and followed Taylor into the free jazz movement, although the main influence on Lacy's evolution as a musician and composer was Theolonius Monk, in whose group he worked for a time. Lacy was now a famous avant-gardist and composer, who lived in Paris since 1970 until in 1992 he was awarded a five year MacArthur "genius" Fellowship. The French government's attempt to tax the money eventually drove him back to the USA. The meeting between the former students of Bechet was not a success. Hadlock told me that he was not impressed by what Lacy showed him on clarinet, and Lacy told me he was surprised that his old friend had changed so little in his music.

Another of our jazz circle was J Norton-Smith, the son of a wealthy industrialist and conscious of his genealogy. The family house on Philadelphia's Main Line was large; outside old stables were converted into a garage to hold many automobiles. Norton played banjo and piano very well, and was then attending Haverford College, the elite men's college often paired in one's mind with nearby Bryn Mawr College for women, Swarthmore making the third of the exclusive colleges in the Philadelphia suburbs. Norton was brilliant, eccentric, learned, impossible. He lead his own traditional jazz band consisting of musicians from good colleges who lived or had been raised in or near Ardmore, and was proud that they included no Jews, until he decided that I had a much better press roll than his usual drummer. I remember once we auditioned for the Paul Whiteman TV show. I had not realized that Whiteman was still alive. Norton was a follower of Ezra Pound and dressed as eccentrically—he had a belted tweed Norfolk jacket that buttoned high in front, a black velvet cape—while making malicious comments on the clothing of others. He had a nervous breakdown directing Sartre's *Huit Clos* in French at Haverford. Between his third and fourth year he went to New York, had a French mistress, and supposedly played professionally at Stuyvesant Casino to support himself. I sometimes wonder could he really have been accepted into the New York musicians union so effortlessly.

After Haverford he studied with C. S. Lewis at Magdalen College, Oxford, and with Lewis's backing moved through various English departments in England and Scotland before being made Professor at Dundee University. He was a famous mediaevalist; besides editing texts he was series editor for several publishers.

He had also become a parody, said to be more British than the British. He did not drive an automobile but had an ex-colonial wife who did. Few academics in England and Scotland seemed to know much about his past except that he had once been a professional jazz musician who played in New York with Sidney Bechet and other greats. When I was a visiting Professor at the University of Stirling (1979) it was thought possibly amusing to have Norton and me meet, but this was not to be. A pity. I really would have like to see him again. We need not have talked about old times. He died during 1988.

Walter kept meeting and introducing me to other musicians from the area and we would often jam at Walter's mother's house in Camden. One, Tony (I can never remember his other name), a pianist from beyond the area, came through the black section of Camden asking if there were any old 78 rpm recordings in the attic or basement he could see and perhaps purchase. Following the tradition of older jazz record collectors he had a black bag filled with bananas to eat when hungry as you could peel the skin and have a clean fruit even if your hands were filthy from the dust. I remember a very good, somewhat older trumpet player from Pennsylvania who introduced me to Surrealism and who claimed to himself be a surrealist. I did not know that there were surrealists in small Pennsylvanian towns.

This was the period when the New Orleans jazz revival had even reached some high schools. We were envious of the Scarsdale High Band led by clarinetist Bob Wilbur. Wilbur studied with Bechet, and the band made commercially sold recordings of their music. There was also a group in Baltimore with a good trombonist, but it was surprising when Walter heard on the radio the Pennsauken Ramblers of

Merchantville, New Jersey, where my Aunt Florence lived, playing "Stumbling" on a Saturday afternoon Horn & Hardart (a quick food chain) talent show. He soon made contact with them. They were indeed high school students, their leader was an accordionist, and I even played drums for them at one of their high school dances. We sometimes borrowed their string bass player, although that required my mother being willing to pick him up and drive us to Walter's.

Walter lived in an old wooden house in what was thought the black part of Camden. His mother was one of those blacks who still voted Republican because "Abraham Lincoln freed the slaves". She was a nurse or medical officer at, was it, RCA Victor or Campbell's Soup, both major employers in Camden. Old fashioned in dress and appearance she had a quiet dignity although Walter once told me that she had twice been raped returning from work. There was an up-right piano in the small living room where we had jam sessions. One time Walter returned with a well-known white revivalist ragtime pianist from the Pacific North West. Portland? We made an acetate duo of "Maple Leaf Rag" on which I played on my wood block and cowbell throughout. I was proud of my six-stroke roll that I had copied from Baby Dodds and which I liked to use behind piano solos. I had long forgotten about the evening until decades later when visiting San Francisco Hadlock played the recording for Adele.

Another acquaintance was "Jazz" Friel, a clarinetist in Philadelphia who held radical libertarian views. We jammed sometimes at his apartment in Philadelphia where through Walter I briefly met Richard T Gibson, the controversial writer who according to some was a black radical and according to others worked for the CIA. He would be cofounder of the Fair Play for Cuba Committee, to which Walter would belong.

When I was at Columbia Friel came to New York where with Walter we sat in one night with the pianist Champion Jack Dupree. I felt honored but Friel complained that Champion Jack was musically illiterate and impossible to play with, as he was inattentive to the length or structure of a composition (his 12 bar blues could be 11 or 13 as he felt like it and he changed chords as erratically).

Drafted into the Army Friel was sent to Japan where he became famous for recordings of his own compositions and married a Japanese woman from an aristocratic family who followed him to Levittown Pennsylvania where he worked for the Army as they were the only employers who would accept someone unwilling to join a union. To make matters worse he demanded that his wife only serve meat and potatoes and he took no interest in her kimonos and tea ceremonies that fascinated their neighbors.

Walter and I would attend films that had scenes with jazz musicians, such as *New Orleans*, a real cornucopia with Louis Armstrong, Billie Holiday, Kid Ory, Zutty Singleton, Woody Herman, Meade Lux Lewis and many others. We would go to movie houses in Camden and Philadelphia that still had touring black swing bands on weekends. We would pretend to be of drinking age and go to jazz clubs in Philadelphia; I remember once meeting Mercer Ellington, Duke Ellington's son. I sat in with banjoist Elmer Snowden's band one afternoon and he told me that similar to many younger drummers I played behind the beat as I was used to practicing to records. He was probably right, but I could have replied that I purposefully entered late as I was imitating Baby Dodds, whose drum rolls filled space between beats and the phrases of the horns.

Baby Dodds was my hero. He had been one of the early jazz musicians and played on classic jazz recordings by King Oliver and Louis Armstrong in the 1920s. He was still playing regularly in New York, Philadelphia and was on some of the Bunk Johnson records made in New Orleans trying to recapture early jazz styles. I had asked him about his tuning of his drums; he had showed me "nerve beats", a technique of holding two drumsticks together so that they rattled as the arm's nerve muscles are made to contact and tremble. I could have cried when he had a stroke and could hardly hold his drumsticks in his hands. I wanted to offer to assemble and carry his drum set. Decades later in Paris I would be stunned by an aging, smiling Art Blakey, going deaf, playing drums badly in the wrong tempo while supposedly leading his Jazz Messengers.

It was partly because of Walter I transferred for my final year of high school to Camden. There were other reasons as well. Most of my Gloucester friends had already graduated. One punched a teacher, was expelled, and joined a travelling circus. During my junior year I had found another family with which I could take refuge, the Salines in Camden.

I do not remember how I was related to or met them, but I became indebted to their friendship and used their address as my home when I changed schools. If I missed a day and the truant office came he would be told I was upstairs ill and could not be disturbed. Through the Salines I met an educated middle-class Jewish community of people my age or a bit older who went to good universities, read books, and were not part of warring gangs. Through the Salines I learned not to always blame "the ref" for the defeat of the team I supported (the father was a "ref"), not to brag about fights and toughness, not to speak like a thug.

Joyce, the younger of the two daughters, was the main influence, and I probably had a crush on her. She was pretty, dark haired, dark eyed, romantically thin and flat breasted. She was a year or two older than me and I looked upon her as civilized when she spoke of Edna St Vincent Millay and the then popular American-Lebanese mystic Kahil Gibran, who wrote *The Prophet*. At a place where her friends congregated we listened to the juke box and she pointed out some passage in a Glenn Miller recording where a trumpet without a break continued a phrase started by a clarinet. She was not an intellectual like my cousin Mimi in Philadelphia, but she was much prettier and closer to my age and I was enchanted by her and the normality of her family, although my mother with unexpected prejudice claimed that they were Southern European Jews of Italian origin.

Joyce's older sister Molly studied speech therapy at Syracuse University and tried to soften some of my rougher edges, including a tendency to lisp as my tongue had not learned to go where it should when making S and th sounds. She told me that she would never say she loved a man before she married him—which I understood as meaning that she kept her virginity. It was possible before the birth control pill for a perfectly normal woman to undergo four years of university education and be engaged to marry without losing her virginity, although her virtue might be tested at one or two orifices. She repeated to me an amusing conversation she had overheard between a newly engaged couple in which the woman and man with conscious humor insisted that they were each always right and that the other should get used to it and stop arguing. It seemed to both of us charming and witty, a level of cleverness with which I had been unfamiliar in Gloucester. I have since imitated it. The Salines sat down as a family and ate together to three course meals. My mother remarked that the father sold automobiles and she kept her distance, but I had found a temporary home. As usual women were the main influences in educating me about the world beyond the streets of Gloucester.

The year at Camden High School was mostly a delight. Camden was populated by Jews, blacks, Polish and Italian. We mixed at school, and the senior class president was black, a good student who played football and won a scholarship to Brown University. Others in my class would go to the University of Virginia, Penn State, and the University of Pennsylvania. Usually one senior would be admitted to Harvard. I was counseled to apply but I had my mind made up that I would go to New York. I applied only to Columbia and NYU. Besides wanting to move to New York, which I thought of as somehow a gateway to Paris or London, I had read an article in *The New Yorker* about some pro-fascists at Harvard during the Second World War. Columbia had a reputation for liberal and left professors.

I had many friends at Camden and while most were Jews they included, besides Walter, Ray Drozdowski, who was Polish. We had a similar sense of humor, made puns, and ethnic jokes, and made fun of pieties although Ray remained close to his mother, the Polish community, and its culture. Ray would miraculously find my address and contact me about a class reunion decades later when I was a professor in

New Zealand. I looked him up during a trip to the States. He was now married, had children, and had done well for himself. I asked his wife how they had met at Penn State and she said it was a blind date after someone had told him that she "put out". He was a lawyer for tobacco companies, defending them against claims that they knowingly sold cancer- causing cigarettes. In support he and his wife smoked heavily. When I telephoned him a few years later, he was recovering from a car accident. He had a stroke, blacked out, smashed into a tree, and needed five by-pass operations. He was retiring from his job and through contacts would be appointed a state judge in Camden. A few years further on I asked for his aid in renewing a driver's license that was expiring before I would return to the USA and he told me that was against the law.

The Jews and Polish students often mixed at dances at a Jewish centre. Italians were less common and blacks, except for Walter, even rarer among my friends. Several Italian families were known to be part of the Mafia. For some reason a very pretty junior, who was a cheer leader, and whom I hardly knew beyond saying hello between classes, developed a crush on me that she publically announced at a football game. The following week I had a verbal warning that it would not be good for my health to follow up the invitation.

Camden, unlike Gloucester, offered many course options and I decided to take Journalism instead of a science subject. (Latin would have been impossible for me.) It might have been the first time Journalism was offered and as I had been editor of the Gloucester High School newspaper I became editor of the one being started in Camden. In Gloucester I primarily wrote articles and chose what was to be featured. At Camden I had other responsibilities, such as lay out, font choice and size, and illustrations. I felt I was going to be a writer and this was encouraging. It was not until I tried out for the *Columbia Spectator* and was told I would be expected to spend weekends not on dates but on duty as an apprentice reporter at a police station that I realized that I had no wish to become an editor at the *New York Times*.

The situation in drama (not a course) was worse. In Gloucester I had a large role—Greek landlord Appopolous,—in the junior play, *My Sister Eileen*. (I did not then know that the original author of the amusing *New Yorker* sketches on which the theatre and film adaptations were made was the wife of Nathaniel West of *Miss Lonely Hearts*, *Day of the Locust* and other dark satires that I had recently read with admiration.) In Camden I tried out for the senior musical, A *Yank at King Arthur's Court*, and did not even become a walk on for crowd scenes.

I would over the years have a conflicted relationship with the theatre that I usually thought inferior to poetry. At Columbia as a freshman I quit the chorus of an ancient Greek tragedy when (once more) I was told that because of rehearsals I should forget about any other social life. Instead I took dates or went by myself to plays on Broadway and in the Village. Then while in Mexico a few years later I found myself in Thornton Wilder's *The Skin of Our Teeth*. Bored by repeating a few lines at rehearsal and maybe three performances, I kept experimenting with my delivery and my position on stage, which upset the blocking, lighting, cues, everyone except me. I was not meant for the stage.

I however wrote a book on John Dryden's plays for my doctorate, started an innovative theatre department in Nigeria, wrote study guides to plays by G. B Shaw and Ibsen, and became co-editor with Adele of the *Modern Dramatists*, a series of some fifty books, and editor of a series of more than a dozen books on 17th and 18th century *English Drama*. I also commissioned and edited books on contemporary American and Commonwealth theatre. I wrote a performance oriented book on Shakespeare's *Coriolanus*. I do not enjoy being on stage, it bores me. I enjoy seeing and listening to plays, but given a choice I probably will decide to go to a film, unless the theatre event is a Baroque opera which is my favorite form of drama.

I began my senior year still a sexual virgin and then I became experienced so rapidly and inconsequentially in the back of my car it made no difference. I had no regular girlfriend until towards the end of the year I began seeing a girl in Philadelphia towards whom I had no romantic feelings but did have sexual ambitions. We kissed, I felt, but her legs remained crossed. I did not believe her when she told me that she had a boyfriend in Philadelphia; then years later I met him and he told me, without knowing who I was, that he had lost his girlfriend to some superior guy with a green Oldsmobile who had gone to Columbia.

# Chapter 3: The Big Apple

*New York and Columbia 1950–54*

By my second year at Columbia Walter and Hadlock had come to New York. Walter at first stayed with an aunt on Sugar Hill and as he explored Harlem I met, regularly sat in with and became friends with blues singer's guitarists Brownie McGhee and his younger brother Sticks at Felton's Lounge on 125th Street. Stick's had recorded "Drinking Wine Spo-dee o dee", an early rocker, for a small black recording company in 1947 and then recorded it, in a more cleaned up version, for Atlantic two years later and it became a best seller, near the top of the pops list. We had sung along to it in Camden High School, especially the refrain "pass that bottle to me".

Sticks was a more modern musician stylistically than Brownie, more attuned to the direction that black and white music were taking, but, while "Drinking Wine" was covered and became a hit for bigger name artists he never had another success. Too many of the follow up recordings had titles that sounded like "Pass that bottle again" or "Drinking more wine" and he began drifting out of the music business. When I met him he still accompanied Brownie for another year or so but he had a day job as, I think, a salesman and his weekends at Felton's were accompanied by another trade.

One night I agreed to drive one of his women home, and innocently we talked for hours. This did not seem unnatural to me as some nights I went back with Brownie to his apartment and had coffee with him and his wife. I do not know what that woman said but when I next saw Sticks he threatened permanently to alter the lower part of my anatomy. Why had I not twigged that Sticks was a part-time pimp? Brownie one night told me when younger he used to transport white women on pay day to black rural workers for blues parties. Although he did not make it as a jumping and jiving singer-guitarist Sticks understood and was willing to share his knowledge of the art, explaining to a white clarinetist I knew such techniques as holding the first note and delay the entry of the next phrase to create swing and excitement.

I had previously heard Brownie on old style blues recordings, and then on the "This is Jazz" radio show with his regular partner harmonica player and singer Sonny Terry. They had recorded on Folkways. Sonny was a folk musician famous for his train, fox hunting, and other imitative sound effects. Although Sonny and Brownie would sometimes play old music they played in Harlem mostly the Chicago blues made famous by Muddy Waters, B B King, Buddy Guy, and Bobby Blue Bland, that had become popular with black audiences. It was louder, faster, and less introspective than the old Mississippi blues from which it evolved when it came north to the cities from the rural south. Brownie had an amplifier to increase his sound but remained basically a piedmont style non-amplified guitarist in an electric era. He had to move with the times and played what was now popular with black audiences. He sometimes played for large dances in northern New Jersey, had a regular drummer for such dates, and I was not asked to sit in. At Felton's I would power along with the broken triplet shuffle rhythm beat common to Chicago and most older blues. Somehow I was deaf to the back beat on the four that had become common to most contemporary black American music. When a soul singer who was sharing the bill in New Jersey asked to use Brownie's drummer he was told that I was a good drummer. One number was enough to show that I was out of my idiom.

The group at Felton's included Bob Gaddy on piano, Bob Harris on tub bass, and a woman called Maybelle (not the famous Big Maybelle of the period). Gaddy had picked up piano in church and the Gospel style was easily transformed into a boogie-woogie left hand. He played standing and as the upright piano was on wheels and the crowd at Felton's pounded on the piano it kept moving this way, that way. I was surprised to learn that Gaddy was still performing until recently. Bob Harris had a good voice as a singer. I don't know why he brought a tub-bass to Felton's as, I seem to remember, that on some recordings he played a normal string bass. One night when he did not show up I decided to play the tub bass and the one string cut my unhardened fingers so badly that they had to be bandaged for weeks. The Barnard girl I had brought was rightly impressed.

After a while I stopped bringing my full drum set into Felton's and limited myself to a snare drum and high hat cymbals. Although Harlem was a place for whites to avoid in the early 1950s, I did not fear violence or being robbed. I parked my car near a street lamp and walked down well-lit streets. This was a decade before the white blues bands famous during the 1960s and I disliked whites in the audience as they often requested folk music and I was once asked not to play drums so that Brownie and Sonny could be tape-recorded without my Caucasian presence. The same would occur in Ibadan Nigeria where I regularly sat in, even was lead drummer, with a local Apala band, and someone from the BBC wanted me to stop playing otherwise, I imagine he felt, the music was not authentic. You can add these to the reasons why I dislike folk music fans.

Although on a few occasions Brownie and other Harlem musicians would at parties join the jazz musicians I knew, the two groups were distinct. Except at Lennox Lounge when we once jammed with Champion Jack Dupree I do not recall Walter sitting in with blues musicians. He sometimes led jam sessions at NOLA Studios where we paid for time as if we were rehearsing. I cannot remember who played beyond Hadlock, Walter and myself. I seem to recall Joe Muranyi who later became clarinetist for Louis Armstrong's All Stars and still later was part of the Louisiana Repertory Jazz Ensemble. Michael Steig, nephew of the famous artist, sometimes played guitar. He was unusually sensitive towards pitch and could take forever tuning up, the model for my repeated joke about some musicians preferring to tune up all night rather than play. Walter seemed to have trouble finding guitarists and bassists, especially as we tried to avoid the dreaded Greenwich Village folkies with their 1930s -1940s protest songs, sentimental ballads, and bouncing ball up and down rhythm. I used to get along with Roger Sprung when I was buying LPS at Sam Goody's where he worked, but when leading the "authentic" folk scene with his banjo I avoided him and his friends. I never liked the famous folk revival groups he inspired such as the Weavers. Then one evening Walter turned up at NOLA's with someone who looked like, called himself, and probably was Carl Sandberg. His repertory distinctly was not jazz and the evening ended when he wanted to play "Come Josephine on my flying machine."

One musician I came to know was Jack Fine who had been in the army, worked at Commodore Records, took advertisements for the recently started *Village Voice*, lived in Brooklyn, was married and had a child, was a card carrying CP member, and played sensitive trumpet and sang in the manner of Hot Lips Page. I would enjoy his "Gee Baby ain't I good to you" and I can still remember his playing the blues one night in a jam session-party at Columbia with a mixed group of musicians including Mike Steig and Brownie McGhee. I thought if playing jazz could always be like this I do not want to do anything else in life. Such large parties were not always peaceful. I remember a fight that Bill Haddad, who had worked in merchant boats, broke up by hurling himself back first through like a cannonball. I no doubt would have tried to reason with the protagonists and been knifed.

Jack would record with many well-known musicians, go through several marriages, spend some years playing in Paris, and move to New Orleans where he started the Jazz Vipers, and still continues to play trumpet, now more forcefully in the local style, and has become famous. You might have seen him in the HBO TV show Treme. But then half of New Orleans at one time or other appeared on Treme.

By my third year at Columbia I was part of a New Orleans style band with a trumpet, Steve Knight on trombone, Bob LaGuardia on clarinet, a pianist, Duffy Givner (banjo). We all were studying in some school at Columbia, and often rehearsed at my AEPi Fraternity House. Duffy was working on a doctorate in Philosophy. I rediscovered him decades later when I was visiting fellow at Calgary University and he was a Professor at the University of Regina, Saskatchewan. He had not published much but his wife Joan Givner was a powerful feminist inspired biographer whom I met when she gave a talk at Calgary. Besides two biographies Joan would later publish a novel, children's books and her autobiography. The Regina faculty included another old friend, Saros Cowasjee, poet, satirist, author of books on Sean O'Casey and editor of many anthologies of writing by and about India. We had studied for doctorates together at the Leeds University, England during the late 1950s. We had once formed with Adele a lawn bowling team, "The Wogs".

Norman, our pianist, had a lively older thin blond girlfriend who was divorced and had a child. She took dance lessons and once told me that she wanted to marry a Jew as they treated women like princesses unlike the Boston Irish among whom she was raised.

Steve Knight was about two years younger than me, had long blond hair which he kept combing, claimed to have had much sexual experience, and I thought of him as an artist. His father taught art at Columbia and the family had a house at Woodstock—an artist colony for decades before the famous 1969 festival. One summer I went with Steve to Woodstock planning to stay, decided that I did not want to spend months washing dishes at a restaurant and returned to New York City. I lost track of Steve, looked him up on Google and found that after I left Columbia he had joined drummer Bob Thompson's The Red Onion Jazz Band. At times Hadlock and Mike Steig also worked with the band. Thompson, already a professional drummer, taught at Columbia while working on his doctorate. He continued leading the Red Onion as he progressed up the academic ladder, founding the Society of NeuroScience, and became a distinguished Professor of Psychology at Hunter University.

Steve later became interested in free form improvisation, and moved on to rock and then on to non-Western music. He played many different instruments including the Oud. As a keyboardist he was one of the original members of Mountain which performed at the 1969 Woodstock Festival and as a bassist he played the odd rhythms and structures of Taksim. He was also twice elected to the Woodstock town council. He knew girls from some of the elite high schools, such as Fieldston, around Manhattan. At a few parties we played everyone seemed to be preparing for a part in a future Woody Allen film. Steve would fall asleep on a bed and stay overnight. I never knew what took place but I envied his easy way with elite women.

Many of the woman I met while at Columbia came from families more financially and socially successful than my own. This was before men's colleges admitted women and our hunting grounds especially were Barnard College, Sarah Lawrence College, and Bennington, as well as Hunter College of the City University (that pun is too easy, but yes it was a mostly female branch of CCNY). I met at Barnard an attractive daughter of a now famous Russian avant-garde composer but never managed to take her out as her weekends were spent on Greyhound busses "discovering America". Sarah Lawrence had a daughter of Jack Warner, the movie mogul, and a Roosevelt. Bill Haddad married her and moved into Democratic Party politics and significant government positions with the Kennedy clan.

At Barnard I met and briefly took out a young woman whose father I seem to remember taught Brazilian literature at NYU. We went to the Village for some reason and as she was a few years older than me and was wearing a leopard print jacket and red slacks I felt insecure and inexperienced No knowing what to do I took her to see Dick Hadlock as they shared a Brazilian connection. A few weeks later they started dating and eventually married. Dick wanted to teach American Indians in California and Ruth worked as reference librarian in children's literature at various schools. When I visited with them decades later Dick had already become a well-established for his jazz record programs on Public Radio stations. Ruth had graying hair and looked like an unsexy reference librarian. It was unfair but I could not help think "I was once scared of that." Later I told the story to a poet friend, Shirley Geok-lin Lim, who remarked that if the woman was wearing a leopard print coat and red slacks she was probably more insecure than I was.

I did join the Columbia marching band as rehearsals were during week nights and there were free trips to football games. I was a sucker for the band's very Ivy League uniform, blue blazer, light grey flannel trousers, blue and white (Columbia's colors) stripped tie, white buck skin shoes. Being a New Orleans style jazz drummer I had no trouble picking up the marching beats and my press rolls were admired. I became the second drummer and no one suspected that I could hardly read music and was simply good at intuiting what was wanted. Then disaster occurred. A small group regularly broadcast live classical concerts on the radio and as the first drummer was ill I was told to take his place. My protests that I did not read music were taken as modesty and suddenly there I was, musically unlettered, trying to read a score by Stravinsky. Let's see, a roll begins here and seems to end there, by which time the music moved on. There seemed nothing to do but improvise. After the broadcast no one said anything to me, not even the criticism that would have allowed me to explain. I did not play with the marching band the next year. I was aware that I was a musical illiterate and began trying to follow the scores of Beethoven piano sonatas and other compositions. A few years later I would start writing news columns and then analytical articles about jazz and other music in such publications as the *Record Changer*, *Jazz Monthly*, and *Jazz Review*. The analysis would sometimes mention precise details, details that I always made certain trained musicians also could hear and confirm what I claimed.

I had not realized that Columbia College formerly had a quota to avoid being overwhelmed by New York city Jews. It was altered to acceptance by regions, one third students from New York, one third from the East coast, another third from the remainder of the country. I was not at first aware of the ethnic and religious make-up of the College until a few nights into my first week I overheard someone say he was the only White Anglo-Saxon Protestant in his class. Such distinctions were there although not usually mentioned.

I was not "rushed" by the few "Jewish" fraternities at Columbia but was invited to join with a small group that was starting a local chapter of AEPi, a large national fraternity. At first we met in a few small rented rooms, by next year we had an apartment with some rooms to live in, (which I did after being tossed out of the University dormitory because of an unnecessary dispute with the dorm counselor) and then, with the aid of the national organization, a real fraternity house in which we were supposed to eat lunch and have our parties. It was useful to me as a place during holidays to have jam sessions, and of course for bringing girlfriends. Sassoon Shahmoon, about two years after me, belonged to both the famous Sassoon and Shahmoon Iraqi families, and was given by his father, a real patriarch in Quebec, seed money to support himself and told to work hard and invest or starve. At night he washed dishes while we studied or played.

The Columbia chapter early won a dispute with the national organization about inducting non-Jewish members, but kept losing a running battle about admitting blacks. I do not know whether we were a little ahead of or behind our times, as racial attitudes were changing rapidly, but our senior class had a black president.

II

While a student at Columbia I had my first trips outside the United States, the first during the summer of 1952 to Europe, the second during the summer of 1953 to Mexico. The summer of 1952 I travelled with Joel, a classmate, to Europe on a Holland American line ship filled with other students. There was a college jazz band, much drinking, and late night hunger. We "tipped" the main cook a dollar for pork chop sandwiches and the like until the captain tried to stop him and was knifed. The final days there were only cold sandwiches for meals. I was supposed to be studying French language and culture in Paris but as usual it did not turn out that way. Being well informed about French literature I tested much higher than my language abilities and was placed in a class I could not follow. The Culture lectures, as was common then, seemed to consist of the names of the mistresses of those who ruled France in the past. I soon drifted away.

I sometimes went to La Mistral, a second hand bookshop opened a few years earlier by George Whitman, who supposedly had spent years riding the rails like a 1930s hobo although an heir to a wealthy industrial family. Whatever the truth about his past he had started the mostly English language bookshop with some inherited money and it was already a center for writers as they could sleep there at no cost provided that they did not mind sharing a bed with his dog and spending a hours watching the books, especially Whitman's prized collection of Olympia Press and still banned books by Henry Miller and others. Among those I met were Alexander Trocchi. Christopher Logue, and Terry Southern who using pseudonyms had written books for Maurice Girodias. The first two were also editors of the new journal *Merlin* and connected to *Zeno* or *Zero* from Tangiers. I remember exploring cave-like tunnels near the bookstore with them. Editions de Minuit had recently published Samuel Beckett's *Molloy* and *Malone Meut*. Although the famous small theatre production of *Waiting for Godot* by Roger Blin was a year later, we were already talking about Beckett.

At night I hung around the cafes on Boulevard St Germain hoping to catch a glimpse of Sartre, Camus, and other literary stars of the period. L'Abbye was a cellar club owned by the black American actor Gordon Heath where he sung with his partner Lee Payant. To avoid neighbors complaining you were told to snap your fingers rather than applaud. There were silent films played at the wrong speed, as was still common, at a crude forerunner of the Cinematique. I danced at the Club St Germain and other jazz clubs. There were some spectacular acrobatic swing dancers whom Simone de Beauvoir wrote about as if they were true existentialists living the non-bourgeois life to which she aspired. In fact they were aspiring professional dancers who soon would tour internationally.

Joel had met Ed Jones, a Sierra Leonean living in London, and I soon accompanied Ed to meet Madelaine Rousseau, editor of *Le Musee Vivant*. She was an expert on African art, had a private collection of her own, and her editorial board included many of the famous writers and intellectuals of the time. She thought of herself as the mother of Africans in Paris, and many would visit her bringing small presents including African masks that she would repair, paint and sell to collectors and museums. A strong eccentric personality she used to brag that when she broke with the Communist Party they termed her the French equivalent of a Nigger lover. She also had quarreled with those around *Presence Africiane*, whom she described as Boulevard St Michel Africans who only knew Paris although I doubt that she had ever lived in Africa. An early teacher of modern and African art to workers she had known Picasso and many of the great painters, and when she needed money she would ask someone to sign, so that she could sell, one of the gift paintings and prints that covered the walls of her small apartment. I would often see Ed Jones and Madame Rousseau in the future. I would stay at Ed's apartment in London. Once after I had taught in Nigeria for many years and had started my own small collection of African art I had when seeing Rousseau's blurted out "But they are fakes", by which I meant modern and had not been used ceremonially. Decades later I told that story to an African art dealer who sold to museums and he looked stunned. He had bought from her and assumed that her reputation was enough providence.

One purpose in going to Paris was the hope of improving my sex life, which would seem even more hopeless after a supposedly erotic show performed by two not so young prostitutes was followed by pleasureless rapid intercourse. This was my fourth such disappointment. Was that what it was all about? Then a few nights before leaving Paris I was with some musicians I knew, beating rhythms on a violin or guitar case, and was picked up by a French woman who I took back to my room. She was married, experienced, energetic, and we kept using different positions. She was also one of the few women I have known who had obvious liquid organisms. I could not stop, indeed after several hours when she wanted to sleep I had to continue. Noon the next day I was still erect and masturbated. So this was it. Technically I had not been a virgin for several years, but I felt this was the night I became experienced.

Let me confess that when I wrote in the last chapter that I had my first sexual experience in the back of my car while at Camden High School I was not telling the truth. There was a homosexual encounter with one of my parent's employees when I was perhaps ten years old. I cannot remember the details, and do not know now how much was enticement, innocence, or temptation, but I did not enjoy, have always been ashamed of, and almost never mention it. It is still hid by a curtain in my mind. The curtain is a fitting image as it took place in the bathroom in the apartment above the pharmacy, where employees also used the toilet. Although I am not prejudged against homosexuals and lesbians that early experience left me embarrassed.

In London Joel and I stayed in Golders Green with a wealthy Jewish family whose son we had met in Paris. It was still the period of rationing and they served elaborate meaty meals and insisted we take trips in the family car. The mother would announce this and that person, about whom I was supposed to know, lived there while I tried to tune her out. I surprised by the son's identification with Israel, an emotion I did not feel. When he asked me where I "would hide when they came to get the Jews" it seemed meaningless as I was an American, but the impact of the German extermination of the Jews remained with many Europeans. A few years later in Heidelberg one of my language teachers started speaking admiringly about my Jewish face, something I would have denied. Soon I understood. In a village I watched the crowd laugh and applaud when a man began to goose-step during a parade, and I would meet Germans who went into exile during the Nazi era.

Travelling on the underground to the jazz club at 100 Oxford Street I met and was attracted to an English girl who was also going there. We "jived" all night to New Orleans style jazz, spent the next evening wandering through Soho, and then the next day went swimming. This was the first of many times when a day at a swimming pool in England left me sick as hell. I spent the next nights not in her arms but wrapped in blankets trying to sweat out the cold while stinking of Vicks. We occasionally wrote to each other for several years, and I would send her some presents next summer from Mexico but by the time we met again in 1954 we had both moved on to other romantic interests. One Hundred Oxford Street would, however, keep being part of my life. I would listen to jazz and dance there in 1954, meet students who would be my first friends at Leeds University, and even more than half a century later at the Palm Court Café in New Orleans I would reminisce with the owner, Nina Buck, about when we used to listen to jazz and dance at 100 Oxford Street. The last I heard of it there was discussion whether to reconvert 100 Oxford Street, which had been a place to dance for many generations to many styles of music, to some other function or make it as national treasure.

## III

Throughout my time at Columbia I was a member of Students for Democratic Action, the student off shoot of the liberal Americans for Democratic Action. One of major issues was our response to the McCarthyist "red" hunting that continued into the early 1950s as several of the university's best known leftist scholars were either denied tenure or somehow pushed out, My own instincts were to defend anyone on the left, but this was thought naïve by Bill Haddad and there was a reluctance to be identified with anyone who might turn out to be a card carrying member of the Communist Party or a "fellow traveler." This was more than an academic debate as the Democrats needed to distinguish themselves from those thought to be acting on behalf of enemies of the USA. I was idealistic and naïve as many of those I read with approval or wanted to defend were proved or confessed latter to being members of the CP. I was learning that words and reality could be very different. Out of curiosity I attended a meeting of a Schactmanite group and listened to a denunciation of both Russian and American imperialism. There were perhaps not more than a dozen attending and I thought that at least half are probably FBI agents. A Chinese friend who was in ROTC (of which I unthinkingly disapproved) explained to me one night the importance of foreign supply bases in time of combat by using chess pieces as examples.

During the 1952 Eisenhower-Stevenson presidential campaign I took the initiative to advertise for volunteers to support Stevenson. The group rapidly grew to many hundreds and we undertook voter registration in Harlem. The group's leadership attracted the attention of the Democratic Party that not only wanted to advise but also hoped to find promising politicians. Those doing degrees in such areas as Law or Sociology were happy for the contacts. I was oblivious to what I had accomplished and, as would happen at other times, blind to what I was being offered.

I returned home to Gloucester between my third and fourth years, was sent to Camden County Hospital to have hemorrhoids removed, picked up a case of crabs and suffered pains in my lower spine (that I continue to have) as a result of a spinal anesthesia. I could not stand my mother braying about how can such a small operation leave me so nervous and I disappeared for a couple of weeks, staying with classmates in New York. I returned determined once more to flee Gloucester and left for Mexico City where another classmate was taking summer courses in Spanish at the American run Mexico City College. We travelled to such places as Morelia and small fishing villages in Michoacán but most of my time was spent hanging around with some older students from Notre Dame University. I generally enjoyed their company and found them surprisingly serious except for Jacques, from Montreal who was studying at Georgetown University, who kept making anti-Semitic comments. I was stunned when we travelled as a group to Acapulco and skipped out of our small hotel without paying. It was not what I would do. In Acapulco I had swam out too far, been dragged by the current some distance, then tossed up on sea urchin covered rocks and passed out. A peasant found me and hours later was still methodically digging out the needles with a pin that he sterilized with the flame from a candle. I was then taken to a local doctor for a tetanus shot. Many needles remained under my skin during the next decade, would suddenly appear and need be helped out with a knife or scissors. I am prone to near drowning experiences and have been pulled out by rip tides and thought I was dying at some of the best beaches in the world.

Mexico City was then inexpensive and it was possible regularly to eat at good restaurants. There was also much violence and guns were often visible. I had a stupid rapid bout with a prostitute, listened to Mariachis bands, went to bull fights, watched and learned how to bet at Jai Lai matches, and ate grilled "worms" at a steak house favored by Mexicans. One night I went to a theatre to listen to Perez Prado then at the height of his fame for his mambo recordings. I was struck by how his female vocalist was over-endowed and the crowd loved her for it. I always enjoyed contemporary Latin music that I thought of as similar to jazz. In New York I would hear Latin groups that featured jazz musicians, and years late in San Francisco Adele and I danced one night to the Fania All Stars.

On the way back from Mexico to New York I stopped for some days in New Orleans where after meeting "brothers" I stayed at the Tulane AEpi house and listened to their girlfriends complaining about how they were expected to give presents to their doctors, a complaint which would still be repeated a half a century later when I regularly travelled to New Orleans to hear jazz each year. The summer of 1953 it was still possible to hear George Lewis and his band with Kid Howard, and I visited Paul Barbarin who was at home to get a copy of the lead sheet for his "Bourbon Street Parade", which had been introduced at the 1949 Mardi Gras. Barbarin was one of the great traditional drummers. He worked during the Depression with Luis Russell, Louis Armstrong and during the trad jazz revival regularly played at a Chicago jazz club. He comes from one of those New Orleans "families" that regularly produce good musicians. The trombonist Lucien Barbarin, who regularly tours with the Preservation Hall Jazz Band and with Harry Connick Jr, is his nephew. New Orleans was still segregated and I was asked if I wanted a "white" or "Negro" taxi. The white taxi would only take me to the boundary of the black area and I had to walk from there.

At an art gallery, a forerunner of Preservation Hall, I listened to a traditional band and talked with Bill Russell about Baby Dodds whom he had recorded talking about his drumming as well as playing on the influential Bunk Johnson recordings that had propelled the New Orleans jazz revival. I thought of Russell as a hero and still do. He had found money and the musicians to record some of the first recordings of earlier New Orleans jazz that he issued and sold in limited edition of perhaps a 100 at a time on his American Music Records. The recordings were on pure vinyl and not available commercially as Russell a Marxist was not willing to pay the government's taxes. I had heard that Russell was well-known experimental composer who had neglected his own compositional career to find and record older New Orleans musicians who still survived. His recordings rapidly became known internationally and when I lived in England many of the musicians had obtained or made copies. Now Bill Russell's papers, manuscript and other documents are the main attraction of the New Orleans Historical Collection.

## IV

My academic life at Columbia was more extremely up and down than middling. An up was Mark Van Doren's course in Poetry from which I would a few years later publish my undergraduate explication of Wallace Stevens' "Metaphors of a Magnifico" to prove to myself that I really at times displayed ability at Columbia.

The downs included Science courses. The College had two years of general education course requirements and some were not to my liking. For Science I choose Astronomy rather than the popular Psychology. I did not know that the University required all professors to teach an undergraduate course and Astronomy although cross-listed as an introductory was intended for doctoral candidates. The professor, who had been awarded a Nobel Prize for his research on the "red shift", soon shed about a hundred for his class by writing impossibly difficult mathematical equations on the black board and leaving. This was not a course in star gazing. After a few weeks there were fewer than a dozen graduate students of Physics and me. Each week I would lose my expensive text book and need to purchase another. As the mid-term examination approached it became clear that I was determined not to drop the course and was assigned the professor's Assistant to get me through, which he somehow did. I passed the final examination and signed up for the second semester which was much easier and I learned such useful information as how to calculate the time in Philadelphia if I were on Mars, or was it the other way around?

Needing a second Science course with a laboratory component I decided in my second year to take Botany in which I proved so inept with microscopes that I was given a "pass" if I promised not to touch another one again. I had broken too many slides. I never was destined to be an experimental scientist.

Even French was a problem as I had trouble learning simple grammar. I wondered how I could pass the College's language requirement until one day a French teacher noticed I was reading Sartre's *Jean Genet Comedien et Martyr*. I explained that I read French but could not remember for tests anything about tenses and the conjugation of verbs. I was allowed to sign up for an advanced reading course in modern French literature and that was how I passed the requirement, otherwise I would probably still be a Columbia student. As I have taught at three French universities and lived in Paris for over a decade this part of my past might seem improbable but it is true and will explain why although I am at ease in France where I shop, read newspapers, go to films, I avoid conversations in French. I still stumble over tenses, grammar, rules.

I think it was during my Senior year Eric Bentley came to Columbia replacing the well-known, charming but middle brow Joseph Wood Krutch as the professor of Drama. I admired Bentley, read his *The Playwright as Thinker* and *In Search of Theatre* and his many anthologies of European plays, but his classes were too early in the morning for me and as I did not take some tests, he marked me down a whole grade for the course despite my final examination and term paper. Thirty years later I was co-editor with Adele of a series of fifty short books on Modern Dramatists published by Macmillan in England and received a letter from Bentley proposing to write on a dramatist whose volume I had already assigned. I felt uncomfortable turning him down as almost everything I knew about modern drama I had learned from reading him.

Although Columbia did not have "majors" there was an equivalent and being undisciplined I only signed up for three of the six courses that formed the history of English Literature before the twentieth century. I took instead courses in art history which interested me. Show me a reproduction of a Cezanne painting and I could instinctively explain how it was a forerunner of all later art movements. I viewed Renaissance and Baroque art as if they were Cubist experiments in form, color, movement and dimension. I was often conscious of the tiny details that my instructors planned to point out. Art appreciation did not require much actual historical knowledge; one of my teacher's remarked "King, don't you ever crack a book?" It was true that I had foolish notions that a superior person should treat learning with insolence and never be caught taking notes or studying. Any reading I did for courses was late at night when I would not be observed. Art history courses were perfect for my vanity; all I needed do was look at pictures or sculptures to see what was significant.

I never thought, however, about becoming an art historian or critic. My future would be in literature or music which in retrospect seems foolish as I still had not learned to read musical scores and I was then not interested in most periods of English literature. I was still following an earlier modernist avant-garde which meant I was passionate about Shakespeare, Seventeen Century metaphysical poetry and Poetry and Drama rather than Prose. I predictably disliked Milton, the Romantics and Victorians. I was distrustful of novels unless they were experimental. In a course on seventeenth century literature I worked out my critical principles while discussing the metaphysical poets, what did I really mean when speaking of "levels of meaning"? I was strongly influenced by William Empson's *Seven Types of Ambiguity*, *Some Versions of the Pastoral*, and the puzzling *Structure of Complex Words*, but their brilliance was clouded by the American New Critics who had their own jargon to discuss the complexities of a literary work.

Although Columbia College with Lionel Trilling, Richard Chase and F. W. Dupee appeared a home for *Partisan Review* intellectuals with their mixture of Marxism and psychoanalysis, many of my teachers, including Trilling, still taught the glories of the Romantic poets and the Victorian novel. There was little explication of texts and class room discussion was likely to be of biography, reputation, society, politics. *PMLA* and other academic journals were still opposing the New Criticism as unlearned and unpublishable, as, curiously, did the *Partisan Review*, which had its own perspective on culture, especially literature, as an expression of the individual's relationship to society. The *Partisan* Review authors although considered did not regard themselves as part of the New Criticism which confusingly had become identified with conservative southerners, some Marxists, and Marxists who turned Roman Catholic. Dupee had moved from Communist Party organizer to a co-founder of *Partisan Review* and literary editor of *New Masses*. While most of his generation of intellectuals were shaped by the Cold War he was later sympathetic to the anti-American anti-imperialism of Edward Said.

As an undergraduate I still needed guidance in reading literary texts. Those who offered it in the form of explication, the New Critics, were still regarded with suspicion or hostility within the academy. There were places that had hired New Critics as English literature professors, but the only one at Columbia was William York Tindall, who taught in the Graduate School, stuffed with mid-westerners in large lecture halls seeking advanced degrees, looked down upon by those who taught in the College with its Ivy League exclusivity and Great Books humanities tradition. In my third year at Columbia I had taken a graduate course with Tindall in the explication of contemporary texts and instead of a final examination I wrote a long essay on Wallace Stevens *The Man with* a *Blue Guitar*. On this basis I had talked my way into Tindall's doctoral seminar on *Finnegans Wake*, where (it now sounds like a joke) I had startled the other eleven students (all at least ten years older than myself) by suggesting two puns on Albanian words. Tindall and the writer, translator, art dealer, and Joyce expert Nathan Halper, who helped with the seminar, left the room to huddle and returned to announce that I was the first to propose Albanian puns in the *Wake* and that I was probably right, as Joyce would have known some Albanian while working in Trieste. My first contribution to literary studies and no doubt as inconsequential and absurd as many which followed. I liked Tindall, he was witty, charming, off hand, and full of gossip about the drinking habits of famous poets and the graffiti that could be found in the toilets of Greenwich Village bars.

At the end of the course needing to examine me Tindall assigned a page he choose at random from the *Wake* that I was supposed to comment upon for two or three hours. Impossible? By chance I had been reading at my then girlfriend's her copy of Joseph Campbell and Henry Morton Robinson's A *Skeleton Guide to Finnegans Wake* (I had some beautiful and very bright girlfriends; I eventually married one ) and as she was interested in art I had looked at the discussion of the *Book Of Kells* in the *Wake*. That is what the page Tindall would choose was mostly about. I could have been faced by some other page and filled my examination booklet with despairing doodles, but I was obviously going to be A + this time. Except that the English Department in the College would not include my two graduate school A grades in my average. I was told that a graduate school professor would assume anyone from the college was an A student.

Through Tindall's course on Contemporary Texts I met Irving Feldman, who was then aggressively a city product with blacken teeth and clothes and speech to match. He explained that his first book of poetry was rejected by forty publishers before he resubmitted to the first one who now accepted it. I would met Irving much later in Deia where he was visiting Robert Graves. He now had a Spanish speaking wife, was the only American I knew who played fronton, had published many volumes of poetry, was a full professor at a major university upstate New York, had whiten teeth and dressed in tweeds. At Columbia we had often passed time chatting in the company of a student who had a sense of humor similar to my own and who taught me the immortal French rhyme:

> Je vous aime,
> Je vous adore,
> Comme mon chien,
> Qui est mort.

Tindall and Lionel Trilling were two sides of Columbia for me. Trilling had spent over a decade teaching at Columbia before getting tenure as it was felt that a Jew had no natural relationship to English literature. Although his book on Matthew Arnold was excellent and he pronounced authoritatively about British manners, morals and sensibility (concepts he often used while teaching) he did not visit England until he was long established as a famous literary critic.

I would read Trilling with respect in *Partisan Review*, but as a teacher he was haughty, pretentious, assumed our interest in the nineteenth century literature he taught. He could suddenly lash out in anger if questioned or contradicted or if he felt he was being upstaged. He would, in defense of Jane Austen, question why people were now against social, class, and other prejudices. Why should not people dislike Jews? It would be foolish to object as this was a trap; Trilling was by origin from a Vienna Jewish family and had himself at Columbia suffered for it. He seemed to feel that life was a challenge of the kind depicted by Balzac in which the end was more important than the means until one was in the position to challenge others who were trying to climb up the social ladder. He was conflicted as he also desired to rebel against acceptance of social convention, to become an artist and live creatively. Many of his insights as literary and cultural critic derived from this conflict within himself that also explains his ambivalence towards Allen Ginsberg (one of his former students) and the Beats (who could be said to have originated among a few Columbia students), the drug culture of the 1960s and others whom he in Arnoldian terms regarded as Barbarians. They were the libido escaping from the suppressing Ego and Super Ego.

Trilling could be surprising. He had known the Marxists who were still part of the intelligentsia and his friends at *Partisan Review* were originally Trotskyites reacting to Stalinism. During a time when defending Alger Hiss seemed necessary for those against McCarthyism he stunned me by proclaiming in class that the hated Hiss. Whitaker Chambers, whom he had known, was an honorable man and would not lie. He proved correct. He also had a vicious tongue and was known for antagonisms. Asked at a party by an equally famous art critic in his circle of friends what he should do to make up a quarrel Trilling was said to have replied "Get rid of your stupid wife".

I once foolishly asked in class for him to explain what he meant by saying something in a Wordsworth poem was "basic" like an episode he alluded to a Faulkner novel was "basic" and was told that if he threw me down the stairs and jumped on me that would be "basic". Several students avoided the class for days after. Yet I admired Trilling dispute the outbreaks of anger. I had previously convinced him into letting me take this Senior course in my Junior Year and while we disagreed about literature and its criticism he would talk about it outside the class room. He scorned the New Critics as minor, hated Kenneth Burke (who I was thinking of studying with for my MA). He said anyone, even R P Blackmur, but not Kenneth Burke. I had not known that *Partisan Review* had a review demolishing Burke as a Marxist and as someone who wrote obscurely. He did tell me to read F R Leavis, who later was an influence on me.

From the threat of throwing me down the stairs I learned something important about life, my literary and cultural heroes could be awful in real life, and it was time I grew up and stopped thinking of myself as unique. That might have been the most important lesson I learned at Columbia. I did not fully appreciate this until later.

At the time I was stunned by Trilling's behavior as I did not know he had a drinking problem until his wife, Diana, wrote about her marriage. I was impressed by the appearance of Diana; they looked smart, a stylish couple in contrast to Tindall and his wife who in my mind is dragging behind him, an Irish washer woman calling him James. Why James? I don't know why my memory made that up, but it seems fitting. Years later I would read Tindall's book on John Bunyan, probably his doctoral dissertation, and be impressed by his scholarship easy wit, and ability to treat the obsessions of others as amusing follies.

My professors at Columbia included Richard Chase with whom I took a semester of his two semester course on American literature and a course of Creative Writing for which I wrote a supposedly Satanic verse drama which included among its parts the Lord's Prayer in reverse order. As I was not religious such attempted blasphemy was in imitation of nineteen century and early twentieth century French writers, although Sartre also made use of such conventions of defying God. I think the main influence was a Canadian I had met in Paris who spoke of killing someone and dumping him in the Seine. At the time I had felt dangerously close to accepting this fantasy. Chase had questioned me about parts of the verse play but otherwise was politely uninterested. He was more interested in that I was thinking of marrying and said that when he was my age no one wanted to marry as it would interfere with one's career. For some reason we disagreed about the actress Tullulah Bankhead who I thought marvelous and he regarded as a tramp. How could you not love a woman who spoke of herself as "pure as the driven slush"? She recently spoke at some Democratic Party event I attended.

The four years at Columbia were a time to discover the world and oneself. Towards the end there was a scramble to choose futures. Joel withdrew from accepting Law School to become a sociologist. A friend who planned to be a research scientist in Physics decided to go to Law School as it would make him more valuable to be a scientist who understood law. I used to go to jam sessions on Monday nights at a well-known jazz club in the Village where professionals mixed with amateurs like myself. One night I was drinking with a trombonist I knew who was complaining that he was often on the road, seldom saw his family and when he did get back from jobs his wife was asleep and she had to wake up early to prepare the children for school. He was a Catholic and thinking of joining a monastery. I knew that I could not live such a life. I needed the close ties of family to make up for what I lacked as a child. I also knew I did not have family money nor wanted to wash dishes to support being a writer. My options were closing and I decided to work for a doctorate in English and a career as a university teacher of English literature.

Originally I had planned to go to Bennington College in New Hampshire to do an MA with Kenneth Burke. Burke suggested I first spend a year studying with Stanley Edgar Hyman—author of *The Armed Vision*—before working directly with him. I arranged to meet Hyman to discuss what I would do. Hyman was charming as was his wife the novelist Shirley Jackson, but as soon as we went into his study he shut the door and began pulling small bottles of rye or some other drink from under pillows of the couches. He was an alcoholic, Jackson was trying to keep him off booze, and he had squirreled it away where he hoped she would not find it. I don't remember much more about our meeting except that I later received a letter from the president of Bennington saying that I could not study for an MA in literature there as it was then a woman's College and the only men admitted for study were in the performing arts, such as dance and theatre.

What to do? Columbia College usually sent the best graduate in English to Claire College, Cambridge but I was not the best in English that year, it was getting late, and I had no interest in going to Cambridge. I wanted to study with G. Wilson Knight at Leeds University whose books on Shakespeare had opened my eyes to the many harmonies between the images, themes, and events in the plays. He had even made me appreciate such Romantic poets as Byron. I thought him a great literary critic, someone who in his own eccentric way was attuned to creativity and great writing. He was known to have influenced T. S. Eliot's poem 'Coriolan', to have created the fashion for abstract scenic design for Shakespearean productions in England for decades until replaced by Brecht's ideas, he was an influence on several contemporary poets, and would, while I was at Leeds, become a major influence on Wole Soyinka. Some Russians would even create an international Wilson Knight Society devoted to the study of the Humanities across national and temporal boundaries. But he was teaching at Leeds University, an industrial red-brick in the North of England, and I was told by one of my teachers at Columbia that if I went to Leeds I could not come back to Columbia. So be it.

Around December of my final year I met through Bill, Fred and Sid Haddad, a girl from Bennington who I would see again in Florida where I spent Christmas and the New Years with Fred who returned home to see his father. I remember we drove all night, non-stop, outraced police cars and made it in 24 hours. On the way back we took our time, stopped in South Carolina for a meal and a scene I thought existed only in movies and novels, as the head of a family ranted at his daughter for the godless, communist ideas she was picking up at the University of South Carolina. He then kneeled over, whether drunk or dead I did not stay to learn.

Fred's father's Lebanese restaurant was my first taste of middle eastern food and of the healthiness of using olive oil instead of butter, of eating pureed pulses, and simply grilling meats. The food was delicious.

I would see the girl from Bennington regularly after I returned to New York and soon she was talking about marriage, children, how we must become Quakers or Unitarians as she needed a religion to replace her Roman Catholicism, even hinting that like her parents we would have affairs, have violent quarrels, and make up. The attractions of mutual masturbation made me accept such fantasies until her mother put her foot down and told her to stop. I was shaken, then relieved that I emerged from such insanity without harm. So many young women then wanted to be married by the time they graduated college that from my first year at Columbia that had been an obstacle course. I had taken a Barnard freshman out two or three times when she told me we could meet no more as she had decided to marry someone from Harvard.

I became wary of getting caught up in such fantasies. My next girlfriend taught me that love and work were the essentials of life and even when I was in England sent me a cook book that I still have as I had to learn how to cook for myself. I, however, remained wary of someone whose parents divorced and who was on probation at Barnard because of poor grades. Too many of the young woman I knew lacked stability. Another Barnard student whom I had taken out was caught cheating in her final examinations, no doubt a sign of insecurity as she was bright enough to pass her courses without hiding notes in a sanitary towel dispenser.

While waiting to leave for England I started an affair with a slightly older woman who had returned from studying in Paris. She was experienced and energetic in three positions on top. She also was or was not (it was never clear) engaged to someone at Yale who her patents did not want her to marry because although he was also a child of Czechoslovakian immigrants his parents were protestant rather than a Catholic, or, I cannot remember, from a different branch of Catholicism. Maybe both families were Catholics and one was Slovak? She was pregnant by him and while she went to Germany to visit with her father who was in the army I found out from Walter's girlfriend that drinking a bottle of gin and taking many aspirins while soaking in an extremely hot bath would bring on an abortion. Which it did. When I saw her again she had decided to marry the Yale boyfriend. The day before I left for Leeds the Bennington girl turned up, implied that she would like to renew our relationship, but by then I was glad to have escaped. I must, however, still been in the mood to marry as on the ship taking us to Europe I met my future wife.

# Chapter 4: Beginner's England

*Leeds, England 1954–56*

When I first arrived at Leeds I felt like crying. Had I exchanged the excitement of New York for such a dark, grimy, damp, chilly, gloomy, sooty future? It appeared a city without glamour. I should telephone home of my arrival, but I did not want to admit the mistake. Leeds would turn out to be marvelous, the social life and University better than my years at Columbia, but this would take time. My years at Leeds had already begun inauspiciously when the English girl I had known in London during the summer of 1952 told me that Leeds was an awful choice. She agreed with my Columbia teachers that I should have gone to Oxford, Cambridge, London or at least Edinburgh University.

I was not used to British food, social classes or humor. The guest house where I first stayed served scones and other foods that were then exotic to me. My first night I listened with amazement while being told that the travelling salesman who I was listening to had attended a famous public school, happily took a third class degree in Agriculture at Oxford, and had a good job as a "representative". The platinum blonde, also a travelling sales person, was impressed and they went out for the evening. The next day I was asked at the university what religion I belong to and when I said none I was told that with so many to choose from in America there must be one that could suit me. There were several new Africans at the Student Union asking about political organizations.

By my second night, I had discovered a local jazz club, sat in on drums, became a "Yank", and was the day after in a van going to Manchester where the band had a gig. We ate at a Chinese restaurant with three kinds of potatoes for the veg. English food was terrible in those days and potatoes were the norm. In London I had met some young people from Leeds at a jazz club, and I soon had friends, a girlfriend, and jazz club to haunt when bored with studying. As usual a large part of my life somehow revolved around jazz music and musicians.

I rented two large rooms in an immense Victorian house that once must have been built for industrialists. It was divided into four apartments, two to a floor, with a shared bath and toilet. The front rooms had small subdivisions for a kitchen. Heating in both rooms was with a chimney and that meant a dirty coal fire which I learned to start by first rolling a page or two of newspaper and then making it into a ball which I lit. It was a dirty affair and even after a long soak in the tub on weekends (the British then bathed once a week if that) I would feel dirty. 1954 was still a time of coal rationing; although meat was no longer rationed the butchers cut pieces precisely to the weight you ordered. The gas for the communal tub and the kitchen fire was paid for by pennies in the slot of a meter. I seldom had enough pennies, found it easy to open the meters, and would leave a shilling or a 2s/6p coin for the equivalent pennies, which the meter man told me was illegal and several times threatened to report me. Like everyone I soon had an electric fire in the bedroom. It was not as warm but it was practical.

Feeding myself was at first a problem. The local green grocer had little green that was not in a, to me, limited range of tins. I thought American bread was bad, I had not previously known British bread, soft, tasteless. Part of the problem was that I did not know my way around. As a child I had learned that there was American cheese and Swiss cheese, when I asked the green grocer for American cheese I was told that he had none, but did have Cheshire, Cheddar and Wendsleydale. Decades later while being a visiting distinguished Research Fellow at Calgary University I would laugh at an American visiting professor who had, I claimed, a nervous breakdown from the cultural shock of tinned foods being in two languages, English and French. In the fall of 1954 I was hardly more sophisticated. On November 5 I travelled to grimy smoky Birmingham with other Leeds students to see a production of a Wilson Knight play and was stunned as grey stunted clothed effigies emerged from alleyways demanding "a penny for the Guy". Before the year was over I could distinguish between a Yorkshire and Mancunian "duk".

If food at the Student Union was British post-World War II stodgy, it could be kept down with a half pint of Ale. The other students seemed more grateful than unhappy with it. Much of what I assumed as normal was not. No one seemed to know of rice. Vinegar was malt. Olive oil was sold in miniscule bottles at a pharmacy. Garlic was something supposedly eaten by the French and Italians. I once heard a female student ask another whether she could eat curry. I was once invited out by fellow students to breakfast, lunch, and "tea"; each meal consisted of omelet, tinned peas, and, inevitably, rissole potatoes. I would sometimes treat myself to a meal at the Metropole Hotel, where the cantaloupe was so hard as to be uneatable despite the sprinkling of ginger, and the exotic was a ham steak cooked with tinned pineapple. It is difficult to recall the terrible food of that period in contrast to the excitement, variety, and availability of good, especially foreign, food in England now.

What was I to do at Leeds as a student? Americans after their general education took specialized courses for their MA and at the start of their Ph.d programs. It was usual for Columbia graduates who went for two years further study in England to do the second and third years of a British Honors degree, for which they would receive a BA Hons, pay at Oxford and Cambridge for their MA and be allowed on return to the USA to proceed to a Ph.d. program. Bonamy Dobree, Professor and Head of Department, thought I might edit an edition of the works of Michael Drayton for my MA. A few nights of reading in Drayton's works convinced me that I was out of my depths. We decided that I would do the complete $2^{nd}$ and $3^{rd}$ years of the Leeds Honors degree in English and decide at the end of the year what to do next. This meant I followed courses in Poetry, Novel and Drama (at Leeds there was normally a choice between them), Chaucer, Milton, a special period (I forget the dates, but they were the late eighteenth century and the early Romantics, I remember studying Blake's poetry and David Hartley's influence on Coleridge), and two Special Authors (normally one but I decided to do both Dobree's seminar on John Dryden and Wilson Knight's Ibsen.) There should be a course in Shakespeare but perhaps he was covered in the two years of Drama lectures. Besides I had studied Shakespeare at Columbia. Between the various courses I wrote perhaps an essay a week although I also had to take final examinations. During 1954–5 Wilson Knight was my tutor and eccentric enough to let me present 30 page typed essays instead of the expected 5–6 hand written pages. As I had years earlier picked up the journalist's habit of printing in pencil rather than carefully writing in ink, I asked him whether I need write the final examinations with a pen and was told I could not use a typewriter as the sound would annoy others,

Wilson Knight was brilliant and even more eccentric than I had expected. I should have realized that he was a spiritualist who thought he spoke with the dead, and whose ideas were first formed by such late nineteenth century thinkers and movements as Rudolf Steiner's Eurythmy. Those brilliant patterns of good and evil and the need to do evil to do good were filtered through Friedrich Nietzsche's *Beyond Good and Evil*, those stylized stage gestures were eurythmic body movements. He had taken now very dated fashions from earlier in the 20th century and turned them into brilliant insights about the ways values and character found expression in literary symbols. As a lecturer Wilson was endearingly hopeless. His lectures on tragedy brought out paradoxes and absurdities; he would start laughing and tears literally rolled down his checks. Lectures on Comedy were tragic, no one laughed or smiled or even found anything amusing except the contradiction between Wilson's topic and the effect. He complained whenever a student failed to return a book or pamphlet to him; future generations of students would suffer from lack of knowledge as a consequence.

Other lecturers included Arnold Kettle, who had written a two volume introduction to the history of the English novel, and who was high in the hierarchy of the Communist Party. When after a lecture I asked Kettle about some Marxist interpretation of James Joyce he was rude and continued to be over the years, warning leftist students, at the time of the controversy over Pasternak's *Doctor Zhivago*, that I was the other face of Allan Dulles as I had written a long essay discussing Pasternak for an issue of *Geste*, a student magazine Adele edited. Kettle was later sent by the British Council to be visiting professor and head of department at an East African university where the English department was in revolt against Western influences. The feeling was that an orthodox Marxist would know how to handle rambunctious lefties. On his return to England he became the first Professor of English at the Open University. I was then teaching in Nigeria and had thought of applying for the position but I kept imagining someone asking me about television and my having to admit that I never watched and knew nothing about it, the wrong background and attitude for someone at a university pioneering in television teaching.

Bonamy Dobree was proud that he had appointed people with diverse critical, political, and spiritual inclinations. Harold Fisch, who taught Milton, was a practicing Jew who would later move to Israel. I remember three lectures on Milton's prose and marriage. I especially enjoyed the lectures on Poetry by Geoffrey Hill (newly appointed) and John Heath-Stubbs (the second Gregory Poetry Fellow). They discussed the details of technique from varied perspectives. Tony Harrison, then a student, would sit in back of the lecture hall writing poems to preset rhyme schemes. He published one later. I especially liked how at Leeds there would be sometimes be "bursts", three radically different lectures on a writer or text, looking at a text, affiliation or historical context from varied angles.

At Columbia my teachers were part of the New York intelligentsia, a group to which I assumed I belonged despite my attraction to the New Critics and their conservative Agrarian and Catholic tendencies. I shared the New Yorkers anti-Stalinism although my liberal idealism at times made me unsuspicious and naïve about how deeply the Communist party had infiltrated the government. Moving to England resulted in a change in perspective, a gradual involvement into other cultural and political assumptions. I would pick up the snobbishness of the British establishment, the assumptions that contacts and friendships were the paths to careers and partly defined oneself, the gossiping about but tolerance of homosexuality, lesbians, and other supposed eccentricities. I remained an American, individualistic, nonconformist, always stunned by how easily the British would put group ahead of individual rights, their tolerance of Communist party members, the respect that they would give to figures of authority in church, politics, and the academic and literary world. The British could be very cultured, but they were seldom openly intellectuals.

At Leeds I blossomed as I adapted the British voice to have a writing persona. I began writing 10, 15, 30 page essays without the jargon that I had learned in the USA but with the insights I had picked up with the terminology. Indeed I was starting to sound like Trilling pontificating about taste, culture, morals, politics and sensibility. Wilson Knight was my tutor but I had come under the influence on Dobree who was a model for writing clearly and with commonsense with a strong foundation in scholarship. My models for prose, George Orwell, Hemingway, later V. S. Naipaul, gained clarity through conciseness, selection, short sentences; Dobree was part of an older tradition, closer to John Dryden in that his writing seemed based on the rhythms and postures of educated colloquial speech. His was the voice of experience but without cynicism or vulgarity. The attraction of Trilling's prose was its assumptions that his readers were highly literate who would agree with his generalizations about modern culture and its discontents. He was a "we" who articulated what the educated and well-read would feel if they engaged in "intellectualization". Dobree wrote as if he were speaking his own mind but which appealed to a consensus of those with experience and common sense. It was tinged with the authority of being part of an upper class that knew the world rather than being an intellectual, academic or literary critic.

His books *Restoration Comedy* and *Restoration Tragedy* were a revelation to me of appreciation and how to write about earlier literature. He ranged widely over authors and periods including Kipling, an admiration he shared with his friend T S Eliot; he even wrote about Joyce's *Finnegans Wake* in *Modern Prose Style*. *English Literature of the Early Eighteen Century*, was the most readable volume of the Oxford History of English Literature. When I was invited to write a volume in a new version of the Oxford History of English Literature I felt both that I represented a new era and how unworthy I and the other new authors were of replacing our elders. We were academics, Dobree had lived. He had been a member of the Bloomsbury Group, was friends with Leonard Woolf, D H Lawrence and Herbert Read. He had been in and out of the Army as an officer, had a lasting interest in Indians and India, and had been first appointed a Professor at Leeds after a short spell as Professor at the University of Cairo. His papers are now at the Harry Ransom Humanities Research Center at the University of Texas.

Dobree would at times appear behind me and whisper in confidence that Strachey or some other famous person in the past had a low voice or beard like mine. In class he would suddenly recall events from his earlier days and nights ("we all undressed but nothing happened"), had surprising respect for his students (he apologized for a shambling guest lecture by William Empson), drank too much brandy with the editor of the Yorkshire Post and would be found sleeping in the English Department's small house by the "char" who would light a fire and prepare a cup of tea. Dobree was said to annoy critics at the University Senate by announcing that he had already discussed the issue with Tom (T. S. Eliot) or, untopable, with the Queen Mother. As Dobree apparently knew everyone and everything I was surprised when he asked me where I read about the comic plot of Drdyen's *Marriage ala Mode* ironically reflecting the heroic plot. It seemed obvious, the sort of parallel found in Shakespeare's plays. I was more proud than annoyed when I found the insight in Dobree's later British Council small Dryden book. If Dobree were using my insights I was no longer a student learning how to read literature. Several years later when putting together a collection of essays, *Dryden's Mind and Art* (1969), I asked Dobree to contribute. Editing the book I suggested to him that he move his conclusion to the start of his essay. Dobree told another English Professor that "when Bruce King starts improving my essays it is time to retire." I always had confidence in my editing of others.

## II

I took advantage of that first year at Leeds to travel to Spain at Christmas, France at Easter and Germany during the summer. Don, the Columbia friend who was studying in Mexico City the previous summer was now in Madrid where along with eating 12 grapes at midnight on New Year's Eve I had twelve drinks and woke up head aching the next day and was later told that I had made anti-Semitic remarks about a man in our group who I had envied as having the attention of a woman I was hoping to attract. I gather that the remarks were not very offensive, but they still shock me as revealing a self-hate for my own Jewishness. I had previously flown with Don to Palma de Mallorca where I was proud of myself for eating for the first time Octopus cooked in its ink (wishing it were Turkey) for Christmas, signed the guest book which included Rita Hayworth and Prince Aly Khan) of a small newly opened restaurant, listened to bad flamenco singers (who were possibly prostitutes), and spent several nights in Valencia as there was then no connection between Palma and Madrid for flight reservations. One flew a segment at a time and hoped for a connection.

On my way back to England I stopped in Paris to see Adele who I knew had been touring Italy with other Americans. We would travel together during Easter holidays, become lovers, spend part of the summer together in Heidelberg and marry the following Christmas in St Louis.

I had travelled to England on the Queen Elizabeth. My mother mysteriously through family or other connections presented me with a ticket for passage in Second Class. My tablemates included a well-known Hollywood and Broadway actor and director being brought to London for play in which he had a part. He talked impressively of the Actor's Studio and a period of sleeping in his automobile while trying to break into films, but his street-wise manner and diction was perhaps too much Method. The British reviews thought his performance as an American was over-done. There was also at our table a quiet lovely Egyptian Jew who said that her summer with her aunt on Long Island was too tiring and she would be happy to return to Egypt. She was travelling on a passport issued by Spain on the basis that her family originated from there centuries ago. The actor-director predicted that she would eventually marry a Parisian Jewish lawyer. I soon began exploring the ship. I had met a seaman at a New York jazz club in which I was playing, but the lower part the Queen Elizabeth proved dispiriting, the crew kept in place by large allotments of beer and rum, like some pirate story from the past.

Peter Mayer, a fraternity brother who was going for a year at Christchurch, Oxford, was in third class where I found a bar filled with students going to Europe including Adele, a recent graduate of the University of Iowa who had a one year scholarship to study French in Paris. I do not know why I was immediately attracted but I was jealous of other men who had her attention. She remembers spilling her beer on her blouse leaning over to listen to my low voice. I arranged for her to have a meal with us in Second Class. We ran into each other briefly in London, kept in touch, and then during the Easter holiday hitchhiked from Paris to Marseille on our way to Corsica.

Hitchhiking sounds naïve and dangerous and I doubt that I would have considered it except for Adele who had a determined innocence and who was living on her scholarship. The actual dangers included French drunken driving (Adele kept putting her hand on the wheel to correct our direction) and the suggestion that we save money by sharing a room with three beds for the night. Instead we rented three rooms, although only two were used. When we reached Marseille a friend of Adele's remarked that she looked different and I was learning how to purchase condoms in France. Corsica was beaches, lovemaking, and a local insisting that King was a Chinese name so either I was not King or I was Chinese.

In Heidelberg I did everything possible to avoid learning much German. I now command such phrases as "heiße Würstchen mit Senf" or "mit Gurke" which I speak in a Southwest German accent and I have a vocabulary that includes "Schmetterling", "Nilpferd", and "Wassermelone". Stanley and Liz, who I knew from Columbia and Barnard, were married and I was best man, which meant writing my American passport number in an ancient book. This was the first time I met their parents and I was amused that Stanley's father insisted that no meal, even a wedding luncheon, could start without soup, so lunch began with hot soup. Stanley and Liz knew each other from Prague where their parents worked, then in London, where Stanley had studied first at the London School of Economics, and then in New York after he transferred to Columbia. He knew many languages and would amuse me by shifting among them if he thought someone was trying to listen to us in restaurants. He was now doing his military service and planned to follow his father into diplomatic service. President Eisenhower however had made the examination for the Foreign Service more egalitarian by changing the information demanded from worldwide to American. Knowing the rivers of Europe was replaced by the rivers of Texas. Needing an alternative career Stanley passed the examination for the CIA and changed his name from Novotny to Norden. He claimed the family name was no big issue, it had been changed to Novotny when his father had been posted to Prague by the British.

## III

A shipboard romance, becoming lovers on Easter day, love making in Paris, Corsica and Heidelberg seems like an old fashioned movie, and indeed I began informing the women in my life that I had made my choice and would be married. There was one problem, Adele was engaged to someone else and felt that she had to see him to explain what had happened and why. She already had accepted a scholarship to study for an MA in French at Washington University, St Louis. For reasons that now make no sense but did at the time I decided to go to the University of Minnesota for a year. My awareness of American geography was never good beyond the Philadelphia-New York area and I really did think that Minnesota was closer to St Louis than New York. Besides I had been told at Columbia that the University of Minnesota was a home of the New Critics.

Joel had a teaching assistantship at the university and we shared an apartment off campus. At Christmas Adele and I were married by a judge Tammany in St Louis. Our best man was supposed to be Bob Koestler who had a second hand record jazz shop, recorded local blues musicians and had a small theatre behind the shop for viewing silent films. He did not turn up at the marriage ceremony as he had a customer who purchased $50 of old 78 rpms, more than Bob would normally sell in a week. Koestler would move his Jazz Record Mart to Chicago where old records were a small part of his expanding business in his large shop. Delmark Records became an international brand, large enough that when I wrote to Koester from Nigeria to sing the praises of High Life music he offered to pay me to record some which I never did and later when I told him that Andre Tanker's band was making a rare trip from Port of Spain, Trinidad, to Boston, he told me that he was too busy to come but Tanker should mention his name to Rounder Records. Nothing came of it. Rounder wanted lively West Indian party music and Tanker wanted to record a suite he had composed. Andre can be stubborn and I doubt that any major label has recorded him and his great band.

Adele resigned after the semester and flew into Minneapolis during a snow storm. Joel could not get his car started and no taxis would travel to the airport where Adele was waiting, feeling abandoned. When she did arrive the bed in our room made so much noise the next couple of days that Joel moved out and we found an apartment near the university. We were a newly married couple learning about each other and experimenting. We tried to master traditional French cooking and reduced an immense amount of beef kidney fat in an attempt to make soufflé potatoes. Only a couple souffléd but our ceiling was dripping with fat. I noticed our neighbor was training a squirrel to follow a train of nuts into her apartment and decided to try it myself. The squirrel once inside panicked, ran all over, tearing up pillows, shitting, before finding its way out the door. The neighbor had observed and accused us of squirrelnapping.

I was disappointed with my classes at Minnesota and decided to return to Leeds. Although the first year students were expected to buy such older New Critical textbooks as the Cleanth Brooks and Robert Penn Warren *Understanding Poetry* and *Understanding Fiction* the graduate courses I had at Minnesota consisted of gossip about Ruskin's shock at finding his wife a woman after his wedding, eccentric discussions of the possible pronunciation of lines from Renaissance plays, and textual problems in Chaucer's *Canterbury Tales*—the final examination asked for ten books on medieval literature to recommend for a small college library. Allen Tate read from index cards occurrences of the same or similar words in Aristotle's *Poetics*, an exercise in methodology that I suspect was repeated every year.

The only course of interest and at professional level was the seminar in American Realism from Twain to Henry James. The professors who changed every quarter cared for their material and the classes included Brom Weber who had edited two books of Hart Crane's works. Philip Young, a visiting professor who had recently published a book about Hemingway, and Weber urged me to publish my essays on Twain (where I claimed that *Pudd'nhead Wilson* was the only place he faced up to black and female sexuality) and Sarah Orne Jewett (I thought sentimental the language of her local color) I just wanted to get away from Minnesota where the doctoral candidates I knew obsessed with mastering trivia about minor authors for their "Prelims", examinations which they had to pass before they could begin their dissertations.

When I left Leeds University at the end of the 1954–5 academic year I was invited to continue for my Master's Degree but I had returned to the USA to marry Adele. I had left the University of Minnesota some credits short of my MA and now had wait while Leeds decided whether I could proceed for a doctorate. The decision was that I should come for a year and ask for my research to be upgraded from MA to Doctorial level. While waiting we were in New York the summer of 1956. I wasted time at Columba University taking courses in German language. I listened to horror stories about the tyrannical arbitrariness of Columbia doctoral committees for English literature in which candidates were told that if they had not read a specific work by Freud there was no point in continuing. If they were lucky the failed candidate could start again at Wisconsin or Michigan University. Adele worked as a secretary for Desclee, a Belgian biblical publisher where she replaced a woman I knew a few years previously who was now leaving for Paris where she would eventually settle, marry, and become the first literary agent for American publishers and authors.

That summer was the last time I saw Peter Mayer who had returned from Oxford and was, perplexing to me, joining the United States Maritime Services. Perhaps it should not have surprised me as the Haddad brothers and others I knew at Columbia had worked on ships before finding their way, but Peter, who I associated with tennis and a winning social manner, seemed an unlikely person for the life of a sailor. He would afterwards have a brilliant career in publishing, rising to be CEO of Penguin Books, Head of Pocket Books, and be awarded many distinguished prizes and titles for supporting independent publishers. The few times our paths crossed we never met but I would hear about him from mutual friends such as David Davidar who headed Penguin India.

I also once more ran into John Fell and John Lincoln Collier who I had known in 1954 when they came to New York from Hamilton College. Fell played clarinet, John trombone, and they co-edited a literary magazine. They had survived by editing several pulp fiction publications and were now in charge of a "tit" magazine that appeared to feature black women.

It was low budget and six photographs of a woman smoking a "reefer" were published facing the opposite direction on the facing page to make a supposed sequence of twelve pictures. They were losing their self-confidence at the time and feeling that pulp paperbacks were their likely future. A prisoner in England had won a major literary competition by copying one of their pulp fiction novels, which rather than encouraging seemed proof of their fate. They eventually crossed the Sloughs of Despond. Fell became a university professor of cinema studies and author of such standard works as *Film and the Narrative Tradition.* The other John became the author of many influential, controversial jazz books including lives of Louis Armstrong and Duke Ellington and, sometimes with his brother as collaborator, prize winning juvenile fiction. *The Rise of Selfishness in America* is another of his many publications.

# Chapter 5: New England

## *Leeds, England 1957–60*

A major change was taking place when I returned to Leeds University. The small almost intimate English Department in its old fashioned row house rapidly became central to English studies in England and throughout the Commonwealth. It initiated or brought together developments that were widely imitated or a model for others. The English Literature and Language departments had been merged into a School of English that in a few months would be headed by A. Norman Jeffares who seven years previously had hardly turned 30 when appointed to the Jury Chair of English at Adelaide University, Australia. "Derry" soon introduced to the Leeds School of English programs in American Literature (then rarely taught in England), Anglo-Irish Literature, and Commonwealth Literature. The latter two were practically his inventions. Later he would add other areas such a Bibliographical Studies, Folklore, Stylistics, and Performance Theatre.

Dobree had moved the Leeds English Hons degree away from an historical approach to more direct engagement with literature through courses in genre and special authors that were spiced with relevant European and classic texts. He had also persuaded the wealthy businessman Eric Gregory to fund nine years of Fellowships in the Creative Arts that had barely started during my first year at Leeds. Jeffares brought the study of English even more directly into the present by including the then influential American, Irish, and rapidly developing Commonwealth literatures. Derry was aware that English studies had to change to keep up with the new cultural and political world that was coming into being with the Cold War and decolonization, although he would have spoken of such matters in practical terms. Not being, like Dobree, part of the social and culture elite, he was concerned with his students' careers and the usefulness of such courses as bibliography and folklore both in expanding his growing empire of English studies and in helping his students find jobs. He believed that the larger the department the more strength it had and the more difficult it was for other's to scrutinize its budget. Derry taught me to increase a department's proposed budget ten percent each year. Let others find reasons to reduce it. Derry did not like staff to become specialists who would teach nothing else and give the same lectures year after year. After three years you were expected to prepare another area of English studies. He could combine the outrageous with the obviously practical. Wanting a pied a terre in London he decided that the School of English needed a place of its own near the British Museum, which was then in Bloomsbury, for purposes of research. He and others would use it when consulting documents, original editions, and meeting publishers.

Before Derry many English departments were stuffed with those who had taken a first class degree at Oxford or Cambridge and felt that they had to do little beyond teach until the class of their undergraduate degree was mentioned in their *London Times* obituary. Derry had no patience with such snobbery. His appointments were expected to have doctorates, publish, and seek advancement, often by teaching abroad. Soon the new universities in England and abroad were filled with Leeds graduates, former Leeds lecturers, and Jeffares was advising the British Council and other behind the scene influences on whom they should appoint, who might make a good Professor, and shifting pieces of his expanding empire around the globe. At times he wrote recommendations for all the candidates for a position and no doubt a private word or two with a Vice-Chancellor or a British Council official influenced the choice of who was appointed.

He had a sense of humor, was amused by outrageous behavior if it made for a good story, and was loyal to those he supported. When I resigned at the start of the academic year at Calgary he recommended me to Molly Mahood at the University College, Ibadan, who already had Martin Banham, a Leeds graduate, teaching drama in her English Department. Derry hoped Molly would join him at Leeds. When I wanted to return to England he recommended me to Frank Kermode who was replacing L. C. Knights at Bristol. Frank had been appointed

Professor at Manchester the same year and at the same age as Derry had at Leeds. Although they were different in their intellectual and cultural tastes, they began with a mutual interest in Yeats, remained in touch, and trusted each other's judgment. When I found it impossible to live on my salary as a lecturer at Bristol it was Derry who suggested to the British Council that as an American I might be appointed to the Professorship of English at the University of Lagos where the previous Professor had departed over a politically sensitive issue.

A few years later I would while staying with Richard Coe, a former lecturer in French at Leeds who had moved to Australia and returned to a Chair at the University of Warwick, be stunned by a late night knock on the cottage door and be presented with an immense doctoral dissertation surveying Commonwealth Literature that needed to be read rapidly and examined the next week at Leeds as Alastair Niven, who had taught at the University of Ghana and was then lecturing at Leeds, whom I had not previously heard of, had been appointed to a lectureship at Stirling University Scotland where Derry was advising the Vice Chancellor on appointments. Niven had written his dissertation under Derry's supervision and I made a point of questioning him for several hours before passing him. It represented an immense amount of work but the approach was different than my own.

It was assumed that Derry was seeking a Vice Chancellorship or some similar high level administrative appointment but he shocked everyone by writing himself a research professorship at Stirling, leaving himself free to write, edit and travel, while his friend Tommy Dunne (who had been a professor in Ghana and Canada) ran the department. At a time when I was finishing a year as visiting professor in Paris and needed something for the next year, Derry arranged for me to come to Stirling as a Visiting Professor to replace Alastair who was leaving to become Director General of the Africa Center in London. Alastair would become Director of Literature at the Arts Council, then the British Council, and greatly improve England's relations with the cultural life of the Commonwealth and with the emerging "minority" communities within England during a time when Margaret Thatcher's government was being accused of little Englandism and racism. Somehow he always found ways to support black, Asian, feminist, and even leftist groups and publications. I see he is now listed in *Derbett's* But I am getting far ahead of my story.

Derry Jeffares began as a Classicist at Trinity College, Dublin; after a short time he realized that English studies were a better road ahead. Descended from the protestant gentry that was falling on hard times after Ireland's independence Derry, having lived through the Great Depression and war years, as well as Ireland's own internal wars, knew he had to earn money to support himself and some of the exotic members of his family. Exile to England or elsewhere was necessary and he would expect those he appointed to have a history of teaching abroad or be willing to teach abroad. He was also always a charmer, persistent, and often got what he wanted. He had gone to the same school as W. B. Yeats and became talked about when as student he persuaded Yeats to send him a poem, "What Then, Sang Plato's Ghost ", for the school magazine that he edited. After Yeats' death he persuaded the widow into allowing him access to his papers, the basis for his first doctorate at Trinity College, Dublin, which was followed, as soon as the war allowed him to move to England, by a second doctorate at Oriel College, Oxford. The two were combined in his *W. B. Yeats*: *Man and Poetry* (*1948*), that along with Richard Ellmann's *Yeats*: *The Man and the Masks* laid the foundation of Yeats studies.

He was not an intellectual, had no deep interests in culture beyond such traditional branches of the fine arts as drawing, landscape, and potting, and he was politically conservative. He was not at ease with the various cultural and literary battles that were then and continued to be the concern of literary critics. Whether preparing a definitive scholarly edition or a study guide for students Derry concentrated on biography, annotation, commentary, and textual problems; his books were for use not to persuade. Others wrote books of literary criticism or edited rare manuscripts; Derry' would publish *Commentary on the Collected Poetry of W B Yeats* (1968) and he would keep his books up to date with new editions such as the *New Commentary on the Collected Poems of Yeats* (1984) which were more useful towards understanding the works than books by literary critics and theorists.

He spent a year teaching at a provincial university in newly liberated Nederland, and was appointed for two years at the University of Edinburgh where he was told that only he or another lecturer would gain tenure. Instead he applied for the Jury Chair and landed in the middle of what would be a long running battle between two camps in the English departments of Australia, the nationalists and those more attuned to the New Critics and F. R. Leavis and *Scrutiny*. As he would do at Leeds, Jeffares avoided taking sides and built his friendships from varied sources, many outside the universities. He bought a second hand Rolls Royce and founded the Vintage Car Club of Southern Australia. At Leeds he rapidly created an active social life by giving dinners, holding parties, and introducing to each other people who otherwise were unlikely to have met. His guests became part of his circle of friends which ranged from talented students in English, school teachers, poets, lecturers and Professors in other departments, artists, business men, and the famous.

We were invited to a meal with Mary McCarthy during which I drank too much and went to sleep in another room. Rather than expelling me from the inner circle it appears to have given me a place. Derry being Irish liked people to eat, drink, and enjoy themselves. He found suspect those too thin, cautious, tight-lipped. When Allen Tate kept trying to get more to drink while staying at the University Guest House it was thought an amusing story. Ever since the wife of an artist pulled me down on top of her at a Jeffares' party I fear drunken women asking me to dance, but Derry (wrongly) told friends that I easily attracted women, which led to an embarrassing party in Paris when a woman who heard the story felt attracted.

Even more rapidly than creating new sections of the ever expanding Leeds School of English Derry launched series of books into a, by comparison, previously sedated British academic and educational establishment. His initial method was to find someone as a point of entry with whom he would share editorship honors and royalties. Robin Lorimer, son of a famous scholar, had recently joined Oliver and Boyd in Scotland. He co-edited the "Mind and Art" series, then the famous "Writers and Critics". After Lorimer left Oliver & Boyd Derry moved on to Macmillan. Macmillan was Yeats' publisher and had not realized the potential of their rights until Derry edited W B *Yeats: Selected Poetry* (1962), after which Derry became an advisor for much that they published in English literature. He would edit the multi-volume Macmillan History of English Literature. Suheil Bushrui, who had written a dissertation on Yeats under the supervision of Frank Prince, became part of Derry's ever widening circle of influential contacts, providing holidays in the Lebanon, and the initial financial backers for the York Notes, later taken on by Longman. Suheil was listed as co-editor and for years received his 5% on each copy.

I mention these series as like others who Jeffares befriended I also wrote and edited books for his series and later, when through his recommendations I became head of English departments, appointed him as visiting external examiner. Although the benefits were mutual and not pressured, they were largely one-sided, I gained much more than I returned. Derry took chances on those offered opportunities. *Dryden's Mind & Art* (1969) consisting of five republished and five newly commissioned essays paid unbelievably generously by recent standards. Each new essayist received 100 pounds, then $500. I cannot conceive the present equivalent. Besides Dobree I asked the American scholar William Frost, and the poets John Heath-Stubbs and Tony Harrison, and contributed an essay myself. Tony had been while an undergraduate at Leeds editor of the now famous student weekly *Poetry & Audience*. He like Derry was originally a classicist, and was writing a doctoral dissertations on translations of Virgil, which he purposefully never finished as he knew that would lead to a career as an academic instead of as a poet, dramatist and translator. We would both teach at universities in Nigeria.

Leeds, which I was advised against while at Columbia, had become a center for the arts as well as a powerful base for English Studies. The Gregory Fellows included the painters Allan Davie and Terry Frost and the sculptor Hubert "Nibs" Dalwood. Davie normally lived in St Ives, played jazz saxophone, painted large gestures that suggested unconscious images and had the mind of a painter; he bragged that he would nip into books wherever he felt and might read snippets from the back before moving towards the front. Nibs, by contrast, was disciplined and articulate about the history of modernism that had influenced his work. I was not surprised when he moved to an American university. Norbert Lynton joined the Leeds School of Art in 1955. His 1949 marriage to Jan was already being tested by exchanges that I wanted to avoid. Jan asked a visiting American psychologist who came to see us whether her children would be disturbed by seeing a different man instead of their father in the morning.

Among Jeffares' close friends were the social anthropologist Fernando Henriques and his attractive wife Rosamond. Henriques, of mixed Jamaican parentage, one of the light skinned Jamaican Catholics anciently of Spanish-Portuguese Jewish origin, had been brought up in England and published several books about the sociology of Jamaica before his *Love in Action: The Sociology of* Sex became a paperback best seller and earned him his Professorship at the new University of Sussex in 1964. Readers did not know that the book was the first part of a serious study of sexuality through history until the present. Many decades later while working on a book about "black and Asian" British writing, *The Internationalization of English Literature*, I interviewed his son, the film director and author Julian Henriques, who turned the meeting around by asking me about his parents' life at Leeds. Fernando had felt much color prejudice in England during the war including being told that as a non-white he could not enlist as an officer; Leeds had been a period of romance according to family mythology.

Another couple who were close friends of the Jeffares were Marietta and J. Peter Nettl. Nettl who was born in Vienna came to England in 1936 and was on his way towards a brilliant career as an historian at St Johns, Oxford, until his father insisted that he come to Leeds to run the family business. He had married Marietta, a dancer in the Hungarian opera's ballet, and it was not exactly a calm relationship. Marietta was thought beautiful, but not as intellectual as the society in which she moved, and still had to report to the local police as an alien. Peter continued his research and writing while at Leeds and suddenly in the mid-1960s resumed his academic career. His two volume *Rosa Luxemberg* (1966) remains a classic and is still in print. On their way to Peter taking up a Professorship at the University of Pennsylvania, the couple quarreled in the private airplane they had chartered and one moved to the back of the airplane. Marietta survived the crash and remained in the USA. There were rumors, with no known foundation, that Peter was involved with Mossad.

There were at first three of us working on our doctorates, Michael Millgate, Saros Cowasjee, and myself. A few years later the School of English would have so many post-graduate students, mostly from the Commonwealth, that it would difficult to keep track of them or which to take seriously. At first, however, the three of us saw each other frequently. Michael Millgate, an Oxford graduate, was five years older than myself, had come to Leeds to do his doctorate after teaching at the University of Michigan, drove a car, and became Derry's right hand man, chaperoning distinguished guests, such as C. P Snow. He was appointed as a lecturer in 1958. "Writers and Critics" started in 1960 with Frank Kermode's *Wallace Stevens*, G. S. Fraser's *Ezra Pound*, J. M. Cohen's *Robert Graves*. Millgate's *William Faulkner* joined Richard Coe's *Ionesco*, Ronald Gray's *Brecht* and Stewart Sanderson's *Hemingway* in 1961. In 1960 Douglas Grant, an eighteenth-century scholar, was made the first Chair of American Literature at Leeds and England. When invited to a Professorship and Head of Department at York University, Toronto, he recommended Michael who after four years moved across to the University of Toronto during 1967 where he became famous, after other Faulkner books, as a scholar of Thomas Hardy. Jane Barr, who was a student at Leeds when they married in

1960, also moved to Toronto University where she became famous for her books on nineteenth-century English literature.

Leeds during late fifties was a nest of singing poets. Tony Harrison, James Simmons, and Wole Soyinka were students, Geoffrey Hill a lecturer, and there were the Gregory Fellows of Poetry John Heath-Stubbs, Thomas Blackburn and then Jon Silkin. There were always parties and dinners. Geoffrey Hill, was the son of a policeman, already a brilliant but difficult poet, puritanical about literature and many other aspects of life. He told me that I enjoyed literature too much to be a literary critic.

Jon settled in Leeds until 1965 when he moved to Newcastle; his house was a center for literary activity and where we often spent time playing bridge and trying to convince him to enjoy food for itself rather than it being "filling". Saros Cowasjee once proposed to make us an Indian meal for lunch. By the time we found the various ingredients he claimed to need and watched him watching tomatoes, onions and garlic and various spices change in the pot to the correct color it was midnight. Saros confessed that he had seldom cooked Indian food. Servants did it in India. Arriving late at the flat we were renting meant being locked out by a landlady who—I am not making this up—insisted that the past was much more genteel, even those delivering coal wore white gloves.

Although Jon pretended to be working class he had attended good schools before doing his military service after which he did indeed for a few years work as a laborer including, he claimed, digging graves. His family were Jewish immigrants who had within three generations became trusted members of the government, and included Baron Silkin of Dulwich (created 1950), and needed an Act of parliament to be the first family to refuse inheriting a peerage. Although Jon rejected his family he often alluded to his relatives with disdain, wanting you to know his connection, and his voice betrayed his upper class schooling. He had rented an apartment in Hampstead from two writers in which he sublet rooms to two other young writers, one of whom, David Mercer, regularly appeared in *Stand*, a literary magazine that

Jon started in 1952 and sold in pubs. I already knew Jon's poetry before he came to Leeds. When it first appeared I had given Adele a copy of his first book of poetry, *The Peaceable Kingdom* (1954).

Jon came to Leeds with Ann Fraser, a nurse with whom he lived for several years until she moved out. Their relationship had a touch of sadomasochism which informs much of his poetry from this period in which the mutual attractions of pain are part of being human. Ann would always love Jon and after she left he would turn up in London saying she should not forgo preparing meat in her cooking which she rightly understood as Jon, who was now married to and a proudly announced vegetarian, wanted to eat some beef or chicken.

Jon wanted to restart *Stand* and it appeared that there would be money from Yorkshire businessmen to relaunch the magazine with an issue or two, but Jon kept putting off going cap in hand for funds. We would need telephone to say he was suddenly ill until one night we called a taxi and pushed him out. *Stand* began again. Jon always had an eye for talent and soon issues carried poetry by many of the better British poets, plays by David Mercer, poetry reviews by pre-theorist Terry Eagleton, as well as writings by Emanuel Litvinoff and translations of Elias Canetti. Although he was early influenced by T S Eliot's verse and still was obsessively enthusiastic about Ezra Pound, Jon was conscious how Jewish authors had been marginalized in England. His publications on the World War One poets began with his writings about Isaac Rosenberg. Canetti was then alive and ignored in England. Jon gave me a volume of Wallace Stevens to review for *Stand* 4.2 (Summer 1960), one of my earliest non-jazz publications, and about a year later I wrote a newsletter about the New York City Ballet, but after that I settled for a free copy of *Stand* until Jon died and the magazine was taken over by others.

During her first year at Leeds Adele took undergraduate courses in the French Department with Richard Coe, Bill Ireland, J C Ireson and others. Coe, Ireland, and Ireson were part of the wave of productive intellectuals filling British universities and who would soon, after a book or two, become professors at one of the new universities or abroad. Coe was especially brilliant as he translated from Russian and German as well as French, seemed tuned into the variety of left politics on the continent and had a Russian wife. We sometimes spent evenings with him and his wife gossiping literature and politics and listening to recordings. He would tell me to listen to the folk melodies in the work of a Russian composer, I would polemicize about the improvisations of Miles Davis and John Coltrane.

When students in the French department decided to start a magazine there was disagreement what direction it should take. After an issue of "What I did last summer in France", the editorship of *Geste* was turned over to Adele and it became one of the several lively student publications characteristic of Leeds since the weekly one penny *Poetry & Audience* was started by a visiting Canadian, Ralph Maude, in 1953. Following the interests of those involved *Geste* rapidly became an international literary magazine, usually peddled alongside P&A at the Student's Union. Francophonic Africa and Boris Pasternak were among the issues devoted to a single topic. At the time there was much discussion in England about the publication and banning of *Lolita*. Copies of the special Nabokov issue of *Geste* were sent to every Member of Parliament (few replied) and many opinion makers. We were told we could be imprisoned, but later the issue and an article I wrote for it, appeared in the bibliography of books tracing Nabokov's reputation, and especially that of *Lolita*, in England.

I knew that Adele was salutatorian in her graduating year at the University of Iowa and no doubt as bright as most of the people we knew but I kept hearing of her exceptionally high grades in her courses. G T Clapton who headed the French department remarked skeptically that he had never seen grades like them and assigned George Hainsworth, one of the more conservative members of the department as her MA supervisor. They might have worked well together as they both were interested in the style and techniques of prose fiction, except that Hainsworth felt French literature had ended with the nineteenth century realists. He was also said to have a bad, at times violent, relationship with his wife. When Adele decided to write her thesis on Albert Camus, Hainsworth looked him up, discovered two articles had already written about him and decided that was enough. The only thing left would be a criticism of why he was not Balzac. Clapton was reluctant to intervene as he believed educating women was a waste of time; they would marry, have children, and were only at university to catch husbands. Three years later Adele submitted a larger than usual MA thesis. Clapton, who replaced Hainsworth at the examination, repeatedly tried to embarrass her by asking why she had written "sex" when she meant "fornication". John Cruickshank, the external examiner and recently author of a book on Camus, more embarrassingly asked why the thesis was not being awarded a doctorate. As it had already been submitted for an MA the only possibility would be to withdraw it, pay an additional year of fees, and resubmit as a doctoral dissertation, which Adele declined. She would later, among her other publications, write two books and edit another on Camus. Five years after we had left Leeds, 1965, Philip Thody, who had written the first book on Camus in 1957, was appointed Professor and Head of the French Department.

I used to enjoy listening to Jeanne Jeffares and Rosamond Henriques gossip, especially about the other women they knew. X took hours to make up, Y never had a "mature orgasm". I admired Jeanne. Born in Belgium (her father was a secret British agent), she had worked in England during the war, and married Derry when they were both young. She was witty, observant, had a talent for drawing, was a good cook, and had decided her role was to support Derry's career. Neither accepted that times had changed from when they started out and that a university lecturer might stay home to care for his ill wife, might take the children to school, or might suddenly leave his wife and England for a student. Men could have affairs but how could you imagine that a new wife would be an improvement on the first? Derry told me that I was my own worst enemy as I would no sooner be offered an interview than I would be asking for a job for Adele. He tried to persuade that if you were lucky enough to have a job, then as you got to know people a few years later something might turn up for your wife. He knew from comments by Richard Coe and others that Adele was exceptional and asked her to write the *Camus* volume in the Writers and Critics series and after that sold extremely well and was translated into several languages the *Proust* volume. I was surprised one day when Jeanne told me that Adele was not only beautiful but was one of the few women who would become more beautiful as she aged. As Jeanne also believed in homeopathy and some other oddities I was skeptical but this turned out to be true

Music remained a major part of my life and that brought me into contact with Leeds' working class and art school culture. For a time I played drums with Eddie O'Donnell's White Eagle Jazz Band, another off-shoot of the New Orleans trad revival. Eddie had been trombonist and remained in contact with Ken Colyer, whose band I sometimes sat in with when I was in London and visited Ken's club. I admired Ken; he was one of the first British musicians to jump ship in New Orleans to play with the rediscovered veterans, and he brought the style back to England where he, Chris Barber and other then famous musicians would start the Crane River Band before going their individual ways. Ken remained the "purist" and would become a legend immortalized for a time by the Ken Colyer Trust, which, until it broke up, had its own imitation Ken Colyer revivalist band and would reissue his recordings and those of others of the period. The Trust's name has been kept going by a British tourist agency offering annual group trips to New Orleans.

Eddie O'Donnell was thin, ironic, spoke with a local accent, led a group of "ravers" who closed pubs and pissed in red telephone boxes afterwards. We sometimes played in pubs in Leeds, and sometimes would travel on snowy Yorkshire nights by train to play in towns with names I can no longer recall. Eddie and his friends had no sense of bourgeois ethics; hired to play for a party they were likely to return with something that "the geezer would not miss". His normal job was as a nude model for Leeds College of Art and he would tell in a toneless voice of female art students asking him to spread his legs to have a clearer view of his anatomy.

The jazz and university scenes blended together in Leeds at that time, or at least they did for me. I remember Wole Soyinka at a pub wanting to sing a Louis Armstrong number with our band. I had no idea that black American movie, radio and musical stars were regarded in Africa as a sign of the unlimited opportunities for blacks in America. To me, with the liberal prejudices I had picked up from the Beboppers, Armstrong's mugging to the audience was Uncle Tom entertaining whites.

We had a large party at Jon Silkin's house to which Jeanne Jeffares brought a chicken liver pate labeled "chaque pate est une geste". I was drumming with Eddie's band when Martin Bolland, a Johnny Dodds influenced clarinetist who was a friend, turned up with some of his football team; soon shouting and a pub-style drunken brawl was in the making. Jeanne smartly brought it to an end by asking the loudest brawler to dance with her. A lesson in female diplomacy.

Leeds was changing from a working class refuge from Oxbridge to a place for those who did not make Oxbridge or some of the better new universities that were starting, such as Sussex; Canterbury, and York. When I first went arrived many of the students I knew had or would be accepted to an Oxford or Cambridge college but came to Leeds for reasons which ranged from having had psychological problems at Cambridge to their family not being able or willing to afford the additional year that they had to wait until their admission to Oxford. Accents were strong as was support for the left of the Labour Party. I would be introduced to older brothers who I was told were officially "bastards", conceived before marriage according to Church registrations at that time. Many of the teachers were consciously working class, such as Frederick May, lecturer, then Professor of Italian, who through translations and student productions began the appreciation of Pirandello in England. Rationing was ending but signs of the war remained, especially in the bombed out sections of London which had great West Indian musicians, such the saxophonist Joe Harriott and trumpeter Dizzy Reece, both from Jamaica, playing modern jazz in racially mixed black dance halls.

By the time I left Leeds the economy was booming, the standard of living rising, and many students were imitating the class system of the south of England. I remember playing drums with a supposed jazz quartet for a sweet something party in which the young men and women were dressed as for a debutante's ball. The industrialist father looked on the dancing, wine and champagne drinking youths with pride and boredom until he saw the cymbals of my drum kit and wanted to know how they were made and the metallic composition. Where there's muck there is money had brought wealth to Yorkshire, financed the university expansion, and was now being displayed at this party.

## II

During our time at Leeds there were many trips to the continent during university holiday periods or because I decided to finish writing my dissertation in Italy or because we felt that this might be our last of Europe. We were often in Paris which will explain why we were at the famous reading by the Beats at George Whitman's Le Mistral Bookshop where Gregory Corso urged Allen Ginsberg and others to undress and have an orgy. After urging the other poets to "put their cock on the table"—his party trick—he found Paris more chilly than California and sheepishly put on his clothes. Interviewed about the reading several days later by an amused but sympathetic Art Buchwald, Ginsberg claimed that he had two big burly bearded friends in the audience who would protect him. This became part of literary history but is false. I was one of the only two beards in the audience and being newly married I was more concerned with protecting my wife than the Beats whom I regarded as the American vulgarity I had left behind. The Beats were wined and dined by the American literary establishment in Paris as long as they entertained with their pretense of being revolutionaries. The Mistral reading however was my "madelaine". Having recalled it for a friend I began writing this autobiography.

Later in Mumbai, while working on my first book about modern Indian poetry, I would be told of an almost identical scene when the Beats arrived. This time it was at a poet's apartment rather than a bookshop. Once more Ginsberg and company had set up a local to read his poetry; then one of them shouted that such literature was dead, and they would seemingly spontaneously explode into reading of their own writings as an illustration of what poetry should be like now.

I was more interested in other more spontaneous shows as when I was having coffee at a small bar across the Seine from Notre Dame and there was a pretty but thin American speaking badly accented French to a good looking guy in what could have been pimp's clothing and some other badly dressed man was somehow involved. I was watching, without knowing it, Jean Seberg, Jean-Paul Belmondo, and Godard discussing a scene for *A bout de soufflé* (*Breathless*).

It was difficult then, as now, to spend much time in Paris without stumbling across some aspect of a movie being made. In the small inexpensive hotel near Le Mistral across from Notre Dame where we were staying there was one shower and it was not working. Instead the proprietor whispered to follow her and I was taking a tub bath in a room newly decorated in pink with pink towels, curtains, soap, even a pink bidet. I was offered no explanation but the situation became clear a few nights later when we found a large crowd in front of the hotel and had to convince the police we were staying there. Henri-George Clouzot was directing *La Verite* with Brigit Bardot whose bathroom I had used. It seemed impossible to avoid her. We were dancing at a hotel restaurant in Seville when suddenly all the dancers moved over as if a ship had tilted. Bardot had entered with a partner.

A close friend in Paris during this period was the New Orleans clarinetist Albert Nicholas who I used to hear in New York and Philadelphia before he left permanently for Europe. "Nick" had played with many of the greats including Louis Armstrong. He felt himself someone who had evolved during the swing period to work with such sophisticated musicians as John Kirby and he resented being ghettoized with the trad jazz revival. He claimed that George Lewis and Bunk

Johnson, were musical illiterates left behind while he and their betters moved on. Although much admired and usually employed in Europe he was lonely. This was a great period for American Bebop musicians in Paris; Kenny Clarke and Bud Powell were among those with regular club jobs. The white French progressive jazz musicians regarded Nick as a traditionalist and while he had many friends he was part of no society except the jazz clubs and down home restaurants where he saw other black Americans. Being around white Americans could be dangerous. One night leaving Club St Germain a white American GI made a racist remark about him and only Adele's intervention prevented a knife fight. I admired Nick's playing and his courage and wrote about him first in Hadlock's *The Record Changer* (1957) and at more length in the British *Jazz Monthly* (1961). In the large room he rented in a left-bank hotel he would cook on a burner New Orleans red beans and rice and we listened to his record collection. Adele thought she might try to use his recipe for a large student union event at Leeds, but the quantities for 200–250 servings were too imposing.

After he was invited to the first Sopot jazz festival Nick asked me to write about jazz in Poland which meant I had to contact many of the leading musicians and intellectuals. For some months during 1958 I regularly contributed "Jazz W Anglii" to *Jazz* (Poland), but stopped after an issue in which my column was accompanied by a photograph of a band playing for a Communist Party rally. I had no idea of what my columns said in translation. This began a period of my life when I was regularly in contact with such writers as Leopold Tyrmand, received translations of the work of the great Bruno Schultz, then unknown in the West, and would receive visitors in Leeds who felt that their careers were blocked as the Russians regarding them as counter-revolutionaries would not give them work visas and the West still saw them as Communists. My "Jazz in Poland" appeared in the March 1958 *Jazz Monthly* and was the first major article on the topic. I was invited to be a judge at the next Sopot Jazz Festival. Unfortunately the organizers could not afford my transportation and suggested that I smuggle a tuba into Poland to pay for my expenses. I also kept hearing stories from my Polish friends and from Richard Coe about Arnold Kettle denouncing people to the Russians.

Somehow I managed to finish my doctoral dissertation, mostly because I left Leeds and Paris for months and lived with Adele and a portable Olivetti typewriter in pensions in Florence, Positano and Viareggio. Richard Coe recommended a pension in the heart of Florence which was fine—we stayed there several times—except for the burnt flour soup that it often served. In Florence we went to a Glen Gould concert and argued with a young American historian, and his attractive wife, about playing Bach on a piano. They belonged to the original instrument school and wanted a harpsichord. The main Italian Dryden scholar already had two books to her credit but would only speak to me through Adele as if she were afraid of contradicting a man. In Positano I wrote almost non-stop from volumes of Dryden's plays forwarded to me from England by my supervisor Douglas Jefferson. New Year's Eve many of the then famous hippie names and models descended on Positano for a drunken and drugged party at a club that reopened for the night on the beach. At midnight people jumped into the by now freezing water. The next day they disappeared and it was time to move on once more, this time to Viareggio for a month or so before returning to London where a disagreeable pimply white faced young customs inspector kept mocking my claim that I was entitled to entry so I could finish my degree. He seemed to think we were either smugglers or intended to stay on illegally and work. He was so busy tearing our luggage apart looking for contraband that he ignored Adele's Ferragamo shoes still packed in their original tissue paper. British custom inspectors have a reputation for such oddness. A friend had naively brought back from the continent many bottles of spirits and scents and when she told the inspector that she had them he thought she was joking and waived her through.

Douglas Jefferson seemed surprised when I presented him with typed bound copies of my dissertation. Douglas had not so far finished a book and the notion of finishing anything seemed to bother him, but Derry laughed as he had himself bypassed his supervisor's caution to finish his doctorate. My external examiner was the poet F T Prince, Professor at the University of Southampton, who could only schedule the examination mid-summer as he had planned to return to South Africa for some months. I went to the USA seeking a job for the next academic year, saw Wole in Paris where he hoped to record when he should have been finishing his MA at Leeds, travelled to Greece where we ran into the young historian now lacking his attractive wife who decided to stay on in Italy, and then returned to England. I travelled with Douglas to Southampton and said little while he and Prince held differing views about what I should do to turn my dissertation into a publishable book. When I returned to London Adele calmly said I knew you would pass and I blew up with months of repressed tension.

I would not see Prince again until the early 1980s when I wanted to leave my Professorship in New Zealand and Brandeis University needed a Professor who could offer a course on Milton to replace Frank whose visa did not allow him to stay longer. I did not get the job. After avoiding the obvious dangers of lecturing on African literature to black American students and discussing the terms of my probable appointment I lost my cool when asked by a student about free will after my Milton lecture. As soon as I said "That is a foolish question" I could see my appointment ending before it had started. I had a bad habit of impressively repressing tension that would unimpressively explode at the wrong time.

# Chapter 6: The Golden Years

*New York Again 1960–61*
*Calgary, Canada 1961–62*
*Ibadan, Nigeria 1962–65*

I was 27 years old when I was awarded my doctorate and I would be 34 when I was first appointed a full professor. It sounds like a success story but the first two years, teaching at universities in New York and Calgary, were a disaster which I would like to forget, except that I cannot skip them as too much happened in New York during 1960–61. To be honest my memory fails to distinguish between that year and other periods in New York around then.

The jazz scene was exciting and the clubs had Miles Davis with John Coltrane and continually changing great drummers, many with names ending in Jones—Philly Joe Jones, Elvin Jones. Charles Mingus with Eric Dolphy was offending his audiences. The Five Spot had Theolonius Monk dancing to his band when not playing piano. Then there was the sensational Ornette Coleman, Don Cherry, Charlie Haden, Eddie Blackwell second engagement at the Five Spot and the beginnings of "free jazz" with its seeming lack of a pre-set harmonic structure, unpredictable changes of direction, motivic based improvisation, strange instruments, and new liberties for the drummer. Free Form divided critical and audience opinion. Many of the Boppers used to more and more advanced harmonic structures said it was not music; Miles Davis pointedly insulted Coleman during an interview and there are rumours of Max Roach trying to punch him.

Many traditional jazz fans like myself welcomed the new music as polyphonic, similar we felt to the old New Orleans jazz before Armstrong and others had made soloing on chord changes characteristic of jazz. Even more curiously the owners of the Five Spot had been persuaded to hire Ornette Coleman in 1959 by Martin Williams, editor of *The Jazz Review*, a publication for which I wrote, and who lived near me in Brooklyn Heights while I was teaching for a year at Brooklyn College.

Martin had been working towards a doctorate in English literature at Columbia University when the opportunity presented itself to become a full time jazz music critic. Whitney Bailliett the *New Yorker* jazz critic had recommended Martin to replace him at some record reviewing publication. With his training in literary criticism Martin brought close detailed analysis, evaluation, and an historical perspective to jazz criticism; there was a new seriousness that had not existed previously. Martin also played clarinet and knew more about chords and harmonies than I, a drummer, did. He would, teacher-like, question you about what you were trying to say and how you expressed it. Although Martin gave attention to the compositional and chordal structure of songs he had been influenced, as many of us were, by Andre Hodeir's *Jazz: its evolution and essence* that explains the music's history in terms of harmonic, thematic and rhythmic variation. Jazz from the earliest known examples until the 1970s could be analyzed by such categories with increased harmonic complexity the motor of change. After about 1970 the story became more complex. Martin was a literary critic not a trained musician and would later be criticized by some professional musicians as a well-meaning amateur, but *The Jazz Review* was a great magazine and Martin's many books about jazz were excellent reading and showed people how to listen to the music, distinguish between the styles and understand why they changed; he showed readers the varied contexts of jazz. His essays and books were always worth reading, but he was transitional; his detailed explanations of music opened the door to the professional analysis of music and the university jazz courses and jazz institutes that followed. One of the contributors to the *Jazz Review* was the professional musician, composer and educator Gunther Schuller whose *Early Jazz* and book on the Swing period made the kind of jazz appreciation and historical speculation I wrote outmoded.

I learned from Martin to appreciate American popular culture. I had assumed, along with many intellectuals, that, outside jazz and other black informed art forms, culture was European. Martin demolished such clichés as a product of leftist anti-Americanism. The American popular culture looked down upon as "commercial" had resulted in good radio and television shows, popular songs, big bands, and other art forms and he tried to find the classics among such mass entertainment. He wanted to understand the production of culture. What happens in a recording studio? Martin would move on to the Smithsonian where for a decade he would organize archives and arrange for concerts and become editor of several selections of jazz music. Afterwards he was in charge of their publications on American culture. The foundations of what became the post-modernist rejection of elitism in favor of mass culture was prepared by his collecting, sponsoring and evaluation of what Americans had actually been creating for the past century. Unlike the theorists and their obscurity he wrote clearly and he brought the same standards to popular as to elite culture.

Brooklyn Heights was becoming an extension of the Village and among my neighbors were the record collectors putting together the Origin Jazz Library, the first anthologies on CDs of previously undocumented early blues musicians like Charlie Patton whose 1929–32 78 rpm recordings were their first release. Although only produced in small pressings, supposedly of 500 which took several years to sell, these had an immense influence as they became better known to blues fans. Until then the great blues singers were the vaudeville tradition represented by Bessie Smith and the modern electric Chicago blues musicians, The Origin reissues were of older country style musicians, jug bands, and singers who had once been popular but fallen outside jazz history and were when rediscovered regarded as folk musicians. I did not know what to make of this new old music but I continued to purchase the Origin CDs after I left New York.

This was a time of "loft" jam sessions. On returning to New York I had telephoned Jack Fine, was told by his wife that I spoke like an English "poofta", and had an embarrassingly bad meal with them while Jack kept trying to pretend he still liked his first wife and child.

He invited me to a jam session at Roswell Rudd's loft near Wall Street. Everyone was there and I saw Dick Hadlock, Bob Thompson and others from the past. Rudd was playing trombone in a strong but odd style with a black alto saxophonist who was probably Archie Shepp or possibly John Tchicai both of whom he would soon record some of the first free jazz records after Ornette Coleman. Reviewing recordings of the New York Art Quartet a few years later in *Jazz Monthly* I claimed he sounded like a cross between Kid Ory and a Stravinsky trombone part. He had in fact started as a Dixieland musician and like his friend Steve Lacy had moved on to the post-bop jazz avant-garde by way of Theolonius Monk's music. At a time when every modern jazz trombonist seemed to imitate J J Johnson's smooth version of bop, Rudd sounded like a circus. Then he disappeared from the jazz scene for decades, playing as a sideman in commercial bands to support his sick wife, until I heard him in Paris playing once more with Steve Lacy. He knocked out everyone including the other musicians with his energy, crude loud sound, and continual witty unexpected quotations from folk and popular music. It was as if the American composer Charles Ives had been resurrected as a jazz trombonist. Rudd's return as a jazz musician was accompanied by acclaim and fame.

    Walter Bowe was also taking new directions. He asked me to meet him at a Village bar where he was reinforced by a group of young blacks with the kind of east African brimless "kofia" cap favored by Jomo Kenyatta. He denounced Bix Beiderbecke, Frankie Traumbauer, and other white musicians, which I found puzzling as I had always been prejudiced in favor of black jazz, and it was Walter along with Dick Hadlock who liked white Chicago jazz. He seemed to be pledging loyalty to some of the black nationalists who were part of the era. Then Martin Williams told me that Walter was now dangerously involved with the Fair Play for Cuba Committee. Walter said Martin was a Republican reporting to the FBI which I gathered Walter thought were the same. (Years later I would question Martin about his politics and he would deny both, but he was always crab-wise when asked of his affiliations.) Next Adele and I were attending a Miles Davis concert, a fundraiser for some hospital in South Africa and Walter and a black woman were outside picketing against it. When they came on stage with their posters while Miles was playing he quit and would not return the remainder of the night and John Coltrane became the only horn on stage. Walter was entering Black Power mythology.

The year at Brooklyn College was a disaster of almost every kind. At Leeds there had been no clear social distinction between students and faculty, the difference blurred with shared interests such as poetry or theatre, or friendships. Whether Derry Jeffares, Douglas Jefferson, or Richard Coe we belonged to a similar environment; even during my first year at Leeds I knew that the Student Union bar was where professors and students talked over a "pint". When I sat down at a table with the head of department and his administrators in the faculty cafeteria at Brooklyn College I made at least two blunders. Junior members of the department were not supposed to be friends with the senior professors. I had also, it was explained, appeared to choose sides. An opposing section of the English department had taken the Head and his friends to court for, from the papers stuffed in my faculty pigeon hole, every crime ever committed by humanity. This was not the usual departmental gossip; hatred flowed as easily as the air. I thought one of my colleagues nutty, bragging about friends in another city who loved her and who would always welcome her to stay. I met another colleague who had written an accepted film script while sitting on the toilet seat each morning before leaving for work. Still another would invite a male student to spend a weekend in bed with him and his wife. That was not the New York cultural life to which I hoped to return.

My experience teaching an adult night class in Modern Drama to the Mirfield Workers Educational Association did not prepare me for Brooklyn College. At Mirfield I lectured and afterwards we chatted over the traditional pint at a pub until it was time for my train back to Leeds. My students were hungry for culture and even if I were reading something as obscure as Djuna Barnes *Nightwood* I had to bring it back to loan someone the next week. Adele and I had had a Sunday lunch at the home with the union organizer who headed the local branch of the WEA and I was impressed by the home-made lemon curd tarts and meat pies, a glimpse of what British food was like before the War and rationing.

By contrast teaching at Brooklyn College was a horror. There were hundreds of first year student "compositions" to correct and mark each week, which left time for little else, although we did go at least once to Harlem to dance. That many of my more aggressive students were obviously Jewish and I had become "British" troubled me. An attractive young lady really did put an arm around my neck, push up tight and said that she would do "anything" for a good mark. Another student had metal boxes of index cards from which she claimed to have written the essay I had given a low grade. When I pointed out the essay was on a different topic than suggested she complained to a high school teacher who complained to the head of my department. Then there were the males who openly announced that reading literature was a waste of time, the aim of life was money, a house in Scarsdale, a wife and two children, several cars and, I forget how many, dogs. Many of the students were bright and sometimes difficult to persuade that they did not have a complete vision of what they read. I remember a student who complained that T S Eliot's "Wasteland" was just the lament of a homosexual over a lost male lover. I do not know from which study guide he had picked up his biographical information but there was some truth in his otherwise limited reading of the poem. Perhaps the worst moment was a student reporting on her reading of D H Lawrence's *Sons and Lovers*, which turned into a long improvised confession about her own sexual experience. I did not know how to stop her without embarrassing her and there was a side of me too fascinated to intervene. The class was being evaluated by the colleague who was loved by her friends in another city, and as soon as she fled out the door I knew I had to leave Brooklyn College for another teaching job in another city, although we had at the start of the year filled our apartment with new furniture.

II

We spent the summer of 1961 in Mexico swimming and watching divers jump off the cliff of a hotel in Acapulco. We bought in an Oaxaca art gallery woven cloths for curtains for the house we would rent in Calgary, Canada. Adele still has a silver necklace from Oaxaca in the shape of a fish skeleton.

The University of Alberta at Calgary had moved into a new building; the landscape and about everything else seemed barren. The winter was freezing, cars had to be plugged into a heater, and when one day the temperature went up to 0 degrees and the sun came out I felt it was spring and did not wear my heavy winter coat. I remember skidding wildly, out of control, in the snow down a hillside in our little Beetle returning from a friendly game of bridge where we left the two professionals arguing with each other why they had done so badly.

My main impression of Calgary at this time was the near universal hypocrisy represented by the police trying to arrest the washing machines at a laundromat for working on Sunday; meanwhile the chief of police was under investigation for live child pornography. Adele was in a local production of Aristophanes' *Lysistrata* which was thought a controversial play; a member of the English Department told his students not to go as it was "dirty" although he was widely known to be part of a group that held orgies and exchanged partners. Then there was my student who complained to authorities that I was teaching a "dirty" poem, Alexander Pope's "The Rape of the Lock". He was arrested downtown by the police a month later for soliciting male sexual partners. There were other kinds of hypocrisy as when the head of my department said he would mark the final examination of one of my students and gave him an A while I might have failed him for answering a question about Wordsworth's poetry by claiming that in our age no one wanted to read about nature and peasants. This was supposedly a "refreshing" point of view; the student was the son of a friend of the Head of Department.

I had two excellent students in my course in Seventeenth Century literature. Michael Lapidge would eventually become Professor of Anglo-Saxon at Cambridge University before moving on to a special university chair at Notre Dame. Michael was quiet and unnoticeable except for his essays unlike Maurice Yacowar who seemed to do everything he could to annoy many in the local community. The first editor of the student newspaper he was soon dismissed for an editorial against celebrating a day sacred to the Canadian war veterans. I had to plead with the Dean not to expel my best student and promise that Yacowar would not do it again; but of course he did a few weeks later under a transparent pseudonym. Yacowar was supposed to write his MA thesis for me the next year and I went away for the summer having, at his suggestion, signed a note to the library saying that they should purchase whatever he needed. I came back to find that he had ordered an immense amount of Restoration erotic literature and I was asked by the librarian was this what I intended. I would soon resign but Yacowar would indeed write his thesis on "imitation" in Restoration erotic literature. He would go on to a doctorate in England and return to start the first Canadian degree in film Studies. His books included Woody Allen's films and film adaptations of Tennessee Williams rather than the Earl of Rochester's poetry.

I might have continued at Calgary except that my schedule for the next year included six mornings of 8 AM classes, the death penalty in an Alberta winter. A colleague whom I regarded as a friend was jealous of my popularity among students. After being told that my schedule would not be changed I immediately applied for a lectureship in English literature at the University of Ibadan, Nigeria. We went to Mexico once more, this time to Puerto Vallarta where I listened to stories about Elizabeth Taylor and Richard Burton's passionate recent affair during the filming of John Huston's *The Night of the Iguana*. I played chess on the beach and at the local chess club, and came down with a throat problem that a doctor claimed was cancer and told me I only had months to live. When I returned to Calgary I had letters asking me whether I was accepting the position at Ibadan and my doctor told me I had pneumonia not cancer. It was a good excuse to resign, reasons of health, but I was told I would never be hired again in Western Canada. We had to sell rapidly the lovely rosewood and teak furniture and Rosenthal china we had bought in New York. I sold my large collection of jazz and classical LPs to the Music Department. Adele sold her signed Picasso and Leger prints to a bookstore whose owner kept saying how could he know whether they were originals. We sent our winter clothes to Harrods in London, the traditional storage place for imperialists taking posts abroad. Decades later, once more out of a steady job, I returned to Calgary as a well-paid distinguished research fellow for a term. The fellowship was arranged by the same person who had assigned me to the six day 8 AM schedule, although we never mentioned it.

## III

Africa then was what was happening. In 1957 Ghana became independent and Chris Barber recorded "Ghana" to celebrate the end of the British Empire and the beginning of the post-colonial era. India had become independent a decade previously but that did not have the same immediacy, perhaps because of the bloodletting that accompanied Partition, as the string of new nations being given or fighting for their independence during the late 1950s and 1960s. An interest in jazz and black American history had led me to an interest in African art, music, dance and ritual. It was a time when there were books investigating the history and range of "Negro" literature and generalizing about "African philosophy". Many of the Africans who had gone to Europe to study after the Second World War were speaking about going "home" and I had been told that as a drummer Africa was my "home".

After some months of waiting in Paris for our visas to Nigeria we were down to our last fifty dollars when the visas and airplane tickets arrived. There was no one to meet us at the Lagos airport; a university celebration meant that no transportation was available. I watched swollen pythons sleeping off their meals while we stayed at the Airport Motel awaiting word from Ibadan. Once there life became better with payment of several months arrears in salary, a salary advance, a loan to purchase a car, promise of an apartment at low rent, and other goodies that the British assumed when taking positions overseas.

The University was still a University College of the University of London and we offered a version of their Honors degree with general lectures and small tutorials. I was soon teaching Thomas Hardy as a special subject, Practical Criticism, and Seventeenth Century Literature. My Nigerian students were mostly from the elite, although a few had prepared for university entrance examinations in small villages by oil lamps. I was invited to parties where the elite of the elite dressed in formal clothes and I was laughed at with embarrassment when I brought cups of coffee and tea to our tutorials; that was the duty of servants. National independence was coming soon and my students were trying to decide whether to be government ministers or librarians.

An idea of how rapidly the university would grow; during my first year we had 7 final year students in English; the next year 12; and soon, as Nigeria and the University, became independent there would be 40. And new universities kept opening in Nigeria as the demand for higher education became wide-spread as it did elsewhere with decolonization. Eventually the graduates would out outnumber the jobs available, but even then, at the start, Nigerians kept a close eye on the tribe of who was employed. Supposedly visas were being held up not from inefficiency, but because Yorubas, who included many graduates, thought that Europeans were taking their jobs, while other tribes felt that it was good to keep Europeans around for a while until they caught up with the Yorubas and Igbos. Even when it was not mentioned there was often a tribal sub-text to life in Nigeria.

The campus was beautifully landscaped, the buildings designed in what I was to recognize as British tropical modernism, the races mixed, the staff was international, and we rapidly had friends and a social life. David and Anne Mobbs were among our first friends. Anne was secretary to the English Department. David was one of several biochemists we came to know and he was unmistakable for his large body, clumsiness and general physical disorder including losing handkerchiefs and an eye that seemed to pop out or was someway unfixed. David spoke of having ended his research towards a doctorate when he backed into his chemical apparatus at Imperial College, London, destroying a year's work. He locked the door and left. He had supported himself while doing his research by gambling and he had not given up some of his tricks. We were often bridge partners and I would say before we started playing that David sometimes cheated and our opponents should keep their cards close especially when David's bent down to pick up a handkerchief or when that eye made a strange motion. "But we have known David longer than you."

David and Anne were defiantly East End working class. We hired Moses, a steward they had fired, trained him to cook French food, serve food correctly, and ordered new clothes including a blue blazer with shining gold colored buttons and white trousers. He looked smart, sharp, and was proud of himself. Anne exclaimed how Moses had changed; they were used to eating large plates of curried meat and rice with a soup spoon. When a Polish economist with whom we sometimes played bridge beat his wife so badly that she did not want to appear in public, David and his working class friends would say something to the effect that while every man had the right to beat his wife there were limits. David and his friends were significantly the Left on the university staff and likely to speak up challenging the Vice Chancellor or anyone in authority. They clearly felt that justice should be universal and kept me informed about what the Nigerian labor unions were doing including being paid off so as not to call strikes. Their politics were socialist not the reformism that Americans call liberal. They scorned trickledown economics, thought individualism and individual rights suspect, and had a lingering affection for the certainties of Communism despite disenchantment with the Soviets, Stalinism, and the Russian occupation of Eastern Europe. Similar to many on the European Left at that time they were appalled by American politics and skeptical of America's positions in the Cold War, but were fans of American culture as found in music, cinema, and big name stars. Anne was involved with the cinema club and kept speaking of American film directors whose names I had heard but about whom I knew little.

We also shared an interest in Nigerian cultures, especially popular music. We would go to a bar to hear a band recommended by students and be welcomed by bouncers saying come on in we have lots of girls for you. Told we were with our wives they would say we have men for them. There was a surprising egalitarianism in Nigeria at such places. A visiting American sociologist decided to interview prostitutes at one dance hall and was locked in a room over night until a senior university administrator arrived promising to pay for the time the women had wasted talking to the man. University administrators were expected to be jacks of all trades. The Vice Chancellor was called to a village to shoot a wild elephant. Nigerians were culturally different. Students complained of having their own individual rooms in the student halls; they were used to sharing and company.

David took up a position at a University in Ghana, then became the first Dean at the Oxford Polytechnic where he created their (and England's first) modular course degree and where I am told a building is named for him. Anne divorced him after he starting being violent. He took up with one of his students, then died of a stroke. We saw Anne in Oxford when Adele was on an exchange with a college for a term; Anne helped immigrants and had become an international organizer for women's groups. She also wrote a book about the lives of ordinary women in Oxford during the past century.

The English Department tried to keep to British university standards. Our graduates went to further degrees at Oxford and Columbia where they were outstanding. If I had to sign an American university form attesting to the competence of a student in English I commented that I wished more American students were as competent. When the University was a college of the University of London we went to England for our final examiners meeting each year, where we were told our standards had become too high and an upper 2.1 was pushed into a "First" by the London external examiner. The new Ibadan degree raised questions about the level for which we were aiming. I was asked by Molly Mahood, Head of department, to look at the first year papers which were based on the close reading of six set texts. There were about 65 percent failures which was clearly unacceptable. Molly said the department had done its best and nothing could be done about it, a view held by others in the department. Over grumbles and complaints I decided that the only way to handle the problem was to add 6 points to each examination result. I then looked at some of the now passed examination scripts and felt I was satisfied. I had changed from the suspect new American lecturer (the only one in the department) to someone to whom one should pay attention.

I respected Molly as a literary critic. I read her books and thought of her as someone who paid attention to details of the text and who argued her position unlike the female Oxford scholars whom she hoped to rejoin. Her literary criticism "had balls". I doubt that she had similar respect for me. I was publishing my first articles in such journals as *Sewanee Review* and *College English* and was proud that editors were telling me how well I wrote. When I read a paper at our departmental seminar about the political significance of a poem by Richard Lovelace she criticized my not mentioning Greek and Latin sources. The essay would often be reprinted without discussion of its classical sources. A paper about the moral complications in *The Adventures of Huckleberry Finn* surprised her; she assumed that Mark Twain's novel was a children's book. The essay would be published and then republished in an anthology about the novel (without asking my permission and without any payment.)

Molly could be sniffy about seeming lapses in taste and assume that people stuck with like. When a visiting external examiner was staying with her I was sickened that she served him roasted chicken and potatoes and cauliflower with a white sauce. Just what everyone needed on a hot tropical night in Nigeria. He was not told in England to take health precautions and died after his return from malaria not the meal. The British warned against eating too much fresh fruit.

Molly assumed we should stay on our arrival and then be invited to meals with a visiting American lecturer whom I disliked. Soon after our arrival we were invited to a meal after which the men adjourned to the garden for a pee and cigarettes while the women went upstairs to the toilet and Adele was told that rank and seniority of the husband determined the order of use.

There were however other sides of Molly unknown or ignored by those who only know that she and Wole Soyinka often disagreed. She had given up a senior position at the University of London to head the department at Ibadan. It was said that half her salary went to a charity in South Africa. Although she seemed often annoyed by the directness of many Yorubas, like Wole, she admitted a preference for Igbos, whom she thought more dignified in their manner. She early told me that Chinua Achebe, who had studied at Ibadan, was a better writer than Soyinka. She even had an affair with one soon to be well-known Nigerian writer when he was at Ibadan, which she treated with discretion; he did not. When some of her preferred students had less than brilliant careers she confessed that her "swans had turned out to be geese." I liked, and still enjoy, her acid wit.

She had prepared a new structure for the English Department's courses once the University began offering its own degrees. The first year in the Faculty required students to take two examinations in each of three subjects. Those doing English took papers in set texts and in language. Specialization started in the second year and included a two paper option, one of which was African literature, taught by John Ramsaran, an Englishman of Caribbean origins. He had published a thirty page booklet on critical approaches to African literature with the Ibadan University press in 1959, which was enough at the time to be an African literature expert. There were also two paper options in American literature and Linguistics. Although the teaching of African literature was only then beginning some of us were aware that a major cultural shift was in process and that the literary cannon would soon be, as we amusingly claimed, Shakespeare to Soyinka.

During 1963 Molly moved to the University at Dar es Salaan for five years before going to a Professorship at the new university of Kent. She was replaced by D E S Maxwell, who was Irish, an Americanist, and a friend of Derry Jeffares. John Ginger, whom I thought even more British than anyone I had ever known in England, resigned to write novels. I remember him praising a book on the British face, which he obviously thought he had, and disagreed on a Practical Criticism exercise I had set; he preferred the opening of J D Salinger's *The Catcher in the Rye* to its obvious source, the opening of *Huckleberry Finn*. The English department was becoming less English. New members included O. R. Dathorne from Guyana and Suheil Bushrui, a Palestinian from the Lebanon. Although both had British doctorates their personalities and interests were different. Ron Dathorne, an amusing novelist, would co-edit the Penguin *Africa in Prose* with Wilfred Feuser of the French Department, an early and influential anthology of African literature. Ron would go on to teach African literature in the America. Bushrui was at the time divided between Irish studies and trying to memorize the *Koran*. A Palestinian who became increasingly anti-Israeli and anti-Jewish he had the problem that faced many of his fellow Baha'i that Islam regarded their faith as heretical and the headquarters were in Haifa, Israel. Bushrui now holds a distinguished professorship sponsored by the Baha'i at a university in America.

Desmond Maxwell was more relaxed and sociable than Molly, more amused and anecdotal like Derry. He had less critical judgment and had written a book on the background of American fiction and a book on *James Gould Cozzens* for the Writers and Critics series, but he was more directly concerned with the needs of members of his department. He would not like Molly ask members of the department to pay for a new cushion for the chair of the secretary to avoid asking the University to pay for it. He would later move on to York University, Toronto, write several good books on Irish literature, and unfortunately go blind.

Central to our social life was the Staff Club with its large swimming pool where we relaxed after teaching, sometimes played bridge, and if wise went home before evening "sundowners" which too often resulted in brutal family quarrels and a few dent fenders. At night we returned for gossip, drinks and sometimes dancing around the well-lighted pool which, I am told, originally attracted snakes. The best dart player was a Nigerian who had returned from Leicester with an impeccable midlands accent. Suheil insisted that the "the tango never dies" or brought Cliff Richards recordings while we tried to play the latest Beatles' hits. There was a continual coming and going of visitors including Americans such as James Meredith whose application for admission to the segregated all white University of Mississippi was a central event in the American civil rights struggle. He was in Ibadan to study for an MA; I was surprised that he seemed conservative and not surprised that he later ran for office as a Republican in New York. An archeologist friend with whom I played chess, Steve Daniels, had studied a central African language so obscure that Oxford discovered he was the only one who could examine himself.

William and Hetta Empson turned up one day. Hetta, who had a strange trick of swimming backwards feet first, explained that William, who appeared an impossibly absent minded professor, went through immigration with safety pins holding his trouser together. As the immigration official hesitated about admitting him, Hetta pinched a baby that started crying and their passports were stamped. Suheil, who had moved to a university house planned a large party that night to honor the Empsons to which we were not invited. Instead we went to the Staff Club and there were the Empsons drinking and gossiping and happy to see us once more. Word had reached them that Suheil would not be serving any alcoholic drinks and they were not going. Eventually Suheil turned up and agreed that some Lebanese he knew would supply the booze and we went off to the party at which we had to listen to Suheil's fulsome praise of Empson, someone known for his economical poetry and unromantic view of the world.

Meanwhile I had become interested in the African popular music I kept hearing. Besides the seemingly traditional Juju music, mostly drumming with a guitar or two, in the poor clubs, and the various Muslim influenced Yoruba drumming bands with their talking drums, the most popular music was High Life, a style that had evolved from European marching, jazz, and dance bands from probably the 1920s. There were versions of such music throughout Africa, the Congo for example had produced some famous bands and South Africa had its versions of American jazz, but West African High Life was what I was hearing and dancing to and I kept asking Africans about its origins. Soon certain names in Sierra Leone and Ghana kept coming up and without doing any real research I had collected a miscellany of information that I published later in *Jazz Monthly*. It turned out to be influential, and real musicologists followed up on the names I mentioned. I would several times be asked if I were the Bruce King who wrote that article. I even had an invitation to contribute to an UNESCO world music project from a famous South African musicologist, which I turned down as I remain musically illiterate and would not have been able to write notation let alone use the appropriate terms in the many Nigerian languages.

Tony Harrison, and Jimmy Simmons were appointed lecturers at the new University of Ahmadu Bello in Zaria, northern Nigeria, and there was international panic as universities searched for those who might represent them at the installation of Ahmadu Bello as Chancellor. Columbia somehow found me and said it would pay my way, so Adele and I hired a car and driver and went north mostly to see Tony and Jimmy once again. They had co-operated on a Nigerian version of *Lysistrata*, and were already thinking of their future directions as writers. Tony married a Czechoslovakian wife, was learning Czech, was planning on moving to Prague University, and hoped to translate Leos Janacek's operas. Jimmy saw himself becoming a folk-poet singer accompanied by guitar on television. They both agreed that poetry should be public, open, accessible unlike the obscurity of modernist high culture. Meanwhile I had to borrow a brightly embroidered university gown and decide what to do about producing an impressive official looking document saluting Ahmadu Bello as Columbia had misdirected whatever parchment it intended to send. Ahmadu Bello was the premier of the Northern Region, hereditary Sardauna of Sokoto, and leader of the party that governed Nigeria. He could have been Nigerian Prime Minister, a role he assigned to another northerner. His robes were said to be made from special silks flown in from Saudi Arabia. The crowds who wanted to watch the ceremony were whipped away by Nigerian guards. An English woman told Adele that white gloves should be worn for the occasion. When later that night I was told that John F Kennedy was assassinated I nervously laughed; it seemed so distant and American.

The American racial scene was of less interest to Nigerians than might be imagined. My students discussing *Huckleberry Finn* would refer to Jim as *a "nigger"* without Twain's irony. *I* had to explain why it would not be wise to do research in Mississippi on William Faulkner's novels. Visiting American radicals seemed out of touch with West Africa. A lecture by Malcolm X concluded with Ron Dathorne, much more black than "red skinned" Malcolm X, angrily disputing his view of whites. Dathorne, from a militant radical Guyanese family, had a white German wife and had been educated in England. Decades later a history of Malcolm X claimed that Dathorne had been a CIA agent, a claim levied against many who disagreed with extremists. Dathorne was upset as he felt it could damage his career, but how could he disprove it? To go to court would only spread the rumor.

Dathorne was one of the many West Indians in Africa interested in finding "roots". He was part of a group of mainly black intellectuals and writers that included Wole Soyinka, the poet Christopher Okigbo, the dramatist J. P. Clark, and the excellent Guyanese novelist, artist and art historian Denis Williams, that at various times edited, contributed to, or founded *Black Orpheus*, *Odu*, and Mbari publications. Ulli Beier who had founded *Black Orpheus* and the Mbari club was teaching for the Extra Mural Department in Osogbo, where he and his first wife Susan Wenger had introduced young Nigerians to using local symbolism and material in their art and became what many consider the start of modern African art. He would return to the campus on Saturday mornings to give a lecture course. These were among the so called golden years at Ibadan when Nigerian literature and modern art were starting to reach international standards and attention. Most of those involved were at some point associated with the University of Ibadan or with the then nearby new University of Ife.

A common mistake is to assume that the literature of these years was primarily about race, and the evils of colonialism and imperialism. The main theme was identity. This was the time of the Ibadan historians who used oral rather than written European sources and who were concerned to assert an African cultural heritage. While the creative writers often used oral sources, Soyinka, Okigbo, and others were more concerned with their mixed European and tribal cultural heritage, or as Clark expressed it, "a man has two hands". The struggle against colonialism was not as fierce or prolonged in West unlike North and East Africa. The British had not settled in West Africa which had been promised and prepared for independence. There was no need for a "Negritude" movement as the British did not proselytize their culture as universal. Edris Makward, part Senegalese, in the French Department was preparing a dissertation on the Negritude movement but Soyinka and the Ibadan intellectuals felt that they had no need to proclaim their Africanity. Nigerian writers and artists did reject the corrupt Africans who were increasingly governing them.

Achebe's first four novels tried to trace the source of this corruption originally from the time that local societies, with their own checks and balances, were being incorporated into an uncomprehending British ruled national state. Achebe wanted to show that African culture was not originally static and unchanging before the coming of the white man and that by bringing various self-governing societies together with the help of native interpreters the British had created a new class that lacked the restraints common in the past. Soyinka's writings mocked idealization of African history and satirized those who had come to power with African independence. Such criticism of local society and of those now in authority was common to the new national literatures until Western intellectuals discovered and reinterpreted them as "postcolonial" documents of anti-imperialism, a process that became established when the Cold War concluded with America's seeming triumph and the New Left began rewriting history and culture according to its own Vietnamese war influenced perspective.

The Mbari Club in Ibadan and the Lebanese restaurant behind it with its court yard, space to dance, performing Apala band, and where plays, such as J. P. Clark's *Song of a Goat*, which included the ritual killing of a goat, were enacted, was a center for social and cultural gatherings. There was a Mbari affiliation to the Congress for Cultural Freedom, an international anti-Communist organization that was later found to receive funding indirectly from the CIA. I did not know of this link at the time, nor did the many writers and intellectuals from the former colonies who indirectly received support. CCF patronage was later thought shameful, but as the Soviets were offering scholarships for studying in Russia and there were writers openly taking Communist money the Americans were amazingly tolerant how their patronage was used, always canalling it through apparently independent cultural and artistic groups, which included those who were openly anti-American.

Behind the restaurant was a library where I could read *Down Beat* and other publications, including the Nigerian and other African writers that Mbari published before they were known abroad. In the courtyard there was usually a Yoruba Apala band in which I started to play the talking drum and eventually would be expected to lead the group although, not knowing Yoruba, I did not know what I was saying. This was possible because the speech of talking drums consists of phrases based on tonal and rhythmic patterns and since I had interiorized the patterns I could easily reproduce the four tones. Apala rhythms were easy in comparison to some tribal rhythms played at student dances at the University.

Yoruba talking drums, with their tension strings, were an import from the Muslim north as were the rhythms. The leader of the group spoke only a few words of English but as we came to know each other, he took me to his house, showed me his amulets against juju, introduced me to traditional drum makers, taught me to look for the now mysterious identifying marks carved into each drum, and surprised me when during the Western Region Crisis he spoke the political slogans of the anti-government opposition about freedom and liberty.

The Western Region Crisis began during 1962 with a split within the Action Group, a Yoruba-based political party, in which one faction joined the national government led by the Northerners. As the two factions quarreled the split intensified and soon an emergency was declared as, the national government claimed, the region had become ungovernable. This articulated the conflict that had long existed between the Muslim dominated North and the more Westernized South and East of Nigeria. Various events that took place during the next years were often violent and I remember driving very carefully through the curving streets of a housing estate directly behind an army truck with its machine gun pointed at me. After a 1965 election in the Western, Yoruba, region, which was thought fraudulent, life became especially dangerous. It was wise when going dancing to decide when choosing a table how to escape if thugs entered and destroyed the club. We once drove to Lagos with a Nigerian hiding in the leg room space in front of the back seat as even ambulances would be stopped and examined by rival gangs. When visiting friends in a small town I was forced to get out of our car after two lorries blocked our going forward or back. They assumed that white people would be supporters of the hated government. As usual the best escape from danger in Nigeria was to stay cool, be amusing, and find which names of people you knew in common were a passport to freedom.

There had been a plan by some supporters of the opposition to hold up the regional radio station in Ibadan and call on the government to resign. When the main person involved chickened out Wole Soyinka took his gun and himself held up the radio station forcing it to play the pre-recorded tape. I had not known what happened when Wole rushed into the University staff club with the gun still showing and wanted to find Jimmy Simmons who had come from Zaria for a conference; he needed a driver to take him where he intended to escape. Jimmy and Wole changed clothing and cars. Wole's voice was immediately recognized and soon the police were searching for him. They also took Jimmy's passport for months. While Wole was in hiding a foolish Leeds visiting professor in Economics, or was it Politics?, was discovered to have entered Nigeria with plans to start a Socialist revolution. If that was not bad enough I was almost arrested on suspicion of smuggling antiquities. Another situation that required a sense of humor.

The Englishman who headed the national museum and who persuaded Nigerians to pass a law against the export of antiquities could not afford trained staff so he hired illiterates who hardly spoke any English to keep an eye on foreigners who might be illegally exporting Nigerian art. It was a hopeless situation since besides individuals having a few pieces as souvenirs there were Francophonic African embassies massively taking antiquities out of Nigeria for sale in Paris. Some Africans with diplomatic immunity had once asked me to buy pieces for them but I wanted nothing to do with it. I did, however, have a small collection and I naïvely asked one of the museum's watchdogs whether he could arrange for my pieces to be taken to Lagos for approval for eventual export when I left Nigeria. Six months later, after I returned from summer holiday, the man turned up with two tall burly homicide detectives and wanted to arrest me for illegally exporting antiquities. I knew that if I could make the detectives amused the situation would somehow be resolved, so I began asking whether or not modern African pieces I had were old and if so how old. Flustered by the detectives laughing at his bewilderment the young man grabbed my favorite statue and said it was an antique, which it was, and he was taking it for investigation. There was no law against having ritual objects, only against their export without approval, but how was I to get it back? This was a time when traders went from door to door trying to sell new and old African "art". The person I had bought the antique from came by with other pieces for sale a week later and I told him what happened. No problem, the chief of police was his "brother", which in Nigeria could mean anything. I stopped at the police headquarters and took my statue back.

Although people discussed what would happen to Wole the assumption was that he would be freed on a technicality. A newspaper described his appearance at the trial as having open trousers (I think they meant shirt), but he was indeed freed on a technicality, no doubt the judge was someone's "brother", perhaps a member of the Ogboni cult to which Wole belongs.

During my time teaching at the University of Ibadan and later Lagos University there was often discussion of such cults. Whites obviously disapproved of them because of the association of cults and blood sacrifice, an association that was still sometimes real. The Ogboni were supposedly the most powerful cult among the Yoruba and it was said that there was still a secret traditional Ogboni cult which practiced the sacrifice of children as well as the reformed Obgonis who used animal blood. While I assume that Wole and Ulli Beier belonged to the reformed branch some Yoruba intellectuals claimed that human sacrifice was normal and only imperialist "decadence" brought it to an end. Wole's own works, especially his plays, offer a personal interpretation of this claim. For Soyinka sacrifice's must be freely chosen, and the sacrificed is a messenger to the gods, participating in the god Ogun's dangerous journey through the abyss of chaos between human life and the divine. It is a journey freely undergone by the creative artist, the actor, and the revolutionary, as well as in ritual sacrifice. Soyinka's own life for decades might be regarded as his own reenactment of this myth. It had characteristics similar to the pattern of heroism that Wilson Knight had described in which breaking the law and moral code was a necessary descent into evil to bring about good.

Wole's short hand allusion to the dangerous journey was the "road", and the road between Ibadan and Lagos, as well as the road between Ibadan and the University, was indeed dangerous. Not being willing to choose between driving on the left like the British or the right like most of the world the Nigerians preferred the center, with the inevitable smash ups. One night driving home we stopped to see if we could help in what was obviously a bloody head on collision between two cars. They had both been driven by Nigerian doctors from the University hospital and both were shouting that once they could stand up they would kill each other.

Although the standard of medical training was high diagnosis could be erratic. A Nigerian who the previous week had fatally misdiagnosed appendicitis as malaria tried to convince Adele that she must have an operation for appendicitis when she had malaria. The British father of a now well-known novelist had a drinking problem and would say that there was no need to take medicines against malaria, it was sufficient to start drinking at sun down and catch the mosquitoes in your hand. On leaving his office you could see him reach into his desk drawer for his bottle. A visiting medical researcher from Johns Hopkins told me that the faculty at the University of Ibadan included some of the most brilliant in the world, except that they were too eccentric to hold jobs at home, and some who were not qualified to be more than school teachers.

Nigeria was interesting whether in learning how to bargain in the market or in the complexities of one's relationship with servants. Each was part of being educated into other cultures. The market women were often more wealthy than they appeared and in Yoruba society they were the money earners unlike their men who devoted themselves to tribal and other politics. There were different styles of bargaining although they all started by naming prices perhaps two or three times more than what was expected. Some traders would expect you to pretend anger and start to walk away before agreeing to a price; there were traders who seemed to agree to a price then start raising the price. I thought this strange, but later while teaching in Israel I found a similar practice; negotiation led to an agreement that later had to be started again. After three years of bargaining in Nigeria it had become habitual; I would see a necktie in the window of a Jermyn Street shop in London and after asking about the price I would start mentioning small flaws until the price came down. The Ibadan market itself was smelly, dirty, exhausting. We went to buy fresh green beans, tomatoes, onions, eggplant, fruit. You could buy alive or dried smoked monkey, the later probably for ritual purposes. The first time Adele bought live chickens they got loose in our Volkswagen and driving home was awful as was the condition of the car. The cloth market was every four days, a Yoruba week.

Just as there were cultural differences in bargaining so there were tribal differences in what constituted corruption and morals. Was it corrupt to give a "gift" after or before a decision by someone in authority? As everything in Nigeria appeared to require "gifts" (usually spoken of as "dash"), including medicine when in hospital, the problem of what was or was not corrupt was perhaps abstract. I had returned from London with an obvious gift for my wife that I had not declared and was asked to meet the customs officer in the men's room where while peeing we bargained over the amount of his "dash". A friend giving a "dash" required by a customs inspector was observed by soldiers who told him to write a statement that could be used against the official. Was this also a shake down?

Moses, our cook, asked Adele to tell his wife that we would not pay for her to visit her village. As Adele did not speak Igbo or understand why she should be involved she tried to avoid whatever was happening by saying a few words in English and hoping that was satisfactory. Months later the wife returned pregnant and Moses asked for some brandy to celebrate the coming baby. His wife had learned that the lack of a child was his fault and returned home to find someone who was not impotent; since whoever it was came from the same village Moses was content and named the girl "Patience".

Treatment of servants often showed national and tribal differences. Americans paid much higher salaries but felt no further obligations. The British paid less but when leaving would try to find future employment for servants. Nigerians paid little and demanded servants be available at any hour, often employing family relatives, but would cover the expenses of marriages, funerals, and other social and religious rituals.

During the summer of 1965 we had gone to Deia, Mallorca, where, we were then staying at Pension Ca'n Quet, I learned that my book, *Dryden's Major Plays* (*1966*), had been accepted for publication. I also was interviewed for, and appointed to, a lectureship at Bristol University. After five years we were returning to England. I felt I would now settle down to a regular job and be a father. Adele had wanted a child for years and as soon as she stopped taking her birth control pills she was pregnant. We returned to Ibadan to give notice of our leaving (Adele's *Camus* was published, she had enrolled to start a doctorate in Paris, and was now a lecturer in French) and after the term sailed on an Elder Dempster line ship to Liverpool. Our immediate neighbor in the apartment across the hall was the famed archeologist Thurston Shaw. We had for some months while his life was in danger from an illness he developed on a dig kept lovely ancient bronzes he had found and he had not had time to turn over to authorities before being rushed to hospital. He murmured something about dropping us in for a drink before we left but by the time we did the farewell party he had secretly assembled in our honor had broken up thinking we would never arrive,

On shipboard a very pregnant Adele lost the final of the table tennis tournament to a Sierra Leonean Army officer who gallantly picked up any balls from the floor. Several months later he led a coup that changed the government of Sierra Leone. I won the ship's chess tournament and a pair of cuff links. The ship stopped for part of a day in the Canary Islands where we barely made it back from our exploration as the gangplank was being raised.

# Chapter 7: Lecturing and Professing

*Bristol, England 1966–68*
*Lagos, Nigeria 1968–70*

I lasted little more than eighteen months at Bristol. I do not think that Frank Kermode, who had replaced L. C. Knights, and who hired me on Derry Jeffares recommendation, lasted much longer. He left for a named chair at the University of London; I left for a professorship few wanted at the University of Lagos, Nigeria.

Bristol began badly. The University knowing that we would need a place to stay arranged a bed-sitter with a landlord who limited use of the shower and bath-tub to a few hours each week. Although we eventually found our own apartment on Buckingham Vale, a small modernized basement flat with an extra room for Nicole, life in Bristol included trying to stave off genteel shabbiness. I was paid poorly by the University and our standard of living dropped. Others I knew who had moved to England received increases in salary as the UK university scale was now higher than the Nigerian. Bristol however appointed me at the bottom of the lecturer's salary and when I discovered that others in the department were paid more though less qualified and with fewer years of university teaching experience I was told that I had accepted the salary when appointed and there nothing to be done about it.

Frank Kermode was often in London and the running of the English department was in the hands of another professor, Henry Gifford, who I did not like and who did not like me. Formerly part of a revolution in English studies that had established the close reading and evaluation of texts the *Scrutiny* critics were now the establishment. It was time for a change, but Gifford thought that judgment stopped with F. R. Leavis's books of literary criticism. I generally agreed with the Leavis canon of significant authors, but felt that it was too limited, even in places constipated. Students enjoyed reading the poems of John Wilmot, Earl of Rochester, and I liked to show that the "line of wit" from John Donne and other early Seventeenth-Century "metaphysical" poets had been carried on into the Restoration by Rochester and from there into the Eighteenth-Century by Swift, Pope, and Samuel Johnson. As Leavis had pronounced Rochester filthy minded or some such puritanical objection there would be no opportunity for students to show familiarity with the poetry on the end of the year examinations.

Gifford kept talking about the department's pastoral duty towards students, but I thought he was encouraging breakdowns. There was a savagely polite diner at his house during which I thought the invited student would either flee or start crying. Knowing I had been in Nigeria the conversation was directed to Gifford's time in India and his wife was called upon to confirm that there were lovely birds following the cows. I resisted saying that they were egrets hoping for seeds in the cow shit. Nigerians sometimes spoke of them as "eagles".

My tormentor questioned whether my tutorial students were given enough attention to the quality of the writing or whether as an American I put more emphasis on themes and meaning, which being American I did. So I was pleased one day to catch him reading a low level study guide before a tutorial. He said he was refreshing his memory. I speak quietly in a low voice and Gifford had hearing problems. He listened outside my lectures hoping for student discontent and, surprised, admitted to me that that he heard no shuffling feet, no whispering. The English Department prided itself on teaching rather than research. I was told that I had a reputation among students as the second best lecturer; Kermode was thought better. This did not surprise me as at Ibadan students had praised my teaching. It was in North America that I had problems with students as I regarded them as uneducated, undisciplined, and spoiled, similar to myself at their age. Oddly, however, I liked to be well thought of by students. When half of my General Degree students disappeared I was upset until I learned that they had returned to the foreign universities from which they had come for a semester.

The English Department regarded itself, or at least some members did, as a place for those who did not quite make Oxford or Cambridge. When interviewing students for places those from public and grammar schools were preferred. Poets, musicians, long haired hippies, the unconventional, were automatically rejected, "not accepted" meant "not acceptable". We were perpetuating a social elite to which few members of the department actually belonged. I was told to "give hell" during interviews to those from the best schools because "they would give you hell later".

Proofs of my *Dryden Major Plays* arrived with a request that I cite first editions rather than the then standard Sir Walter Scott-George Saintsbury collection of Dryden I had used for my doctorate. The Bristol University library did not possess the microfilms of the plays and was unwilling to request a loan from other British universities. I was told to go to Oxford during holidays. The solution was easy but not what the Bristol librarian expected. The University of Ibadan, which was then trying to be a place for research, sent its microfilm copies. Somehow I was already becoming known among scholars and was surprised to be invited by Maynard Mack to compile *Twentieth Century Interpretations of All for Love* (1968); such series were then coming into fashion and to edit a volume was a step up the ladder for a young scholar, especially if you were outside the series editor's circle.

John Wilders, who soon advised the Royal Shakespeare Company's recordings of Shakespeare's plays, and would be literary consultant to films based on the plays, shared my interest in the Seventeenth-Century. We invited each other and our wives to meals, but we were on different wave lengths. He said that he wanted to act his age but it was 1966 and I wanted to be part of the age. The poet Charles Tomlinson was different. He had been influenced by American poetry, had lived in America, and was friendly. I was annoyed when he and his wife came to our apartment for diner and left early because they lived in the country, but I was told by others that I had been favored, Charles and his wife seldom came to social events. Knights had discouraged the department from entertaining which he thought a form of superficiality. Kermode came for a meal with John Wain after too many drinks at a pub, but generally he was in London or in the house he bought in the country where his lonely wife, she told us, offered sherry to the cleaning lady so she would not leave her without company. The marriage eventually broke up.

Loneliness seemed common to those we knew at Bristol. Kermode was different, he was active, commuted to London where he edited *Encounter* (another of those Congress for Cultural Freedom offshoots), and was reading the new French theorists. When we had a party at our apartment for staff and students he recognized and laughed at the Mick Jagger poster I had on a door. Another member of the English department asked what I had meant by "bring pop records" for dancing.

A couple with upper class sounding names that attended upper class schools lived in an apartment near us and were lonely. They appeared to have no social life beyond family. They came twice for a meal and we once had a meal with them. We both had one child and that was the main source of our friendship. When I resigned and we were leaving Bristol they told us that we were their best friends.

Adele gave birth to Nicole in May and we decided that the baby should keep our hours. When Nicole fell from her high chair when being fed we had to explain to the doctor why she was up at midnight. Adele also started to write her *Proust* (1968). I went to the American embassy in London to register Nicole for American citizenship and a passport and was made to swear that she had not voted in a foreign election, had not served in a foreign army and was not a prostitute. When I looked at the embassy official to see if he was smiling I was told that this was a serious matter. I could not yet swear the color of her eyes, they might change. While in London I saw the literary editor of the *New Statesman* about reviewing and was several times asked "who sent you?" As no one had it was obvious that this was a dead end.

At first there were a few large parties given by people we did not know to which we were invited through a young American who had been brought by Kermode from Manchester. After May we stopped going or being invited and made few new friends. New York, Nigeria and Leeds remained significant. Terry Brindley, who had edited *Minotauer*, a student literary magazine in Leeds, was doing research at Bristol and ranting about England being a police state. He was unhappy and verbally critical of his wife along with most of existence the night when we had a meal with them. The next time I saw him he was upset that she had thrown a bowl of cauliflower soup at him; for some reason the cauliflower especially troubled him. I was not surprised the next time to learn that she had started an affair with someone else whom we had known at Leeds.

Another couple we had known in Leeds visited and they now had children and Singaporean citizenship as they did not plan to live any more in England. Ann Fraser came to see us and we began to stay with her whenever we were in London. A Nigerian lawyer told her that he could not marry her, but he would show her that as a nurse she was entitled to a low interest mortgage for her house and she could rent a room to immigrants or Irish laborers who would pay in cash. Her love life was confused by Jon sometimes reappearing and until Sarita was born there was the question of who was the father. A few years later Ann would marry Simon, return the night of wedding with another man, and tell me that "she could sleep with any man but Simon". The marriage broke up but she now had a house in the country with her mother living nearby. Eventually she remarried, now to a respectable lawyer, and became a country lady, her wilder days and nights now I imagine part of a glorious past, which I salute.

The Leeds connection kept returning and I could not help contrast my experience of the two cities. People told me that I was lucky to be living in Bristol, that it was the better university and that the English Department was highly regarded, but Leeds had been fun, tolerant, exciting and Bristol was not. Similar to many Americans I had a driver's license by the time I was sixteen, and felt that an automobile was an extension of my body. In Nigeria I had easily, without offering "dash", passed the absurd test of backing around many oil barrels without touching any. Tested in Bristol I was failed for not looking often enough at the rear view mirror although I thought I had avoided a certain accident when a car ran through a stop sign. A friend laughed; he had stalled at an intersection, released the clutch badly and caused the driving inspector to hurt his head on the window—and passed. Adele was so pregnant that she could not turn the steering wheel to back into the space to park, but she had passed. I should have, like Adele, taken a few lessons at a favored driving school and used their automobile for the examination and I would have passed. This was worse than Nigeria.

Although I began publishing in *Jazz Monthly* when I was studying at Leeds I had not met Albert McCarthy, the editor, until he and his Canadian wife visited Bristol. I knew he also edited an anarchist magazine from St Ives where he lived but I had not thought much about his politics until he told me two stories. Invited to the USA to write about jazz he needed a visa from the Embassy in London. When asked if he belonged to any subversive party he hesitated before saying that he edited the *Anarchist* and was told they knew that but if he had not admitted it he would not have had the visa. He told me that during the Second World War he was imprisoned as a conscientious objector but as he was a good cricket batsman he played for the local army side. On weekends they would dress him as an officer, salute him, and after the match would march him back and he had to change once again into his prisoner's clothing.

I began reviewing records for *Jazz Monthly* and among the recording was an Ambrose Campbell's High Life LP. Campbell was a Nigerian whose band had played at London clubs and dances and was highly regarded in jazz circles. I assumed that the LP was recorded because of my *Jazz Monthly* article "Introducing the High Life", but later learned that Campbell was already known from 78 rpms recording of his West African Rhythm Brothers, supposedly the first black band in England. He is now seen as one of the pioneers of Nigerian High Life.

Martin Williams wrote that Walter Bowe had tried to blow up the Statue of Liberty, was imprisoned in Sing Sing for five years and suggested I write to him. I knew nothing about this but soon learned that Walter had become part of a group apparently brought together by a FBI agent that planned to bomb American monuments. It seemed a clear example of entrapment as it is possible none of the group would have done anything without the agent's encouragement, but Walter had indeed argued for and investigated the use of explosives against the Statue of Liberty. What do you write when one of your best friends has turned into a dangerous extremist? Decades later I started to wonder what had happened to him. Hadlock heard that after Walter left Sing Sing he had for a time started a band "Walter Bowe and his Georgia Bobos" or a name like that. Through on-line searches I found that he had been married to a leading black radical (perhaps the woman who had picketed the Miles Davis concert) and that they had a daughter who was a well-known black actress but who had little knowledge of her father beyond his having been a trumpet player. There were references to him among the Black Liberation Front, the Black Arts Movement, and even a flattering picture for sale of him as a trumpet player, but having become part of American Black Power mythology he had disappeared. There was also a "rent-a-beatnik" photograph of him playing with older style jazz musicians for a loft party in 1959 staged by *Holiday* magazine that suggested nothing of the radicalism of the 1960s. I remember when Walter first moved to New York from Camden and was often in the Village he became friendly with some white progressives from Oberlin College. When he brought a female from Oberlin back to Camden his mother, one of the of black Americans who still voted Republican because that was the party of Abraham Lincoln, ordered "get that white trash out of my house". He had travelled a long way.

On returning to England I asked Harrods to take our clothing from storage. I also ordered a suit made to order and was given a snobbish hard time until I pulled out my Harrods' credit card that I had never used previously. I ordered some shirts from Harvey & Hudson.

In Bristol the delicatessen on White Ladies Lane would cut us a pound from the center of the Cheddar cheese. One of my colleagues in the English Department, younger than me, and also appointed low on the salary scale, was married and had a child, about a month older than Nicole. They regularly ate one poached egg each on fried toast and saved to have a bottle of Sherry over the Christmas holidays. While England was booming they were still living as if the War had recently ended. I could not on my salary eat and dress to the standard I expected and I could not afford to teach at Bristol. There was no likely University position for Adele and if she wanted to teach in a school we were told that she would need an Education diploma. I had informal offers from the new University of California—Davis, University of California—Berkeley, and Western Ontario in Canada, but I accepted an invitation from the Lagos University, Nigeria, to Head the English Division as there was also a lectureship in French for Adele. I went to a fashionable tailor on Jermyn Street to have an extremely lightweight suit made for professorial occasions and was quizzed about who I was and was my picture in the newspaper last week. Even paying was annoying as their practice was to present an annual bill in December. I told them to send it to me in Lagos.

Ben Obumselu, the first Nigerian appointed to the English Department at Ibadan, came through Bristol with his wife and told me that Christopher Okigbo was flying arms to the East and a civil war was likely. I knew the close relationship between the Nigerian literary world and the political opposition so I was not surprised when the fighting started. Kermode talked to the British Council and said that I could change my resignation for a leave of absence, but I did not want to return to the Bristol English Department where there was concern who would teach Old Norse after Susie Tucker retired. Susie was friendly, one of the few in the department to show interest about Nigeria, but I did not care about the future of Old Norse and Old English. I admired Christopher Ricks' book on Milton and he would replace Kermode, but it was time to move on. I was not surprised to learn that the Department had set the wrong examination for the General students. The Seventeenth and Eighteenth centuries alternated with the Romantics and Victorians and this was the year the earlier period was taught. The examination was set for the later period and no one noticed until the students objected. Although I had been one of their lecturers no one had asked me for examination questions.

We once more sold our new, this time London bought, apartment furnishing and the hi-fi set I had used to review records. A lecturer from Spain married an English woman and Adele said we should make them a good price as a wedding gift. As they never paid us it was a real gift.

## II

Adele and Nicole were the only two females on the airplane taking us to Lagos. Indeed they were the only females to go to, rather than flee, Lagos for some time. The Americans and British were advising their nationals to leave, but I trusted the person in the British Council who said that the situation was changing and come, so we did. The first week in the university guest house Adele heard a loud noise and said come to the window there is a bright light in the sky. I said go back to sleep you are hearing the air conditioning. The next day the newspapers were filled with stories and pictures about the air-raid.

It was the only one. There was more danger of being shot by drunken or sleepy Nigerian soldiers. One night I had tried to take a back road from a restaurant into Lagos and blundered into a side entrance of the airport and immediately was blinded by spotlights manned by armed soldiers. I could see machine guns trained on us. That took some explaining. Another time visiting a friend a bullet went over my head. I turned to see a sleepy soldier looking foolishly at his rifle. Returning to the University from dancing a French friend sped through a check point before the Lagos Bridge and I had to lie to the drunken machine gunner that I did not know the driver of the car in front while hoping that he would not take out his anger and humiliation on us. But in general life in Lagos during the Nigeria-Biafra war was safe, if you were not an Igbo. The massacres in Northern Nigeria that had caused the Igbos to flee and declare Biafra independent from Nigeria continued in Lagos where Igbos were hunted down and killed. Those who could departed for Biafra, those who could not, or were willing to take their chances, tried to pass as tribes near to, but different from the Igbos. We did not know that our baby nurse and steward were Igbos until someone told us that Nicole was speaking baby Igbo.

There were other surprises. When I first met the Vice-Chancellor he told me that a government minister was arriving from the Congo which needed a Professor of English. When I refused to go he told me to pick a member of my department. A committee decided that the university needed an entrance examination in English and I was told that my department should mark it. I have a visceral reaction to orders and I resigned on the spot and left the meeting. It was then agreed that the marking would be shared among departments and the markers would be paid. I must have gained respect as I was appointed to the Publications Committee, a committee for funding research, a committee for appointments to the University Hospital, and asked to join the University Orator in writing Honorary Degrees.

The committee for funding research listened to a report concerning evaporation from holes in the heat. There was no pattern to the result and the fault was blamed on the differences of the holes. Funds were asked to prepare concrete holes in the expectation that they would produce better results. The geologist had published a comic novel, but this was unintentionally amusing. As at least one Honorary Degree would be awarded to a Russian whom the Soviets would decide at the last minute to send I had to fill the citation with vague nonsense to which could be attached a name and a few facts. The Publications Committee was more interesting as the members were hoping for proposals that had some relationship to teaching. The Dean of Law proposed commissioning essays for a book that could be used to teach Nigerian Law. I suggested a collection of essays on Nigerian literature and said that a Nigerian should edit it, but was told that it would be more "objective" if I were the editor.

I did not like the way African literature was then usually treated as if there were no differences between regions of Africa and as if poetry, novels, and drama were extensions of anthropology. There appeared to be no clear line between African Studies and the teaching of African literature. It had once been useful to assert an African culture in response to European cultural imperialism, but the time of "Negritude" was over. Now there was a need for a canon of significant national authors and for introductions to some of the main tribal literatures and their forms. I also wanted a book that would include the young Nigerians whom I had taught at Ibadan, like Dan Izevbaye, and those in my Lagos department, such as Abodun Abetugbo. This turned out to be less straight forward than I hoped as several I would like to have asked were now in Biafra. J. P. Clark, told me that the young American writer Paul Theroux had known Christopher Okigbo and could write on him. The Canadian Douglas Killam who had applied for the Professorship at Lagos and who was in my department and who was, unlike me, a specialist in African literature, was asked to write an essay. Eldred Jones was a Sierra Leonean and editor of the journal *African Literature*. Ron Dathorne had introduced me to Amos Tutuola. At the end of the day, which was after I had left Lagos, the published book was the first on any single national African literature and the first to include vernacular literatures along with English language writers. I was also on my way to being an authority of the New English Literatures although, as with *Introduction to Nigerian Literature* (1971), I was self-educating myself by commissioning, writing the "Introduction", and editing the publication. Someone later told me that the reason my comments sounded right was that I was reflecting the views and contrasting opinions of the writers, many of whom I knew.

There was truth in that. Even the decision to plan the book during a time that Nigeria was fighting a terrible war against the declaration of independence by Biafra was an almost unconscious decision. The writers I had known at Ibadan had thought of themselves as Nigerians, and now at Lagos J. P. Clark and Ken Saro-Wiwa were Nigerians who were from tribes in Eastern Nigeria that felt oppressed by the Igbo of Biafra. It was impossible not to feel sympathy for the Igbo after the massacres in the North, but similar to many other oppressed peoples their natural desire for national independence ignored the rights of those who shared the territory they claimed. As the Nigerian government would not surrender the Delta oil fields that Biafra also needed to survive there was no obvious compromise.

Wole Soyinka had tried to find a compromise and was imprisoned as a result; the government thought he intended to lead Yoruba intellectuals into siding with the Igbo. It was true that Wole, correctly, suspected the Northern Muslims of being anti-Igbo, and he was part of the Chief Awolowo Action Group who originally led the Yoruba out of the national coalition and began the Western Region Crisis. The government kept Soyinka in solitary confinement for two years and tried to break his spirit and health, but it feared that his death would unleash a violent uprising among the Yoruba: besides being a famous writer and intellectual he belonged to an important Abeokuta Yoruba Christian family. I kept hearing remarks at the University of Lagos about his condition. Many Yorubas were informed about what the government was doing and how he was resisting. Although denied paper or ink somehow Soyinka managed to write the notes that were later published as *The Man Died*. I wrote to some British magazines about his imprisonment, but I was a small voice among a worldwide choir of important literary and cultural figures.

It is difficult now to imagine the bitter divisions between Nigerian writers during the Nigeria-Biafra war. J. P. Clark had been close friends with Christopher Okigbo who died for Biafra and Clark also quarreled with Achebe. He especially seemed to do what he could to harm Soyinka's reputation. One day Ken Saro-Wiwa turned up in my office announcing that he, an Ogoni, had escaped from Biafra. I knew him from Ibadan and immediately asked the Vice-Chancellor to let me appoint him to a vacant junior lectureship in English. Soon Ken was appointed as Civilian Administrator for Bonny in the Delta. Which was great for him except that as he refused to give up his University of Lagos housing I could not appoint anyone else to replace him. Saro-Wiwa had always been very intelligent, strong willed, and difficult. At Ibadan he had decided to take French; when in his first class Adele told him to stop talking he walked out and never returned. Dathorne had said in a class that a then popular Nigerian novel proved that "pidgin" could only be used for comic purposes. In reply Saro-Wiwa eventually wrote his own *Sozaboy*: A *Novel in Rotten English* about corruption and the Biafran war.

He was brave. Later in life, when he had become a wealthy businessman who was publishing his own highly successful radio and television plays he wrote to me that he was sending his children to be educated in England as he feared what might happen to them in Nigeria. He was close enough to those in power to know that when he challenged the government's military occupation of the oil producing regions of the Delta during the early 1990s he would be imprisoned, although I doubt that he could have suspected the government of framing him and others for murder and hanging them two years later.

With Soyinka imprisoned and Ulli Beier no longer in Nigeria, J. P. Clark took on the editorship of *Black Orpheus* and asked me whether I had anything for the next issue. I had started to write poetry once again when Nicole was born (Adele commented that I also lost weight while she was giving birth). I had published a few in American magazines. I was influenced by the new manner introduced by Robert Bly and James Wright in the late 1950s in which ordinary scenes, followed by surreal imagery, were used to tap otherwise inarticulate emotions. I had also published a review of Wole's novel *The Interpreters* in an earlier issue of *Black Orpheus* so I did not feel that J.P. was necessarily flattering his Head of Department, but I did start joking about renaming the magazine "Black and White Orpheus". It turned out to be a powerful issue with some of Okigbo's final poetry, and an impassioned lament by Clark about Chris's death.

Despite the war we enjoyed a strong social life. The university supplied us with a house on campus that had been screened against mosquitoes and supplied with air conditioners by the Canadians for a former professor who had fled; when he eventually returned we kept the house but offered to give back the sand pile that the Canadians had imported for his children. We often played bridge in a weekly duplicate competition and sometimes socially with Peter Connell from the British Council. Besides J P, our social companions within the English Department included Killam and Abodun Adetugbo. Adetugbo's history interested me. He and his wife married in the USA and had to live in Lagos as they were from two different groups of Yoruba that did not approve of marrying each other. Cities such as Ibadan in Africa originally came into being similarly to the cities of Europe, partly as markets but also as free areas for dissidents, the banished, and others who needed to escape the oppressions of community. I would later hear a similar story when I taught in Zaria when an East African married a Nigerian student and were threatened with death by their communities; they moved to Singapore.

We employed a cook, "house boy", baby nurse and shared a night guard. The first baby nurse, a large woman in her mid-twenties, immediately began to steal from Adele who went to the police who found the jewelry but not the money. I told Adele to drop it as in Nigeria the thief might bribe a judge and the victim would be imprisoned. That is almost what happened. Adele went to observe the trial where the woman's lawyer produced a document claiming that she was underage and Adele was the guilty one for employing her.

There was no staff club on campus but we dined and danced at various clubs, the music ranged from High Life to American soul and twist. At a posh club in Lagos there were weighty older men in Nigerian robes, attractive prostitutes and rumors that men had drunk and danced themselves to death. Frank Cawson of the British Council often had good parties and played recent British recordings including the Beatles. But the best was Fela Ransome Kuti's group in the basement of a Lebanese owned hotel, was it the Empire?, playing their mixture of high life, jazz, James Brown, and funk. Fela was a first cousin of

Wole. Such Christian families produced many leading artists and intellectuals as their children created a modern culture by blending past and present.

The club was so crowded it was difficult to do more than move your body, the dance floor became covered with sweat and condensation, and to outsiders it appeared, I was told by visitors we brought, a scene from Hell. J. P. may have originally told us about it as he did about several of the places we danced. Although we sometimes had meals at each other's house, J. P. was often out on the town, and his wife Ebun would visit us—our houses were within walking distance on campus—as he felt she was secure with us.

Lagos was closer to King's Road than Bristol seemed. Adele wore mini-skirts and even a see-through knit dress and I had long hair, facts commented upon by a local newspaper columnist who wrote "I never thought I would see a Professor of English ....". A couple we knew, he was American, she British, seemed the epitome of the swinging sixties, with their LSD, pot, King's Road clothing, and tales of "so I told her to join us in bed, it would be fun."

In general, however, we did not know "official" Americans, those employed by the American government. They were told to avoid Americans on direct hire to Nigerian institutions; we might be Communists. Our friend, marrying his British girlfriend, had to sign an undated letter of resignation in case she later were found to have undesirable associations. Our daughter could not get the injections and other medical aid offered the children of "official" Americans, nor go to their entertainments. I was once invited to a private dinner at the American embassy when they assumed that as a Professor of English I was British. I tried to keep my distance from the American professors on campus; several were from teacher's colleges and thought we should be giving instruction on black board techniques. Black Americans assumed that they were Africans, although being from no tribe that was a dead end. I felt most of the Americans had no idea that they were in Nigeria and did not look around them; their world remained Bob Dylan and Martin Luther King. The worst was someone who came to me claiming that we must offer extra-mural courses to the "many of thousands of bored suburban house wives" with university education who needed something to do. In Nigeria any woman likely to graduate with a university degree had been proposed to by males seeking a wife who would be paid well in a good job. Women with university degrees were still a rarity in Nigeria.

It was true that many of the stray Americans in African universities had leftist affiliations. When Nigerian university salaries fell below those of in England the British and French governments started to supplement the salaries of their nationals. Someone from the Rockefeller Foundation called together the Americans at the University of Ibadan to tell us that Washington would not permit such supplementation as it feared that some of us were communists. At that meeting many were surprised to find that others assumed to be British were Americans.

At Columbia I knew Sam Astracan who wrote his first novel while still an undergraduate. His older brother was a *New York Times* journalist in Lagos and through him we met Marian and Arnold Zeitland. Arnold was also a journalist. Marian taught a French course in the evening. Marian's family had been Christian missionaries in China where she was raised. As Protestants they warned her against marrying a Catholic; instead she married a Jew. She and Arnold were interested in Nigeria and had commissioned Twin Seven-Seven, a famous Osogbo artist to cover one of their living room walls with a mural. They also adopted a Nigerian child as well as Marian giving birth to a child in a Yoruba village which she thought would be the best way to learn the language through complete emergence. As she was used to the four tones of Mandarian she had little trouble picking up the four tones of Yoruba. She claimed that the village women were always complaining about the lack of sexual stamina of their men; another myth about Africans demolished.

One day a man came into my office asking if I could employ as a secretary his American girlfriend, who would soon arrive in Nigeria. The answer was no, but we soon became friends. He claimed to be a

French-Polish Count with a doctorate in Chemistry who had been sent by the United Nations to help develop local products. He told stories about a noble grandfather so addicted to gambling that he would wager his wife, losing, then winning her back. He was full of money-making schemes. Holding a UN passport he was exempt from the money controls others faced and he would exchange Nigeria naira for dollars at a large profit, then buy dollars with the Naira at an official subsidized rate for the UN, and repeat the process many times, earning a fortune. He invested in the businesses he helped create although that was illegal.

He smuggled his girlfriend into Nigeria and sent her full time through the villages in a car purchasing antiquities that he hid and would show no one. After he had about five hundred he smuggled them to America by way of a Polish freighter. The girlfriend earned her doctorate and a university position and published two books studying the antiquities that she claimed came from a previously unknown Polish exploration of West Africa during the nineteenth century. They had been housed at a Polish museum destroyed during the last war and he had brought them to America. With such a providence some of his pieces won awards at antiquity shows, increased tremendously in value, and when I read about them I did not know whether to warn Africanists about their actual history or laugh. Then he made a major mistake. After he ditched his girlfriend for another woman, she shot him dead. When I telephoned her university for confirmation of what happened they did not want to discuss it.

Although Adele was appointed only at the Lecturer level she was Head of the French Division of the Faculty. Perhaps because of her position we were expected to entertain the Rector of the University of Abidjan on his visit to Lagos. We picked him up at the airport, took him home for a meal and got along well, and were soon invited to visit his University. Thus began a long courtship with the University of Abidjan that was never fully consummated. On our way to the Ivory Coast we spent a disillusioning night in Ghana where the well know

High Life band at our hotel was sleepy; the country still followed Eastern European exchange controls. On entering Ghana you not only had to pay for a visa you had to exchange a large sum of money for local currency. Leaving the country you could try to exchange the remaining money but the currencies available would be Eastern European that could not be exchanged elsewhere.

Abidjan was a clean modern city, laid out and looking like part of the French Riviera. The luxury hotel where we were put up had oysters, lettuce, milk, and wine flown directly from France. The market could be in any provincial French city. All this was a pleasant contrast to the dirt, disorder, and violence of former British West Africa. Even at the University there was a mixture of European and African students, you might have been in France. That was the trouble. The African literature course consisted of three texts, one each by L S Senghor, Aimé Cesaire and J P Sartre's *Preface to* Senghor's 1948 *Anthology de la nouvelle poesie negre.* Only Senghor was African and the syllabus was only concerned with the Negritude movement and had no recent Francophonic African texts. The students of English took us out to a jazz club that played Louis Armstrong recordings and looked like a provincial copy of something metropolitan French. As we left I asked about the drumming I heard and was told in effect that was African bush music of no interest. These Africans had become French in their cultural sensibilities. Even Negritude was essentially a romantic cultural movement that the French were happy to support as it kept their former colonies from nationalist and Marxist politics.

Corruption was always a problem in Nigeria and especially at the University. Someone, supposedly a student, distributed an open letter accusing many of the better known faculty of corruption, especially demanding sexual favors from female students. Although the actual author was unknown the letter was thought to have some connection with the Yoruba faction with which Soyinka was associated and a Professor of History at the University of Ibadan was said to be involved. At the next meeting of the University Senate some Nigerians wanted to close the university until the culprit was caught. They claimed that this was the traditional Nigerian way, the village is made to suffer until it turns over the guilty. I was among those, all non-Nigerians, who claimed that it was absurd to punish so many for the fault of one who might not even be a student. It was a heated debate and left hard feelings, especially as those named in the letter claimed those who wanted to keep the University open believed the accusations, which might have been true. It was difficult to live long in Nigeria without suspecting that some of the letter was correct. J. P. Clark, who was among those named, suddenly went from friend to enemy and disrupted the workings of the department. Although a high level committee was formed that reprimanded Clark I knew that nothing would change as J. P.'s brother was Minister of Education. I had idealistic notions that I should resign in protest rather than permit the situation to continue under my Headship.

I immediately sought another Professorship, managed to get to Vancouver during a Nigerian Airway pilots' strike when an American Pan Am pilot taking pity allowed me to fly on his plane to New York where I arrived with my Lagos-London-Vancouver ticket, and had to talk my way on to a flight to Vancouver. I arrived sleep deprived, jet lagged and very ill from a half cooked meal served after a stop in Dakar. I did not know I was expected to give a lecture about Andrew Marvell's poetry as the letter had been sent by ordinary post and would not reach me until after I returned to Lagos. My confused attempt to improvise a lecture was disastrous. Instead of moving from Lagos to Canada we would spend the next year in Mallorca while I wrote my Marvell book and applied for professorships.

# Chapter 8: Swinging Sixties, Drugs and Graves

*Deia, Mallorca 1964–86*

In Palma de Mallorca Nicole was surprised to see a black couple. In Nigeria she had not made a distinction between white and black. We also learned that Nicole had been given coconuts to amuse herself while Bassey, the boyfriend of her baby nurse, was "playing with Mary's bottom". We rented the main floor of a house in St Augustine on the outskirts of Palma as it was supposedly one of the more sunny areas in winter. Nicole went to kindergarten in Spanish at a school run by nuns and made friends with a Canadian girl whose mother lived near us. The father worked for an oil company in Africa and returned for a week each month. Our house was warmed with a log fire started with pine cones, very different from the dreadful Leeds coal fires. At Christmas I gave up attempting to open the oysters and Adele said she would try. With her left thumb almost falling off we asked the couple in the apartment downstairs for help going to a hospital; he said he had studied medicine before deciding to run his own business. He heavily bandaged Adele's hand after covering it with a powder which we later learned had been banned and might have killed her. She also escaped, with only a burnt eye brow, from the explosion of the butane gas oven. A few friends kept visiting us from Deia, especially Stephanie who was teaching English at the soon-to-be new university. It was then unheard of for foreigners to be hired by the Spanish, but Stephanie had a course in medieval Catalan at Oxford and could speak Majorcan. There was a small but lively Majorcan culture still alive in Palma although it had been banned under Franco.

We started going to Deia the summer of 1964. Adele had a lectureship in French at Ibadan during the 1963–4 academic year and was told by Henri Evans, the new Professor of French, to take the course on language teaching at the University of Besançon during the summer. We had spent a month in Paris, then found our Besançon hotel full of bedbugs and roaches. Henri was staying at a monastery and was not interested in our problem. I think he was happy any place to avoid his wife Jacqueline. Their marriage was not tranquil and she would eventually leave him. Once we had been asked over for an evening of bridge with her and a Frenchman who taught in East Africa. He was being paid by the French government and was seeking another university to which he could be appointed. During the game Jacqueline made a stupid play and he calmly said "Madame, vous êtes une idiote." That should have been enough to end his hopes for Ibadan, but that summer in Paris he seemed to feel the appointment was his if he wanted it. He took us to an expensive meal before going to see his gangster friends in Montmartre.

Not trying to avoid Jacqueline, with whom I had more sympathy than I had with Henri, a couple of nights of bugs was too much for me and we fled south seeking an inexpensive place for the remainder of summer. Barcelona led to Palma de Mallorca and then I remembered that Robert Graves lived in Deia. A taxi took us to Ca'n Quet, a pension with good food and a magnificent terrace just outside the village, where the first night we were questioned by a French butcher "who was Robert Grav-es", whose son William he had slugged. It is expected to protect the honor of your woman, isn't it? Our contract with the University of Ibadan paid our fare to Europe during the summer and we stayed at Ca'n Quet the following summer when we met Jan Lynton who had a house in Deia village and who we knew in Leeds. From then on for about a further decade Deia became our summer home, while I taught in Lagos, Canada and elsewhere and it seemed natural to go there for a self-paid sabbatical after the University of Lagos. As Deia is cold and damp during the winter we spent that period in St Augustine. I bought a very old French car from a departing American and we could travel between St Augustine, Valldemossa, Deia, Fornalutx and other places on the island where we had friends.

One of our friends was Stephanie Sweet who had attended university with Graves' daughter Lucia. The Oxford police charged her under a new law that made landlords responsible for drug use on their property. Stephanie had rented her cottage to some students during the previous summer and became the first person charged under the new law. Although she had not used drugs and had no knowledge of their use in her cottage she was banned from teaching. Robert thought it monstrous and invited her to Deia while he used his contacts to have the House of Lords repeal the law. We became friends with her in Deia and about the time she moved to a stone farm house in Valldemossa we were in St Augustine. Graves' library had many concordances, commentaries on the Bible, and other books that I borrowed through Stephanie to work on my Andrew Marvell manuscript, which partly explains the eccentricities of its sources, notes and information.

I believed that Marvell wrote his best lyrics while working as a tutor to the daughter of Lord Thomas Fairfax and that they contained many allusions and symbols that created a double perspective altering their significance from the secular to the religious. I knew that Fairfax had patronized commentators on the Bible and was part of a social network that included other poets, such as Thomas Stanley, who I thought wrote obliquely about politics and religion during the "winter" after the execution of Charles I. The obvious scholarly way to prove my argument would have been to consult rare manuscripts in Oxford, Leeds, and other libraries in England. Without the resources or inclination to do such leg work I tried to read the poems with the aid of whatever biblical and ancient sources I could find, a methodology sanctioned by New Criticism, biblical commentary, and by the psychedelic, let your mind expand, orthodoxy of the late 1960s.

Deia was certainly a place for aids to mind expanding. An invitation to come over for a drink might lead to someone offering "homemade LSD". There was "Blind George" whose cane was used for smuggling illegal substances. David Solomon, editor of *The Marajuana Papers* (1964) and editor of *LSD: the Consciousness Expanding Drug* (1965) had brought his wife and two daughters; their quarrels could be heard through the Deia nights. Solomon was not a selfless propagandist for flower power; he was a friend of Timothy Leary and was later discovered to have come to Europe to create a network for LSD manufacture and distribution. Solomon had been literary editor of

*Play Boy* when it was publishing writers favorable to drugs and an editor of the American jazz magazine *Metronome* but perhaps ironically because of his preaching drugs, perhaps as a disguise, was often referred to by Paul Arnabaldi and others as "the rabbi". (There was also a famous scholar of Jewish studies named David Solomon.) Paul Arnabaldi, with whom I sometimes played chess, was thought to be mainly interested in young girls and his source of money variously explained before he also was sought as the head of the international ring of drug dealers that included Solomon. He owned the house in Wales where LSD was being manufactured for Europe. Arnabaldi was the only one who escaped capture and was thought to have been tipped off about the "Operation Julie" bust; after having his face slightly modified in Thailand he would return to telephone and see old friends. He lives in Mexico and sometimes appears in Deia.

I was cautious. I depended on my mind to survive and I knew some of those who had not. A young German painter had died by stepping off a moving train. I liked his work and had enjoyed talking to him in Deia. It was, however, almost impossible to avoid having your mind altered. A dinner buffet during a large party at Jakov Lind's "tower" had suspicious looking chopped spices in the salad and the punch was no doubt spiked with LSD. I had to pick up Nicole and one of her friends before midnight but as I drove through the winding narrow roads I was shouting "wheee!!!!!" and felt I was flying. Somehow we made it home without going over a cliff.

Jakov Lind was the second most famous writer in Deia after Robert Graves and possibly as great a writer. An Austrian Jew who was sent to Holland when he was eleven years old he decided his best chance for survival was to pass as Dutch in Nazi Germany. He even cooked for an SS officer who after he realized Lind was a Jew feared denouncing him without implicating himself. After the war Lind emigrated to Israel where he married, fathered a child, returned to Europe and eventually divided his life between Hampstead in London, the Village in New York, and Deia. When I met him he was already becoming internationally famous for his first book of short stories and first novel, published in German, (1962–3), about five years before he started writing in English. Although Lind was thought a Magic Realist similar to Gunther Grass his theme was the absurdity of life as represented especially by the Germans he met during the Holocaust. They were as much his theme as the Europe of the Holocaust.

As his own survival demonstrated, life was bizarre, meaningless; survival, sexual pleasure, and writing were what counted. He had a sense of humor and enjoyed quirks of character. He was one of the Deia chess players, more interested in the personalities of his opponents than technique. He took pleasure in talking you out of a winning position, and commenting upon your behavior. He was also part of the LSD set. Some years after Jakov's wife Faith died from throat cancer his son by his first wife came to Deia from Israel and we would dance in Port of Soller. Unfortunately he seemed to know only Israeli folk dances which would rapidly clear the floor of the discotheque and bring the set to a premature ending.

Eva Figes, a friend of Jakov and Faith, was also a Deia regular. Although we became friends and would often take her to restaurants in Deia and the Port of Soller and stay over with her in London I thought her obsession with her literary career tiresome. Her conversation continually turned to reviews of her novels and which writers, publishers, and reviewers she met at London parties. A Jewish refugee from the Nazis she was weirdly nostalgic for Germany, had an affair with Günter Grass, and hung in the main room of her apartment a blue cloth he had given her. When we brought her a blue cloth from India she pointedly told us that it was not the right color of blue, it did not match Günter's.

She was not a Magic Realist or someone who used surrealism for political allegory as did Jakov, Günter Grass, and others at that time. She was part of a group of British writers, most notably B S Johnson, influenced by the experiments of the French *Nouveau Roman*. She became famous as a feminist with the publication of *Patriarchical Attitudes*. Eva used to foolishly claim that no one previously had questioned Milton's views on women; and I would pedantically ask how could she write such a book without having read Sir Robert Filmer's *Patriarcha*.

Such literary tensions reflected a difference of personalities. People I respected spoke of her as intelligent, but I thought Eva self-centered, emotional, messy, unable to see beyond her own needs. She had gall stones and little interest in food. If I took her out in London to a Chinese meal she would insist on going to one she knew and order an expense meal that I knew she would not eat. Once she insisted we go to an Italian restaurant in Soho that years before had been a favorite among her friends. It turned out to be now patronized by soccer fans happily singing their chants in the other room. Eva kept claiming that they were hooligans and that we leave immediately without paying although we had already eaten the first course. When I had to go back to get my raincoat, she said I was "bourgeois" and had left my rain coat purposely because I wanted to pay the bill. When we went to Joan Littlewood's Theatre Royal Stratford East production of *Oh What a Lovely War* she was afraid of the black youths in the neighborhood. One night after midnight she telephoned a mutual friend and told me to leave her apartment and go as she could not sleep while I was there. I think I missed several obvious signs that she wanted me to make a pass or more. But I would never have become involved with Eva as I thought her dangerous; I had listened to her after an affair broke up saying she was going to telephone the man's wife demanding that he choose. To anyone else but Eva it would be obviously that he had chosen and she was seeking drama.

Adele asked Eva to co-edit the Women Writers Series with her and felt she had made the right choice when Eva's agent had their advances doubled. But Eva did little for the series beyond turn down manuscripts that expressed different opinions than her own. I think she saw co-editing the series as a way to move up in literary circles but when those she hoped to court did not bite she lost interest. As the publishers did not fancy a fight with Eva's agent the series eventually died, but not before there were about thirty titles mostly commissioned and edited by Adele.

There were other British in Deia who we sometimes saw in London and who were friends of others we knew. Jan Lynton led us from Ca'n Quet to the village. Her husband Nobert had become a well-known art critic and university professor. Although they had separated and she had the two boys, Oliver and Jeremy, he would not for years allow her to divorce him as he wanted to avoid marrying the woman with whom he was living. This meant that Jan's few earnings as a part time art teacher were taxed as if they were additional to his income. Eventually she would work her way up to be head of the School of Art at Goldsmith's College, University of London, and he would remarry, but those years of penny pinching left her cautious in her life, and always worried about money. Oliver and Jeremy at times seemed like older brothers to Nicole, once, when she was still little, rescuing her from a motorbike that had fallen on her.

Among Jan's friends were Rosalie de Meric (the second wife of the poet Tom Blackburn) and her daughter, now the writer, Julia Blackburn. They would often be in Deia, lived near people we knew and sometimes saw in London. Their lives were an extreme illustration of British bohemia. Tom was addicted to some medicine that left him confused, which he combined with too much alcohol, making him at times nearly insane. Rosalie was a painter intent on sexual satisfaction with often much younger men. Julia and her mother were openly in competition for the same men, and it was said that Tom, although primarily heterosexual, shared the same man with both women. Rosalie could be crudely obscene and said to a still young Julia in Deia that men would someday be trying to get into her cunt. We visited Rosalie's studio in London where she was supporting an effeminate toy boy who claimed to be unable to survive without her aid. Julia Blackburn's books describe some of the horrors of her childhood and youth.

There were several creative writing teachers in Deia during the summers from a branch of an American university and they thought the village would be a great place for a writers' school during the winter. The trouble began with the American students expecting a choice of what to eat at their meals; fish was difficult to obtain in winter and local cheese was much more expensive than meat. The winters in Deia are depressingly cold and damp, and there was little else to do beyond getting high. Before the first semester was over a Dean arrived from America and closed the college after finding the students and their teachers all experiencing expanded consciousness.

One of the writers was George Cockcroft, "Luke Rheinhart", the author of *The Dice Man* (1971). Adele was the woman in the other apartment that the Dice Man thought of raping. I never regarded him as a good writer but he was lucky. By the late sixties everyone was experimenting with I Ching and throwing dice as a means of divination when a sharp young British agent visiting Deia liked Cockcroft's outline of the book and sold it to publishers and Hollywood for a fortune. It became a cult classic. I thought him foolish. He had rented a sail boat on which he and his family disappeared in the Mediterranean for days until they were rescued.

Because of Graves and Deia's reputation as a center of the arts there were often agents and editors looking for undiscovered talents, but Deia was a place where once brilliant people might write a book and spend the next decade or two in a drugged and drunken haze trying to write their second book. Alfred Dhurssen met an American editor at the main bar; the editor published Alfred's *Memoirs of an Aged Child* which received good reviews. Many years later Alfred sent the same publisher without any covering note explaining his past association his second semi-fictionized autobiography which received a form rejection. It would eventually be self-published as *Difficult Women*. Alfred, who is extremely cold and self-centered, indeed had difficult relationships with women, especially Bunty, a former champion swimmer with whom he lived. Their continual arguing was heard by most of Deia. Their daughter married into English royalty, to someone in line to the throne, a musician, fashionable enough to be often photographed in British publications. The basis of the marriage was believed to be the mutual affection for drugs. They later divorced.

Writers in Deia included Jim McKinley, who taught Creative Writing at the University of Missouri, Kansas City, and edited *New Letters*.

He had a house and wrote a biography of Robert Graves that he told me was going to be an alternate choice for the Book of the Month. Then suddenly the book disappeared. According to one rumor Graves had disapproved of it, according to another Jim had decided he had too much wanted Graves' approval when writing it. He was to read a portion of the unpublished book in 1977 on an audiobook, the same year that he wrote an "Introduction" to a new edition of Grave's *Collected Poems*; McKinley also published *Assassination in America* the same year, but basically he was the author of short stories and magazine articles.

It was undoubtedly McKinley who obtained the commission for writing an article on tracing Hemingway's footsteps in Spain for a sports magazine. Its' co-author was Nick Arnold; it was Nick's only legitimate publication, although for a year or so he was one of several aspiring writers in Deia who wrote pornography on a sadistic book a month contract. Nick was a Harvard graduate who inherited a million dollars, and after military service went to France and then Deia hoping to be a writer. Three wives and a wasted million dollars later he had published little, drank far too much, and would later move to Goa where he could survive on a dollar a day until he died of cancer. I often played chess against him and we had many of the same friends. After he died his daughter Annie, the only one in the family with a fixed mailing address, went to Goa planning to bury his ashes in Paris. After a long struggle with the Indian and French bureaucrats she succeeded but her two druggie half-sisters claimed that she had taken for herself some money Nick had left and during an argument in Paris had knocked out one of Ann's teeth. I was surprised when Ann continued to see her half-sisters, but she said they are two thirds "of the only family I have."

The Americans and some of their Australian friends seemed to set the tone of Deia during the 1960s and 70s, perhaps because of the protest against the Vietnam War, perhaps because the conflict over the use of drugs was intense in the United States. I would often speak to and sometimes dance at parties with Ellen Shirer, who had published several excellent illustrated books for children, and who had been a dancer for Balanchine's New York City Ballet. She seemed to me loudly opinionated and self-centered. She now lives in France and is a painter. I sometimes noticed letters she would write to *The Herald Tribune* screaming against American foreign policies; the editor of the page told me that he had to limit the number he printed by her. Alice Meyer Wallace was another illustrator of children's books who still continues internationally as a painter and art teacher. I always found her more relaxed and amusing then husband Bruce Wallace who was more prominent in Deia, but who became a monk and a hermit. Alice was the talented one.

Several well-known journalists who also wrote books had houses in Deia. I used to listen to Fred Grunfeld's "Music Magazine" on WQXR when I was at Columbia University. He and his parents were Jewish refugees from Germany. After graduating from university in America he became an international cultural journalist and writer. His books ranged from a history of guitar to a biography of Rodin. He moved to Deia in 1960, had a house on a hill outside of the village, and was often travelling on assignments. He once noticed that I liked playing the Grunfeld Defence and told me that it was named after one of his relatives who had made it famous.

I never felt I knew Grunfeld who appeared to live in his own busy world unlike John de St Jorre whom I met soon after he came from Africa to Deia. We shared African stories and John was to write a book about the Nigerian Civil War. He had worked for the British Foreign Service in East Africa, resigned to become the foreign correspondent for the *London Observer*. Driving across Africa to Nigeria he was detained in the Congo and made to squat for hours in a hut while soldiers attempted to understand his passport and decide whether he should be shot or freed, a typical story about Africa, especially the Congo.

For some years John moved back and forth between his house in Deia and New York, where he was advising foundations, and editing their books, on world politics as well as being a journalist and writer of books. I ran into him much later in Paris where he was giving a reading from *The Good Ship Venus* (1994), a book he had written about Olympia Press. Its director Maurice Girodias had published Henry Miller, Nabokov's *Lolita*, and many once banned classics along with "dirty" books. During his research John had interviewed a now aging Dominique Auric who confessed to have written *Histoire* d' O (1954) as gift of love for the writer and editor Jean Paulham. During the mid-1960s a translation was published in English by Olympia Press and Grove Press. John explained that for tax reasons, Richard Seaver an editor at Grove Press, was unwilling to confirm he was the translator. Unsatisfied with the advances for the British and American editions of his book, John had lunch with Tina Brown, whom he had known in England and who was then editing *The New Yorker,* and who published his history of the *Story of* O for more than he had received for the two editions of his book. His screen play of the book has been optioned for a movie.

We used to employ a baby-sitter who our daughter loved, the girlfriend of the saxophonist Didier Malherbe. As soon as she arrived the crayons and colored pencils would come out and when we returned there would be glorious psychedelic drawings. Unfortunately the baby sitter also turned up at 4 or 5 in the morning trying to remember whether she left her aids to consciousness with us and in her altered state begin searching the large house that we rented. Another acquaintance suggested that we employ her daughter as babysitter, but as soon as we returned the first evening it was clear that something was wrong. The girl claimed that Nicole had fallen down the stairs. It seemed that the mother, a rich divorced woman with a young lover, had often hit her daughter who took her anger out on our daughter. After that Nicole either came with us or stayed with Francesco, the Majorcan woman from whom we rented the house. Nicole would watch bull fights on television and thought that the same bull returned for the next fight.

Although it is now said to be its golden period or "Heydays" I had mixed feelings about Deia in the sixties. There were too many claims of being unique and rebellious by people who dressed alike and shared the same opinions. Seeing ten young people in identical long white Indian cotton shirts congratulating themselves for not being in "uniforms" was disillusioning. Many of those now listed as writers or poets published little. A young couple I liked presented another incongruity. During most of the year she was a prostitute in Marseilles, he was her pimp. For several summers they cleansed themselves in Deia through meditation, vegetarianism, and of course wearing the long white cotton Indian shirts.

People kept coming through who knew someone who came to Deia during the summers and it was surprising how often there was some previous association. A former student of Adele's when she taught part time at two private schools in New York during 1960–61 told her that the boys would purposefully drop pencils and books to get a better look down her loose fitting blouses. A pianist's boyfriend said he was a jeweler and using stones we had brought from Nigeria did indeed make Adele some spectacular necklaces and rings, but unless you are dancing nude they are almost unwearable. The last time I convinced Adele to wear one of the larger necklaces someone told her that he was himself a jeweler who had designed similar pieces during the sixties.

After 1968 Daevid Allen, Kevin Ayer, and Robert Wyatt of Soft Machine settled in Deia; by 1970 the group was starting to split and those in Deia became Gong, with Didier Malherbe on reeds and Allen's girlfriend Gilli Smith, on weird voices. I thought of them as Deia's psychedelic party band. They were great for wild dancing. The pixie like Gong act with which Daevid Allen now tours is nothing like the out of your mind music the band played when, as one of their record albums claimed, Deia was "the center of the universe". Malherbe's Tibetan horns and other unlikely instruments that have made him a Parisian cult distract from his being an excellent jazz saxophonist.

Bill Waldren was among the American ex-GIs who after studying painting in Paris came to Deia. He built a large California style open plan house for his family. When "action painting" went out of fashion and regular payments from an American gallery ended he tried his hand at various jobs including decorating night clubs in Palma. Bill enjoyed drinking, was powerful, and was known to have fights while drunk with the Guardia Civil that ended with them hugging each other as friends rather than with him in jail or hospital. One of his buddies was Steve Kopper who had married an American woman who inherited a Deia house. Steve was studying to be an archeologist and Bill would go on digs during which they discovered an ancient, previously unknown, culture in the caves of Mallorca. With Robert Graves' support Bill built a museum for the finds; the museum was also used for concerts by musicians, especially by a seemingly endless flow of good guitarists. The museum attracted outside funds and Earthworks sent Waldren short term would be archeologists who paid well to dig in the caves; being a "digger" in Deia meant being one of Bill's crew. Jackie Waldren supervised the cooking and the large house became a hostel for diggers. Graves built other houses and the museum expanded into a research center.

Bill had learned from Steve how to prepare articles on their discoveries for archeological journals. These came to the attention of Oxford University archeologists who visited Deia to see what was happening and were surprised by the discoveries. They proposed Bill pull together his research and write an Oxford University doctorate. There was one problem. Bill not only did not have a university degree, he had left high school before graduation to pursue a career as a figure skater.

He was invited to Oxford for a drunken weekend, granted a B. Phil, and allowed to work on the doctorate. He thought originally "Doctor Bill" would be a good advertisement for his digs, but he wrote several books on his research, was asked to lecture at Oxford for a term each year and Jackie, during that time, would supervise one of the college halls of residence. She eventually wrote up her research on the inter-action between local people and foreigners and earned a doctorate in anthropology and became, Dr. Jacqueline Waldren, an Oxford lecturer, an ethnographer, author of a book and academic articles on how foreigners and locals change each other.

Central to my social life in Deia was a continual chess game that moved from the cafes to the beach to private houses to restaurants to wherever pieces and a board could be set up. I had never studied chess openings, only played during holidays, and would go home with headaches rethinking moves. Although the standard in Deia was high, and several who had played on their national teams passed through, I was determined to hold my own, which meant that I was excruciatingly slow as I had to think through every move; I had little knowledge or experience upon which to fall back. I did have an instinct for how a game was likely to develop. If in the opening or early game I did this the ending was likely be pawns pushing there or an attack on the king if my opponent, as likely, castled on that side. I was especially weak on the end game and needed an extra couple of pawns or an extra piece to compensate for my lack of technique. Anyone willing to put up a dogged fight at the end was likely at least to tie. I always appreciated those who did not know my weakness resigning early after I somehow conjured an impressive position or threatened to win a piece. I was amazed when two players from national teams repeatedly resigned early to me. If they were Deia regulars they would have known better.

One of the most dedicated players was Paul Crotto, a New Yorker one of those who came to Deia in the summer from Paris where they had settled after studying art on the GI bill. Paul was a handsome man who attracted woman despite his continual complaints about his health. Nick Arnold, who knew him when they first went to Paris, claimed that Paul could sell anything, which I found curious as he was not articulate in a lettered way, his conversation on anything sounded like overheard sound bites. He was, however, a good painter in the increasingly out of date Parisian School. He was aware that he had avoided following those he knew towards action painting, pop art, minimalism, and other postmodernist styles and that perhaps explained his obsession with chess rather than his own paintings which over the decades would sell poorly and sporadically. Every so many years he would be rediscovered and promoted for a few shows by some gallery and then there would be a wait for the next rediscovery. Meanwhile he played more and more chess, painted less and less, and his wives left him. He was proud that his niece was one of the top women chess players in America, and he feared that insanity ran in the family as his mother had been institutionalized. One of his sons would also have mental problems.

Karen, a beautiful and sensuous, Swedish woman was Paul's second wife. She enjoyed fantasizing. She told me at various times that she was Russian, a former ballerina, a former journalist; she claimed to be a chaste wife although Paul talked about their open marriage. She worked as a secretary in the Swedish embassy in Paris and had married Paul when it was possible to rent an inexpensive run down large apartment in Montmartre, fix it up, use it as a studio, and eventually buy it. Now the price of his paintings were no longer quoted in art journals, and Karen was tired of returning from work to find that he had played chess all day instead of painting. She also had an older Swedish multi-millionaire art dealer in love with her.

When she left Paul for the older man and houses in Sweden, Versailles, and the Riviera, and Picassos, Matisses, and African art deposited with Swiss banks, it, surprisingly, was a crisis. She insisted that her new husband buy her a house in Deia, took to pills, then became an alcoholic. She also became financially penny pinching when around Deia friends. We were supposed to meet her and Jan Lynton for lunch at the Drug Store in Paris. Karen announced she would pay for the set price meal but any additional wine we had to pay for ourselves, which I thought offensive. No one expected her to pay for lunch but if she decided to do so why not for wine. Paul stayed at the Deia house during summers, continued to play chess, and soon had a third wife, an English woman, who eventually would leave him for spending too much time on chess and not enough painting.

When Karen's rich husband died she took up with, and married, Jacques Lutz, another fantasist, whom we knew from Deia where he had been Stephanie Sweet's lover. He was also dangerous. He liked to humiliate people, used to beat Stephanie, and had a grossly inflated image of himself based on a short newspaper piece he once wrote, a photocopy of which he showed me when he first arrived in Deia. He did have a talent as a satiric painter, in the manner of George Grosz, which, after a few small exhibitions he did not pursue.

Jacques knew how to use Karen's weaknesses including a streak of masochism, and as she was increasingly institutionalized for alcoholism, he persuaded her to transfer to him her property and money. He would insult her two sons by Paul, calling them "Jew boys", and Karen did nothing about it. She adopted a large cross and believed that she was part of the local Versailles French Catholic aristocracy. Eventually when she died Jacques inherited most of the property except the Versailles house that she left to her sons and the Deia house which she left to Paul. Jacques soon died and the fortune and other property passed to his French family.

For some years we have been fans of William Christie's Art Florissant and sometimes went to their concerts at Versailles, but as it was difficult returning to Paris we arranged to stay for a few days with Karen and Jacques. After being shown the art collection we were fed large oysters and pates from hampers while Karen and Jacques increasingly drank far too much and started to lose their dignity. Karen changed into a bathrobe and then disappeared. Jacques lasted a bit longer; he adopted his speaking to inferiors tone when asking Adele about her "petit thesis", her doctoral dissertation about Paul Nizan that had been published as a book. He still had not written anything beyond that early newspaper article but felt that he was an intellectual and talked about what he planned to write. We left early the next day.

Although he was the better player I usually ended up even with Paul at the end of the summer. I could hold my own or more often won with other regulars but Walter Biemel usually won our games and sometimes slaughtered me. He was remarkable. He had a bad leg but played table tennis at championship level, was an excellent swimmer, played viola, and wrote a book each year which would be translated from German into twenty-some languages. I do not know how he kept up with so much but when he brought me a two page review in German from a Swiss newspaper of my *New English Literatures* (1980) I was flattered and amazed.

Biemel was born in Budapest, studied in Romania and then began work towards his doctorate with Martin Heidegger. To avoid Nazi Germany he lived the final years of the Second World War in Belgium, then returned to Germany where, he told me, he was upset to find that many of his university colleagues were still unrepentant fascists. As he and his wife Marly had both studied with Heidegger they became world authorities on his philosophy. Biemel's interests were, however, much broader than German thought and he published books on a great variety of subjects such as J. P. Sartre and Picasso. Although his specialties were phenomenology and aesthetics he was a philosopher in the older European sense of someone whose field was culture inclusive of social and political aspects as well as explicating ideas. He was also highly disciplined. He would drive in his Peugeot every morning to the nearby village of Fornalutx for a swim (we were often invited to accompany him), the afternoon would be spent writing his next year's lectures which would be published as a new book, and the evenings would be free for chess, music, and socializing. Looking forward to retirement Walter and Marly bought a large house in Deia which they modernized—it had a magnificent large kitchen-dining area,—and then willed it to a local family whom they hired to care for it and themselves.

Bob Bradbury loved to play chess but was a push over and often retreated to his old stone house, with its out house, to paint and brood over his humiliation at losing and was mocked by the stronger egos of the café crowd, especially the painter George Sheridan whose loud voice seemed more a weapon than for conversation. George liked to prove that he could have won his games by restoring the pieces to an earlier stage and restarting as if the game had not already ended with him losing.

Bob and his wife Dorothy came to Deia from California to escape the brutalities of American life. They would devote themselves to the arts, especially to painting. Bob was a skilled cabinet maker and whenever they ran low on funds he would go to New York where for a few months he practiced his expensive skills, then would return to Deia, Dorothy, and painting. It was always possible to drop in late at night for a chess game, chat, and a drink. Both he and Dorothy liked colorful, slightly smeary, paintings that seemed totally out of touch with the contemporary art market. For years Bob painted imitations of Persian rugs from photographs he had found in a book. I was surprised recently to see that their work was now being sold by a British gallery on-line.

Besides being a center for chess players and painters, the Bradbury's house was also a place for music. Their daughter Suzie spent her youth practicing piano at which she became extremely proficient without at the time much sense of musical structure. She would be expected in her teens to perform before an adult audience at their house an evening of Beethoven's *Hammerklavier Sonata* and *Diabelli Variations*, a demanding task for any pianist, but as her fingers rushed over the keys her memory would betray her and sections would be played out of order. Eventually she travelled to England, then Germany, to study music, had a lover, married another man, was divorced, and music made sense. She became an international performer and a piano teacher, part of the classical music scene including festivals that developed in Deia, Palma and Soller.

Not all of our Deia acquaintances were artists and writers. Sam Kaye was a solicitor, his wife Kim was in Special Education. He felt it a duty to defend those accused of being criminals and as they had no children he willed his practice to an orphan whom he had brought into his firm. They would arrive in Deia with a large cold store of British meats and sausages which they claimed were much better than any they could buy locally. We enjoyed their company and would sometimes stay with them in their modern house outside of Luton. They introduced us to a wealthy businessman who told me that it was foolish to expect to become rich though hard work. It was who you knew in the local council that was important. Find out in advance the land the Council planned to purchase. Kim told me a story about an orthodox Jewish woman who married a Canadian and gave him the choice of moving to England or converting to Judaism. He wanted to remain in Canada and choose the later; in this forties he was circumcised and unable to consummate his marriage for many months. Kim suffered from cancer and Sam showed signs of Alzheimer; they apparently died in a double suicide. The executor of their estate was vague and I did not want to know the details.

During 1977 while staying with Sam and Kim we heard on television an announcement of the bust of an international drug ring and of Paul Arnabaldi's escape. After Paul's long relationship to what seems to have been the European wing of the Brotherhood of Eternal Love became known I asked myself how I and others could have been deceived by him and his stories that throughout his life he felt inferior to his older brother who was now a prosecutor in Connecticut, his claim to be living off a pension as a result of hurting his back working for an international aid agency, and that money also came from investments in the Japanese Yen. He had always seemed quiet, one of the many chess players and party goers in Deia. Adele thought him good looking, friendly; he had put a flower in her hair at a party. When Paul bought a small boat and began playing cards at Pedro's restaurant on the beach with three rich Americans rumors circulated that they were running drugs from Thailand. But as far as I know he was never thought to be the head of a drug gang that had a billion dollars stashed away in Swiss and German bank accounts. Paul Crotto and Nick Arnold were his close friends in Deia and Paris and they had telephone calls from him after his escape. Crotto told me that he had seen him in Deia and that despite the Thai operation Arnabaldi looked as he had in the past. Might they have known more about him? Was Arnabaldi even his real name? Some accounts spell it Arnaboldi. There were so many hustlers among the bohemia in Deia, but Paul seemed quiet, almost apologetic, in comparison.

About our third summer in Deia we met James and Norleen Smith, an Irish couple who lived as if they and their friends were cartoon Irish. James taught English at a good teacher's college in London at which John Heath-Stubbs was now also a member of the staff. He knew many of the literary world I knew such as Tom Blackburn. Norleen, protestant Irish, had inherited and was a director of an engineering firm. I had promised Desmond Maxwell an essay for a book on W, B. Yeats and I was reading through the prose. James demonstrated to me, by reading passages aloud at a Deia cafe, the conscious "Irishy", coat dragging in the persona Yeats adopted in many essays. Maxwell claimed that I was the first to write about this aspect of Yeats' prose but I had picked up the insight from James. As if wanting to be cartoon "Irish" he had somehow smashed the car door on his hand while leaving Deia.

I enjoyed their company and would visit them in London when I was staying with Ann Fraser. Such nights began with perhaps a dozen Irish, ranging from taxi drivers to famous actors, laughing at stories about drinking too much -- "and after he presented his credentials he fell dead drunk at the King's feet." As the numbers grew and bottles were emptied the stories became aggressive ("I told him he was the first and last Englishman to enter my door") and I had to keep my head down when, as sometimes happened, there were let's fight rants against American foreign policy or Israel. Adele who was in Paris asked me if I had brought the Smiths a bottle of duty free whiskey and I had to explain that we usually each drank a bottle and a half a night. James claimed that he had offered to join the English army during the War after he failed at enlisting in the Italian army; he had wanted to get out of Ireland. As if to ensure that his son, who was going to the elite St Paul's School, would never become an assimilated Englishman he had followed the Irish Catholic practice of making "Mary" one of the boy's legal names. Then there was the children's party that went on for days while the children tried to waken their drunken parents so that they could leave. The Smith's were fun.

Robert Graves had first come to Deia in 1929 with Laura Riding after her attempted suicide, and they stayed along with a few friends until 1936 when the British insisted on evacuating them. The details can be found in my short biography *Robert Graves* (2008) or at more length in the biographies by Miranda Seymour and Richard Perceval Graves. He returned in 1946 with Beryl, the unmarried mother of his three children. They were not married until 1950. He was one of the first British allowed into post-war Spain, a country severely ruled by Franco's pro-Axis government with the support of the Roman Catholic Church, the only legally recognized religion. Marriage between Roman Catholics and others was not legally allowed.

For an English Protestant with a family and weird spiritual views it was a strange situation. People in Deia still whispered about the mass murder of those with leftist views, including anyone with eye glasses as they were suspected intellectuals, by being pushed over a cliff. Guilt of being Republican was passed on to the next generation. Anyone thought of as related to Republicans could at best be employed in a degrading job. Somehow Graves was allowed to return and settle. His daughter went to a Roman Catholic school in which she was told she was damned. He made important contacts in the government. I remember buying a now almost impossible to find essay by him praising the Franco regime and that in Spain servants were still available and inexpensive unlike in post-war England. He was said to have kept Deia primitive as he remembered it.

When we first came to Deia in the early 1960s there was still no regular supply of electricity. You had to have your own generator and the bulbs were of such a low voltage they were made by hand. There was no paved road to the beach. It was a long hot walk on a dirt road Graves built during his years with Laura Riding. Graves wanted to keep Deia free from tourists.

Spain began to change. Instead of being a closed society and visas being impossible to obtain, the government desired tourists and their money. During the 1950s some Beats and some ex-GIs studying art in Paris came to Deia and managed to buy houses. By the 1960s Deia had become a hippie paradise in part because of Graves and his writings about the White Goddess, mythology, drugs, and other themes that appealed to the flower power, tune out generation. Graves both welcomed the new adoration of himself as a cultural prophet but, now famous, claimed to detest the worshippers, many of whom had not read his poetry or much of his prose but knew of him as someone who claimed drug culture had roots in ancient ritual.

It was impossible to spend many summers in Deia without seeing and coming into contact with Robert. He would unexpectedly sit down at your table in the main café and speak unintelligibly about something with which you were supposed to agree. He owned the tall tower-like house across from the apartments where we sometimes stayed and it was occupied by his friends and muses. Once Toby Grunfeld's long tailed Brazilian monkey escaped into the tree in front of the house and Robert began throwing stones at it, claiming it was damaging the tree. Adele had to run out and warn him that Nicole and other children were playing near the tree. Another time he came knocking on our door in the morning after we had a large late night party. He wanted to know the whereabouts of his current muse who had come to and left the party with a young man to whom Robert had introduced her. I suspected him of often purposefully or foolishly idolizing his muses and then pushing them towards lovers as he needed to write poetry in which he was the adoring victim of his White Goddess. He was a strong powerful man and even after he had lost his mind he could be seen climbing the cliffs over the small Deia beach as if he were a mountaineer in his twenties.

We were not part of Graves inner circle which consisted of family, visiting writers, and a gaggle of camp followers, mostly from New York, many of whom were part of the drug scene. There were also several British writers with whom he had friendships and sometimes competed for his muse. I knew his two younger sons, Juan and Tomas, along with Antonia Dalton, (the daughter of Catherine Dalton, Graves' daughter from his first marriage), Stephanie Sweet, and a few others who were always gossiping about Robert, his eccentricities, and friends. It was impossible not to know his current muse as she belonged rather to our generation than his. The many visiting writers and actors included Kingsley Amis, David Storey, Maggie Smith, almost anyone you can name. Antony Burgess and his wife came and lived for a year in Deia.

Among the New York crowd I liked the painter Esteban Frances who had fled the Spanish Civil War for Paris where he became part of the Surrealists before, in New York, creating costumes and backdrops for George Balanchine. He had starting coming to Deia during the 1950s and took up with Anne Heather, an American living in Deia, whom he often mistreated. We would find her late at night out locked out of their house, her clothing thrown from a window. She was used to it and seemed to find it funny. He could be charming, amusing, and he drew a beautiful sexy donkey for Nicole to be used for "pin the tail on the donkey", a children's game. We kept it for many years until the paper started to crack and it seemed difficult to preserve.

I have always tried to live near water. Without a source of water nearby I feel landlocked, dry. I gravitate towards streams, rivers, lakes. Even in Paris we live in an apartment on the bank of the Canal St Martin, one of the few waterways that cuts through the city. Even more than living near water I like living on islands. St Lucia, Trinidad, Hvar, and Mallorca are among the places I felt at home. I especially like islands when there are hills or mountains in the background, the contrast between the water and solid height feels right, the way the world should be. I enjoyed Mallorca. The mountains of Deia felt solid and gave me peace in contrast to the loud braying of the self-asserting drunks who filled the cafes and restaurants. Even the road from Valldemossa was filled with history, it had an archduke's palace, and had been painted by Gustave Dore as one of his scenes of hell. Valldemossa had its history of Chopin and George Sand but when the tourist buses would depart it once more became a small quiet town. It also had an amusing history. Whereas until recently young men in the villages travelled to other countries to learn a trade and make money before returning home to be married, Valldemossa exported its young ladies to Uruguay as prostitutes, which explained all the letters postmarked Uruguay awaiting replies in the post office. Many of lovely villages had a secret history not included in tourist guides. Fornalutx was said to have had a fight with the Catholic Church which for a century resulted in no priests being assigned to the village and none permitted within.

We had been coming to Deia more than a dozen years. It was "home" while we moved between England, Nigeria, Canada, Scotland, and France. When we went to New Zealand the summer of 1979 our life broke in two as it was years before we returned to Deia. I had also taught at the University of North Alabama for two years and been in India for sixteen months.

Now everyone we knew appeared old and gave private diners instead of holding open parties. The late night dancing that the Guardia Civil unsuccessfully tried to stop, sometimes pointing guns at us, under Franco had ended; the Socialists who now governed Deia passed a law against late night noise and music had to end before midnight. A new generation was taking over the cafes and restaurants. Our daughter told us over lunch about an incident the previous night when a man smashed his old car into the expensive new car of someone with whom his wife was threatening to leave, but I knew the wife and also had been flirted with. She liked to make her husband jealous. Our daughter had a one-night stand with the son of a friend; to be cured afterwards, in Paris, she had to have injections.

Jakov's daughter Oona was married at his Tower to an American TV star and Joy Ryder, an American who sang in German bars, told me about having gone to Muscle Shoals Alabama to record and learn that the soul sound that had made the recordings studio's reputation was now past and it was now aiming for sweet country music. Didn't she want to record with a dulcimer? She felt quarantined in the studio's housing to keep her away from the locals. Having myself gone to Alabama hoping to be near the studio's famous rhythm section I could have warned her; I had also been disappointed.

A few nights later a jam session in Deia broke up when the drunken bongo player began threatening to fight any and every one. As Adele and myself started to walk home an elevated Joy Ryder sat down in the middle of the only road through the town, pleading "don't go". Deia suddenly belonged to the past. We did not return. I did later look up Joy Ryder on-line and found that she was a much better known singer than I thought at the time.

Paul Crotto once told me that Deia was the center of his life. As Paul Arnaboldi disappeared, Bill Waldren, Nick Arnold and others died, he felt the world he knew was dissolving. Perhaps because I was a generation younger, perhaps because Adele and I were wandering scholars, perhaps because I often looked on Deia wryly, I felt I had other lives that were more important to me.

# Chapter 9: Wandering Scholars

*Windsor, Canada 1970–73*
*Zaria, Nigeria 1973–76*
*Columbia, Missouri 1976–77*

After the year in Mallorca our three years in Windsor, Canada (1970-73) seem mostly featureless both in regard to landscape and personalities. That we lived in a rented house scheduled for demolition, owned by Windsor University, shows how little urge we had to settle; emotionally our bags were packed. The street we lived on was close to the University, and could have been a cross section of the city. In the afternoons there were children locked out of their house waiting for their parents to return from work. There were criminal families. At the end of the block a family returned from a night out to find their neighbors burgling their house. After an argument the neighbors agreed to return most of what they had taken, but a few items remained in dispute. There was also a house rented by several male students I knew, as one of them was taking one of my courses, in which they smoked pot and discussed how after graduation they could stick together, take odd jobs and live on unemployment insurance, and continue to enjoy leisure and pot. They thought they should invite a woman or two into their planned commune to cook and clean the house. There were also three academics, one of whom was now permanently ill. Next to us lived a young couple; he was a Canadian in Linguistics, she an American. The other academic was a Dutchman, Adrian Van Den Hoven, married to a Canadian. He taught French, did research on Jean-Paul Sartre, and was the only person who seemed aware that Adele had published books on French literature. As he had intellectual interests—he later became President of an international society devoted to Sartre's works—his house was one of the few at which we spent time. We met decades later in Paris and he had changed little, but his wife Monica had become more independent, assertive, and had done post-graduate work, specializing in a field of nursing. She also noticeably had kept herself in shape and I was not surprised to learn that she and Adrian had in the past separated for some years.

The first year we lived in Windsor Nicole attended a Montessori school. The problem next year was that the public school said that at five she was too young for first grade but this would have been her third year in kindergarten. The French Catholic School was however willing to take her into first grade, so for two years she was expected to play, read and speak in French, and believe in God in school and disbelieve at home. I taught her to read and count in English.

During those three years we became aquatinted with others including a Nigerian with an American girlfriend in the English Department, and an American homosexual who was hired by the department, resented me and Adele and eventually died an alcoholic; he had a habit of being beaten by the rough trade he apparently enjoyed for his intimate moments. No one, however, stands out in my mind except Joyce Carol Oates and her husband Ray Smith. Joyce was always using what she found of interest for material and even before we actually came to Windsor she had used parts of my and Adele's CVs in a short story, based on one of the teachers of English, about a Professor who is carried away by the 1960s, rebels against and pickets his university. Adele would appear again in her fiction, this time in the Parisian "hot pants" and checked stocking, she was wearing when Joyce and Ray came to diner. Adele had just returned from being examined for her doctorate on Paul Nizan (published in 1976). Then there was Joyce's party where, no doubt asking us over in return, we and a guitar playing priest were the only non-psychiatrists. Joyce was writing a book in which there was a psychiatrist.

Her life was organized around her writing. Ray did the cooking, typing, took care of what was needed. His own research as an eighteenth-century English literature scholar, while not abandoned, took second place. She had a condition that made her hyper-active and she only needed four hours of sleep each night. Her creative writing class was mostly a one to one tutorial and she liked to interview her students about their past, which might turn up in her fiction as further examples of the small town American gothic of which Joyce is a master. A disposition towards such subject matter came from her own childhood; she was raised in an apartment above a store in a small town. Such material seemed far from her own interests in the classics of modernism, prestigious little magazines, and literary high culture. She could be generous. When a student of mine wrote an excellent MA thesis on Nadine Gordimer's fiction, Joyce said she would recommend it for publication if chapters were added on other books, which the student being lazy or cautious (perhaps thinking of enlarging the thesis for a doctorate) decided against. She also sent me word praising my ability as a thesis supervisor, although this student did not need supervision. He later published an essay in the book I edited, *The Later Fiction of Nadine Gordimer* (1993)

There were a few excellent students at Windsor, but many were terrible and aggressively brought down the level of classes. I was told by a student studying drama that he hated reading and only cared for what he felt like when performing. He wanted to perform before the class instead of writing essays. Another said that she was only interested in contemporary literature, "not eighteenth-century poets like T S Eliot". One of my better students spoke of the pressure she felt from others not to participate in the class, and she eventually dropped the course. Such sinking of educational standards resulted from the rapid transformation of what was formerly a small Catholic college to a public non-denominational university supported by the Provincial government after 1962. In the next decade the University grew from 1,500 to 8,000 students. As funds were allocated on the basis of the number of students, departments sought whoever they could find. We were encouraged to speak in schools, hand out flyers at shopping malls, do whatever might increase course enrollments. A popular lecturer whom the department usually sneered at for his eccentric opinions was promoted to a full professor on the basis of what he termed "desk drawer publications", meaning his notes. Even worse was the power gained by filling the quota for student representatives on such committees as Promotion and Tenure. What had seemed a liberal idea during the 1960s of giving students a voice in their teaching became a farce as no one voted for their representatives and student representatives became self-appointed. A handful of students would vote for each other knowing that faculty would grant them high marks, scholarships, assistantships, and other goodies in return for their vote on committees. Few faculty members were willing to stand up to someone claiming that student opinion had not been taken into account while making a decision.

The lowering of standards in the English department was part of a general attempt to bring in more students throughout the University. Until then the only Canadian degree in Asian Studies was given by the University of Toronto and required learning two Asian languages. The Canadian Ministry of Foreign Service now had to treat as equal to the Toronto degree the new Windsor Asian Studies major with its popular courses in Asian cooking and Asian erotica. I asked one of the students what she was learning and was told about "parsley" and other herbs.

We went to Toronto for an Irish Studies conference as many of our friends from Leeds and Africa would be there. Derry Jeffares was founder and co-chair of the International Association for the Study of Anglo-Irish Literature; a few years later, in 1973, he became Honorary Life President. There was lots of free drink being offered by whiskey makers, and W. H. Auden, showed me that his walking stick was a hollow cane filled with whiskey to help him though tedious academic lectures. Michael and Jane Millgate, now both professors at the University of Toronto, gave a dinner party at their apartment and at midnight tapped each of us on the shoulder to say that our taxi was waiting. No drunken all night Irish doings for them. We also went to Wayne State University in Detroit another time to listen to Derry lecture. I do not remember the topic but it, like many of his guest appearances, had no relationship to anything the students would have known. Afterwards

I overheard them, as I usually did, talk admiringly of how charming he was.

On our way to another conference in Toronto one of our tires blew out and the car turned over into a ditch. I still do not know how many days I was unconscious but I awoke to Adele sitting by my bed telling me that she had escaped with little damage, and my mother was taking care of Nicole in Windsor. I was heavily sedated and thought the hospital a paradise. I had never eaten such good food, seen such beautiful women as the nurses, been entertained by such great television shows, and enjoyed "space checkers" with a young Chinese patient in the next bed. Then the drugs were withdrawn and I went crazy and demanded to leave immediately. Only the doctor could sign me out and he would not be there until the next day so I telephoned one of my more rebellious students and asked him to meet me with his car at midnight at the back entrance of the hospital and we fled back to Windsor. I telephoned the hospital the next day and was told I would need see a doctor to remove the plaster cast and do exercises for my broken elbow. After Adele was unable to turn up for a few days Wayne State replaced her as the teacher for her French course, and Windsor rehired someone in Renaissance studies whom I had replaced. He was promised a permanent job in future and I found myself increasingly elbowed out of the Seventeenth Century into other areas of English studies, especially the modern period which brought me into confrontation with those who should not be in universities and who wanted A grades for doing no work.

By some quirk of political geography Windsor was on a strip of Canada that was south of Detroit, and was really a commuting suburb of the American city. Traffic went in both directions. Executives in some Detroit corporations lived in the part of Windsor where Joyce Carol Oates lived; the houses were modern, the air free of Detroit smog, it lacked American racial conflict, and Windsor was far less violent. Going to Detroit for an evening of entertainment had its risks. One of my more muscular students returned from a night out in his underpants; even his socks had been stolen. We once had a flat tire and were not allowed into the nearest gas station to telephone for aid as the attendant had been robbed and pistol-whipped some weeks previously. When I could not find our car in a parking garage I was certain it had been stolen until Adele remembered that it was on a different floor.

We were invited to meet the daughter of one of Adele's former university teachers at a Christmas day party in a distinctly white wealthy enclave within black Detroit. In an attempt to discourage burglars there were timed lights flashing on and off in different rooms during the evening and night. The parents told their daughter everything she desired was permitted, but instead of bringing home Sydney Poitier she was living in her room with a drug dealing semi-literate redneck and the parents did not have the courage to throw him out. The only commonality I could find among those invited was that they were ideologically committed to legalizing drugs, but they were not now, and perhaps had never been, drug users, although one of them spent a few years in prison on a related offense. A young man turned up with excellent hash that was generally refused, but if only in protest against our hosts we enjoyed.

Crossing back and forth from Windsor to Detroit was probably easy for those the border guards and customs officials knew. When I had to go to Detroit for an airplane as Windsor airport did not have the necessary equipment during a snow storm, the taxi was not stopped. If we crossed with our Nigerian friend and his white girlfriend the car was searched by the Americans for drugs, as it was when there was a large rock festival in Michigan. Joyce and Ray had two cars; if they travelled in their Beetle their passports would be checked, in their larger car they would be waved through. A Canadian friend with a strong Scots like Maritime accent was accused of being a Russian with false identification. The Canadian officials seemed only interested in those bringing back from America bottles of wine and spirits which were much less expensive than in the government controlled liquor stores of Ontario. A student whose father was one of the custom officials claimed that the confiscated bottles became their property and was either drunk or sold by them.

I am uncertain how my research and writing time was spent in Windsor but I remember being still at work on *Marvell's Allegorical Poetry* (1977) and being obsessed with Graham Greene's *Brighton Rock* and his other "Catholic" novels of that period of his fiction. I endlessly sought symbols, echoes, and patterns of images alluding to the Catholic mass, the Christ story, and Dante's *Divine Comedy*. There was, and still is, no doubt in my mind that Greene had seen Pinkie's story, and those of his other heroes such as Scobie in *The Heart of the Matter*, allegorically as twisted incarnations of the Christ story, with the sinner closer to God than the innocent or non-believer, but there was always something wrong with my details, something that had to be forced as otherwise it was lacking. After a decade of carrying a rough draft of the book around the world with me I eventually threw it away rather than be tempted into wasting more time. I had earlier published in *Etudes Anglais* during 1968 a long article on the Dante allusions.

While at Windsor I offered my first course in African literature in contrast to the occasional lectures I had given in the past. It was the start of my moving away from traditional to new English studies and I began commissioning the essays that formed *Literatures of the World in English* (1974, 1985). Contributors included some of the famous names of the period including Jeffares (Ireland), William Walsh (England) and Balachandra Rajan (India). When the essayist on the United States backed out late I foolishly tried to fill in. The book belongs to a time when we sought unique characteristics in a national literature, such as British manners and morals, American romance.

## II  Zaria

I had thought that we would return once more to a position in England but when offered a Professorship for myself and a Senior Lectureship in French for Adele at Ahmadu Bello University in Zaria, Nigeria I accepted. I was told that to avoid nepotism Adele would be responsible for the French section of an English and Modern Languages Department that I would Head. Later she was made a Reader (Associate Professor or lower level full Professor) but while I was there French officially remained my remit although plans, which fell through, were for it to become a separate department along with other potential departments that began emerging within English and Modern Languages, such as Theater, Hausa, and Arabic. At Ahmadu Bello I discovered a talent that I had not shown since the Eisenhower-Stevenson election when I was at Columbia University, for leadership, planning, organization, and building where there had been little previously.

Although Ahmadu Bello had been a university since 1962 the continuity of its departments in the Faculty of Arts and Social Sciences had not been straight forward. Depending mostly on foreign staff, the faculty would scatter when political events such as the Nigeria-Biafra war, or a major military coup d'etat, brought violence. Our appointments were part of a reconstitution of the faculty under the leadership of the Northern Christians who were now head of the national government and the army as well as an expanding university. Several lecturers I knew at the University of Ibadan, some of whom had taught since elsewhere in Africa, were now professors and heads of their departments. James O'Connell, a Catholic priest, was in charge of Politics and Robert Gavin led History, the two main departments in the faculty except for Sociology which had a visiting American Professor and seemed filled with militants who wanted the blood of everyone, especially whites. A low profiled Indian was in charge of a low profiled Geography department.

When we first arrived at Ahmadu Bello University we spent a few nights at a house on the Agricultural compound and were told that the snakes that had escaped were unlikely to bother us. We inherited a large house formerly occupied by lonely Geoffrey Walton who surprisingly had been Professor of English at ABU. Surprising because he had been one of the contributors to *Scrutiny*, someone whose name I knew well, and because I had earlier noticed an announcement of his appointment as a lowly lecturer at a British university and assumed it was a stop gap job taken by those who fled upheavals in the recently independent nations. The house had two stories, three air conditioned bed rooms and land on three sides, especially in back where I planted a vegetable garden and even put in wild strawberry plants. Knowing that there would be water shortages I went to a German firm in Kaduna that built oil trucks and asked them to make a five hundred gallon water tank for our house. Getting it from Kaduna to Zaria was hell, but it proved necessary for the garden, and drinking, bathing, and washing water for ourselves and those we employed including a cook-steward, house boy, gardener (appropriately named Hyacinth), and night guard.

The night guard was small effeminate, wore lipstick, and was the "wife" of a group of Tuareg, who, being nomadic, use one of themselves for that purpose while away from their homes. This meant that we usually had several guards in our garage keeping our guard company. It was said that the Tuareg paid the local criminals to avoid the houses they guarded. They seemed more a community than individuals. About once a week the Tuareg would gather on the grounds to one side of our house to share and cook the meat they bought while vultures circled in the sky and perched on trees. Our dog, Sesame, part Dachshund, would stay in the house.

Besides Sesame we had two cats. The three were raised and on cool nights slept together. Tigran was indeed named after the world champion chess player Tigran Petrosian, a name we had considered for Nicole if she had been a boy, and would prowl at night, come home with wounds from fights, and despite his name was hopeless at catching birds. I would watch him cautiously stalking birds in our back garden; it would take him so long the birds appeared to yawn before taking flight. When we left Zaria we gave him to a colleague who took him to the USA and then to France where years later a now old, normally bored stand-offish, Tigran must have recognized Adele's smell, as he emerged from his usual place to sit in her lap. He was also the father of Peppermint's kittens. He had to be as she seldom left the house even when she defecated she returned from outside to use the litter box. After Peppermint's second litter of kittens Adele had Tigran fixed and I felt threatened. Peppermint once dragged a kitten up the stairs and settled on a pile of clothes in Nicole's room before giving birth to her second litter. She probably needed to be suckled to start. How had she learned that trick?

Nicole also had a bowl of small tropical fish which we brought from Harrods, London, in a plastic bag large enough that they had air for several days. The Nigerian customs official probably thought us insane. She had a baby chicken which she raised on her veranda and which imprinted with her would follow her around until given away, no doubt to be eaten by others. We did raise chickens for their eggs in a coop in back of the house. They each usually produced an egg a day. If the numbers fell it was time to eat a chicken; those remaining knew the quota was an egg a day once more. Then we tried raising rabbits, but rabbits contrary to the expression do not breed like rabbits as a friend in Soller, Mallorca, who tried to raise rabbits for a living soon learned. Ours needed a much larger rabbit, which we borrowed from a friend. Nicole and a school friend watched the rapid wham bam no thanks mam display. Nicole named one of the rabbits Hyacinth after the gardener and it was her pet. One night it was eaten for dinner by mistake and Nicole became a vegetarian.

We lived on a professorial row. Our neighbor on one side was a medical doctor who had worked in the middle-east and used a water cooler in place of air-conditioning or a ceiling fan. On the other side was a visiting American professor and Head of Sociology who felt that undergraduates and discussion of standards were of no interest; students only became interesting at Master's Degree level. Immediately across from us was a Professor of Morbid Anatomy from the University Hospital. He had a large talking parrot that said "George is in London". His wife would offer to put our meats and dairy products into the Hospital morgue during prolonged electricity failures.

Lalage Bown lived near them. She was one of the ex-Ibadan group, now a UNESCO backed leader in bringing Adult Education to Africa though distance learning. She was one of those who disapproved of the post-WWII expansion of education in England and its former colonies as elitist meritocracy that reinforced the class system. Their assumption was that adult education and something like the Open University in England would transform society. I liked Lalage, did not have similar views, and sometimes asked her for favors as when a new member of our department lived in her garage-guest house while awaiting university housing, always a problem at Nigerian universities. I should mention Lalage's many books or her renown, but what I remember most about her is that she was the slowest eater I have ever known. By the time others were eating dessert she was still thoroughly chewing her first course. She also published a story by my old London Sierra Leonean friend Ed Jones in an anthology of African writing. Ed was always trying to persuade someone to "type" his writings, which meant rewriting them. I seem to remember that someone had translated a story into German and Lalage translated it back into English.

The combined Department of English and French was filled with personalities. Ken Evans, a Senior Lecturer that I inherited, had an isolated life and was obsessed with money. He possessed one knife, one fork, one spoon, and invited no one to meals. He had a reputation of bringing home Nigerian prostitutes, and their nights would loudly conclude when he did not pay whatever she thought had been agreed. He published very little and was often after me to put him forward for promotion to Reader. He was hopeless. I suggested it would improve his chances if he took on some departmental responsibilities, which he was reluctant to do, and then when a university committee would not promote him he wrote letters to the Dean claiming that I misled him. After he retired and returned to England he died alone and was not discovered for some time.

I first met Gaye Shortland at a small reception given when I visited Leeds on my way to Nigeria. She leaned on a table bringing down all the food and drink and wailed "why does this always happen to me!" When I got to Zaira she was already there, a new member of my department. Single young attractive white women are in demand in Africa, and soon Gaye had an older mooning admirer in the French section. She however preferred her Tuareg night guard and went off with him to Niger where she had several children by two Tuareg men, lived as part of their community, opened a restaurant, taught at the local university, and after becoming very ill returned to Ireland where she became famous for her two best-selling autobiographies. She also wrote novels and plays about life in Cork, and became editor of a publishing house. In Zaria she had been thought flaky, but with her return to Ireland, and as a result of her writings, she became a hero to many especially feminists who admired her for defying conventions to create her own life.

One year Anna Rutherford and Kirsten Petersen (who would stay for a second year) arrived in a land rover having crossed the Sahara. Anna was Australian, Professor of English at the University of Aarhus in Denmark, and a well-known organizer of the Association of Commonwealth Literature and Language Studies, founding editor of the journal *Kunapipi* and much else. She was instinctively an empire builder who spoke openly of life as using people and being used. As an Australian she could raise money from Australia, which wanted to project its image abroad, and from several Commonwealth bodies in London. She knew how to fund conferences, publications, visiting writers. Many university appointments in Commonwealth and English literature required her recommendation. And she was often consulted behind the scenes.

Anna may have been the only lesbian I have known who was actually built like a man. She was athletic and tough. I would not liked to have met her in a dark alley or on a wrestling match. She was also without scruples. If invited to give a lecture she would put in for expenses with both the inviting university and at ABU. When there was, as there often was, a long delay in being repaid she invited those who had to sign for the money to a party with some Nigerian female secretaries and much drink. She stayed at Bob Gavin's house for the year as he was on a sabbatical; she used a pick to chop ice from his expensive imported deep-freeze, which he returned to find ruined. Years later I was in Europe at a conference with Anna and Kirsten, when Anna bashed into another car, backed up, and fled with Kirsten saying "Anna, you have gone too far this time". When the other driver caught up with Anna she pretended that she did not know there was an accident and talked her way out of the situation. For someone who appeared and normally sounded like a rough Ozzie from the outbacks, she had a surprising ability to go all feminine, flutter her eye lashes, seem soft, act innocent, and win.

There were at the time only two Nigerians lecturers in English at ABU: Aliyu Mohammed in Linguistics and Kola Ogungbesan in English African literature. Mohammed was shy, kept to himself, and I was told converted later to one of those evangelic Christian churches that keep increasing in Africa. Kola had lived in America, was more sociable, and was friends with both the militants in Sociology and History and with their departmental Heads, Professors Gavin and O'Connell. Kola was bright, subtle, ambitious, wrote well, had literary standards, and was the obvious future Professor and Head of English. I felt the Department should be represented at the international FESTAC 1977 (Second Festival of African Arts and Culture) planned for Lagos in 1976. We co-edited *Celebration of African and Black Writing*, published by the new Ahmadu Bello University Press in co-operation with Oxford University Press. This was the only project ready for the Congress as originally planned; delays caused the Congress to be postponed for a year. Kola did little work on *Celebration* which led to my not wanting him to co-sign the "Introduction" until the Director of the Press insisted. Kola subsequently wrote several books about African literature and after his death his essays were collected and published.

The French department or division had its own characters as if to prove that African universities were staffed by eccentric whites. Don Phillips, an Oxford D. Phil., arrived the same year we did. The problem was that while he was an expert concerning historical French styles he, because of some psychological flaw, would not speak French. What do you do with a French lecturer who will not speak French? We invented a required course in writing French to go with the translation course, at which he was brilliant. When I wanted to write to L. S. Senghor, the famous poet and President of Senegal, to ask him for a "Preface" to a planned collection of essays on African literature, Phillips was more than useful at finding the precise tone and expressions. Senghor declined for the present; he did not mention that his country and Nigeria were at odds on major international issues, but he did praise the style of the letter. Phillips was not always an asset. Lonely, sentimental, he sometimes drank too much and one night at the staff club made a racist remark when a Nigerian kicked a cat. I was under pressure to fire him. I felt that if Nigerians wanted to beat him up or have him deported that was their business but I would not fire someone on the basis of a reported remark.

Rodney Harris came to us highly recommended as an expert on Aimé Césaire (about whom he had written a book in French), but none of his recommendations mentioned his drinking problem. He would carry around a bottle of "water" and before the morning was over would be asleep. His French wife, who also taught, managed to cover his duties. She was very good.

Norman Stokle was an Englishman with a strong interest in theatre. Before coming to ABU he had published translations of French Nineteen-century farces, a book on Camus, and taught at American universities. He had also studied at RADA (Royal Academy of Dramatic Arts) in London, had been in several theatre companies, and had a second beautiful wife, a former student whom he had taught in France. The first beautiful wife, an American, left him. He was not profound, but he was the sort of person who if left on Mars would rapidly learn how to survive, even flourish. He was always talkative and could be clever, inventive, informative, amusing, brash, boring, wrong. He had three children and we spent time with him in Zaria and later in France where he was in charge of the Skidmore University abroad program.

He was excellent at getting to know people and getting things done; other university abroad directors came to him for advice and contacts. He was, however, indiscrete. He told his lovely wife that she was putting on weight and should exercise. She took up tennis, divorced Norman, and married her instructor. Norman soon had an attractive Chinese woman keeping house for him and warming his bed. She divorced her husband in China to marry Norman and soon after obtaining the right to live in France left Norman for a French artist,

His main interest remained in theatre and he can be seen in many television plays. Once while we were vacationing in Turkey we looked up from our meal to the TV set and there was Norman as a German officer in a drama with Joan Collins. His performances are listed on The International Movie Base (IMB).

The Theatre section was led by Michael Etherton. Michael and his friend Andy Horn, an American married to an East African, were deported from Zambia because their plan for theatre in local languages was thought subversive of the government's desire to keep English as a means of holding the many tribes together in one nation. Michael Crowder, an English historian who from early in his career had become a name on the Nigerian intellectual scene, brought Michael and Andy to start a performance group in ABU's new Nigerian cultural center. I objected to what had become a tradition of exhibiting African culture as exotic and primitive. Michael unwillingly found himself in the English Department and kept threatening to resign while I began asking people, including Richard Coe who was now at Warwick University, what we should be doing to create a performance oriented Drama section. It was clear that imitating RADA training was wrong, and that whatever emerged had to be from Nigerian conditions. Michael had his way in building a mud theatre of which he was proud. He had his students improvise in Nigerian languages and use local references in plays based on the plots of European classics, but he also saw that his students wanted to write their own scripts and that there was a great potential in television. Michael's many books include *The*

*Development of African Drama* and, for my Modern Dramatists, *Contemporary Irish Dramatists*.

Michael was helped by Jan Cohen, an American with a Master's Degree in Theatre Administration, who was the wife of Neal Cohen in the Economics Department. Jan taught the students during 1973–75 such basics as promotion, ticket sales, numbered seats, financial records. The Cohens became part of our social life in Zaria as they explored what was available that we had not, such as the swimming pool used by the Air Force. They kept up with what was going on in the USA and I could discuss with them the best-selling *Hite Report on Female Sexuality*. We would see and hear from the Cohens off and on over the years. When Adele was a visiting Professor of French at the University of Missouri we saw them in St Louis where I was fascinated by Jan's sister with an "avant-garde" life style; she performed in porno films during the 1960s. Neal soon began working for USAID or a branch of the American government, employment that kept them for long periods in Nepal and South Africa. Jan always was part of a theatre group as actress, director, and management. Even when Neal was diagnosed as having brain cancer and likely to die he kept taking on such dangerous and improbable assignments that I wondered if this ex-Marxist had become affiliated with the CIA.

Because of the rapid development of Theatre the Vice Chancellor gave me the task of reforming Hausa, in both the sense of restarting the teaching of a language that had not been taught at ABU for some years and rethinking how it should be taught and what should be covered. After I was told it was an impossible task because of political sensitivities I consulted with a broad range of those involved with Hausa. A Malam explained that he and other traditional teachers only taught word lists and memorization. They did not know the grammar. I learned that the grammar had been most thoroughly analyzed by F. W. Parsons, a British District Commissioner during the colonial period who was now a Reader at SOAS (the School of Oriental and African Studies, University of London). I asked him to suggest a syllabus which

I then showed to Philip Jagger and Russell Schuh, both then at the Center for the Study of Nigerian Languages in Kano. I remember discussing the syllabus with someone from a Christian missionary group that also had experience with teaching Hausa. The three year syllabus that I eventually submitted to the ABU Senate was passed without a dissenting comment. By concentrating on linguistics I had unknowingly put off the vexed matter of Hausa literature until a third year option. Vexed because much of the political history of Northern Nigeria, and therefore of status and authority, was based on papers that may be forged. A Muslim school of Nigerian history, challenging the prevailing cultural nationalism of the Ibadan historians, was founded on these documents which are usually inaccessible.

Nothing breeds trouble like success. I was next asked to prepare a syllabus for the teaching of Arabic. I thought this a tiresome task made worse by my being of Jewish origin. I might not think of myself as a Jew but Bushrui at Ibadan early on started to question me, and since he thought I was one then whoever was sent to advise my committee would also think it. There was an Israeli at ABU, probably in Agriculture, who coached the national Nigerian chess team, but he kept an extremely low profile and chess was not "God's Holy Language." What would any good husband do in such a situation but pass the disagreeable task to his wife, an attractive forty year old blond Christian American with great legs who wore mini-skirts. Adele, however, had an interest in language, and an ability to keep calm and calm down others. The first task was to find out more about Arabic teaching. We invited to Zaria a Nigerian novelist with a doctorate in Arabic from a British university, and we learned that classical Arabic had changed little over the centuries and could be read by educated Muslims, but the spoken language had radically changed and would be unintelligible when moving from country to country. With his doctorate in classical Arabic he could not order breakfast from a menu in Cairo. So there was the question of what was the most widely accepted dialect and how to teach it. Tunisia had already faced this problem and had language tapes available for the purpose. An invited Arabic scholar told Adele that she was an infidel who did not understand that God's Holy Language was in the Koran and should not be taught through tapes, but the committee was impressed by Adele's pragmatism.

Soon there was serious talk about dividing the Faculty of Arts and Social Sciences into two separate faculties, an Arts Faculty with departments of English, French, Performing Arts, Nigerian Languages, and Arabic. And a Faculty of Social Sciences that would include History. I did not like the way I was becoming a full time administrator and did not want to become a Dean. I had little time for reading and no time for my own writing. I knew I was out of touch with research in areas that interested me. Increasingly people were coming to me with problems that I hoped would go away or which others could handle. The family whose daughter was caught submitting an essay she had plagiarized and the professor who was thought to be her lover were told to see the lecturer who had marked in red the parallels between the essay and a published paper. The American official who thought I should participate in celebrating an American holiday, although Nicole could not get the inoculations and go to films offered to other American children, would need speak to my secretary. I began having my secretary bring my mail and sending anyone who had to see me to our house. I would sneak into my office at night to finish what needed to be done.

My fears of being a Dean ended the evening of February 13 when much of what I and others had built during the past three years suddenly crumbled. Adele thought it would be fun to have a large February Friday 13th party and weeks before we started making preparations by inviting perhaps a hundred people, buying drinks and food, arranging for music. A tobacco company was to send attractive hostesses to hand out cigarettes. During the evening there was a radio announcement of an attempted coup d'état in Lagos but people should continue as normal, which we did. I had lived through several such turmoil's in Nigeria and made a rule for myself to go about life as if there was nothing to fear. Some people failed to appear, but enough did that those missing were not noticeable. Then the next day we learned that General Murtala, who ruled Nigeria, had been killed in an attempted coup by other army officers and after initial indecisiveness the army backed the Murtala generals against the plotters. All hell broke loose, many in the army and an untold number of civilians were executed, although exactly what was happening and why was never clear as the new government was reluctant to explain who was involved. Those in power were said to believe that the Northern Christians and the British and perhaps the Americans were involved, although the only reasons for such suspicions were that Nigeria had in recent years moved closer to the Soviet Union, General Gowon who had been deposed by an Army coup a year previously had fled to England, and one of the plotters had sent Gowon a puzzling telegram.

The ABU campus became unsafe. Some white lecturers and their family were attacked, non-Muslim Nigerians went into hiding. Dr. Ishaya Audu, a Hausa Christian who had been Vice-Chancellor since 1966, was replaced by a Muslim professor who proved indecisive and incompetent and was himself replaced a year or two later. Both James O'Connell and Robert Gavin were said to have been advisors and speech writers for General Gowon and were told to leave the country. A strange alliance of Marxists and militant Hausa nationalists took over the Faculty of Arts and Social Sciences and formed a committee to rid the university of foreigners, end such new developments as Drama (we were told that theatre was not Nigerian, which was ignorant of Northern as well as of Southern Nigerian history), and generally inspect what was taught. The History Department especially underwent an ideological revolution as the course in the Soviet Revolution would now concentrate on the early years and ignore Stalin's tyranny. A course in the Industrial Revolution was replaced by a course on Medieval Europe presumably to show the universality of clerical rule. I refused to acknowledge the authority of the committee over my Department and was censured. There seemed to me two choices. Wait until the terror started eating its own which would be followed by a return to something more like the past or, my choice, move on.

Adele some months previously was invited to the University of Missouri to be interviewed on campus for an Associate Professorship in French but as Missouri would only pay her expenses within the USA she declined. The man they appointed had since been awarded a Guggenheim Fellowship for the coming year and Adele was asked to be a visiting Associate Professor. We would go to Columbia Missouri while deciding what to do next.

I was sorry to leave ABU. We had a good social life. Besides the continual parties and dinners there was swimming in the afternoons at the staff club and films, dances, and sometimes duplicate bridge at the Lebanese Club. There were several people whose company I regularly enjoyed including the historians Jane and Ian Linden. Their children and Nicole played together and they had a pink pig. Ian with Jane's aid would write several books about Catholicism in Africa, then go to Bhutto, and more recently Ian became Director of Policy at the Tony Blair Faith Foundation. Michael and Cheryl Dash, two West Indians, often stayed in our spare room and we would stay with them when we visited Kano where they taught. Michael contributed an excellent essay to the *Celebration* collection and both wrote for the book I edited on *West Indian Literature* (1979, revised and enlarged 1995).

That book had started when I was wondering what we might teach next within the English Department. We now had Nigerian and American literature and we had a small postgraduate seminar on South African literature as I thought Nigerians should be aware of what was happening outside West African. I wanted the politics, history and literature of each nation to be seen as distinct, but I also retained an older sense of a pan-black anti-colonial struggle. There were many West Indians teaching in Nigeria and I began discussing with them the contents for a book of essays.

I had enjoyed the monthly Arts evenings that the poet John Haynes, who taught in the School of Basic Studies, organized for the English Department. It included poetry readings, sketches, music, displays of local and modern crafts, even once a snake charmer and his snakes. John was married to a Hausa woman who spoke little English and he had a broader range of contacts throughout the university and native arts communities than I could ever hope to have. He was and is an excellent poet from whom I learned much about verse structure and versification. I once showed him a Robert Graves lyric that seemed perfect but I could not understand how Graves had put it together rhythmically. John went through it line by line showing me the many changes in the mode of versification. Until then I thought I knew something about poetry. John may have been the only poet I knew who turned down an offer to publish his second volume of verse as he felt he was not ready. More recently he has been awarded many poetry prizes including the Costa and being short listed for the T. S. Eliot.

Our most significant accomplishment was the exchange of language students with the University of Abidjan. Nigerian students doing a degree in French were expected to study for a year at a French university, but while we were at the University of Lagos that became a problem as France was backing Biafra, French government scholarships became hard to obtain, and many Nigerians objected to students studying in France. The obvious solution was to arrange an exchange with a Francophone African university that would take our students and send us their students of English for a year. Abidjan, where we had contacts, seemed just what we were seeking. Unfortunately the Ivory Coast students found Nigerian food terrible and complained so badly that we had to let them do their own cooking. Ken Evans said he could not understand why they were unhappy, the British ate Nigerian food.

Ahmadu Bello University had been interesting. I would miss the crusty British medical professor who said she married her husband in sickness and health "but not for lunch." No doubt the medical and science people hardly would be affected by the radicalization of the Faculty of Arts and Social Science, but I would. Nigerianisation was coming to the faculty late and was needed, but I would be against many of the ideological developments likely to accompany it. We asked for a leave of absence in case Zaria looked better next year but I did not think we would return.

## III  Columbia, Missouri

On our way to Missouri we stopped over in Philadelphia to see my mother who was living near Cherry Hill, New Jersey. Nicole had never previously heard the Philadelphia accent, a cross between Brooklyn and Southern, and asked what language are they speaking. The United States confused her. Seeing black Americans she would assume that they were Nigerians and be disappointed. "They were only Americans!" The school she attended in Columbia was meant for the children of the faculty and even offered Beginning French, but the teacher's French was so poor that Nicole kept reporting back to us the teacher's errors. Nicole would return to the USA for her undergraduate degree but, unlike Adele and myself, she never felt that she was American. The continual uprooting and travel while young made her emotionally stateless.

Adele was in a Romance Language Department chaired by the famous translator Margaret Sayers Peden. "Petch" appeared relaxed, happy, full of humor, ready to be amused, and liked to refer to herself as a backwoods girl who had made good, and would go into a sort of hillbilly accent to show how she could woo Missouri politicians who distrusted universities and especially foreign languages as, well, foreign. She entertained, liked to be entertained, taught a high level seminar on translation, found time to translate over sixty books from Spanish, mostly by authors from the Americas, and kept winning national prizes for her translations. Although she has been called the "best translator ever" her seminar, in which she showed how to turn a translation into a publishable work of art, was regarded with suspicion by some in her department who thought scholars should dig for facts and not be concerned about such matters as an author's "voice" and "tone". Petch's husband William Peden was a creative writer who did not know Spanish and he was her first audience. She was enthusiastic about the modern Latin American novels she was discovering and began by translating one for him to read.

Another of our friends in the department was Donna Kuizenga, who combined being a feminist with eighteen-century French studies. She had somehow picked up an affected New York Jewish accent as if to prove her credentials as an intellectual. During the year we often saw her and remained in contact for several years afterwards when she returned to the Eastern USA to become a university administrator first as head of a department, subsequently as a Dean. The nice but less interesting boyfriend we knew was displaced by the returning Guggenheim whom Adele had replaced, and he broke Donna's heart for a while. During the summer Donna found us the nice apartment we rented in Paris the next year.

Adele taught advanced courses in Modern French and Francophone African literature along with the usual American undergraduate courses in French Language and Literature. She began research towards a book on the Guinean novelist Camara Laye. I wrote two of the three York Notes study guides (under the co-editorship of Derry Jeffares and Suheil Bushrui) I reluctantly agreed to do and found them unexpectedly interesting work. When you had to explain to students the significance of every scene in a play there could be no faking, no blurring over what you did not understand yourself. Providing short essays as examples also proved more interesting than I had thought. Every word counted, there was no space for academic mumbling. Study guides are frowned upon in the academic world, but I started to feel that every teacher should be made to write good ones instead of long futile studies that had little use.

I had started the year trying to write a book on Samuel Johnson. After about three chapters I had the feeling that I had nothing new to say, I was arguing with others about matters of interpretation and biography. I really wanted to write a book that would help me understand what had happened at Ahmadu Bello University.

I applied for a Rockefeller Fellowship for support during the coming year while I worked on this book. Then unexpectedly, totally out of the blue, I was one of the twelve scholars and artists appointed by the President of France to teach at French universities during the coming year. Michel Fabre, the French expert on African literature in English, wanted a sabbatical, and a French professor I had met at a conference recommended me as Fabre's replacement at the University of Paris III (the New Sorbonne). The Rockefeller Foundation agreed to my teaching in Paris while holding their Fellowship. Then more mana from heaven, Adele was awarded a post-doctoral scholarship by the American Association of University Women to help towards researching her book on Camara Laye.

We were going to Paris for the coming academic year, but first I had a summer grant from the William Clark Library in Los Angeles, and we would go to Paris by way of the Caribbean and Portugal. In Los Angeles we rented a house in Santa Monica Canyon near a great beach and the pier, but the water was too cold for swimming. Our rent-a-wreck had to share the streets with all our neighbors' Porsches. At the Clark Library I was supposed to be reading late Seventeenth- and early Eighteenth-century plays for evidence about acting styles, but I was offered a contract for the West Indian volume of essays and that took precedent. Everyone seemed to have a Hollywood connection. The woman who taught Nicole's drama class for children had been in films. The University Music Department suggested a young guitar teacher who taught her so well that Nicole soon gave family concerts including Bach and the Beatles. John Povey—who I had met in Lagos—as well as teaching African literature at UCLA sold real estate (as did many people we met) which enabled him and his wife to have a magnificent house on a terrace above Malibu with a swimming pool. He told me about the movie stars living near him whose houses slid away forever in the mud during heavy rains. In his will he left a substantial sum to start a center for English Studies in South Africa where he had emigrated from England as a working class youth. Travelling from one end of Los Angeles to another was as much trouble as visiting a foreign county. Russell Schuh, who taught Hausa at UCLA, lived so far away that it would have taken hours to visit him or he to visit us. Max Novak, UCLA's Defoe scholar, and his wife should have warned us that they did not eat pig products; they would not have been served ham with melon, followed by stuffed pork chops, and would have had a more filling meal.

We drove to Santa Barbara to see Garth St Omer who published four remarkable novels about St Lucia. He had lung problems and I was then chain smoking so our interview took place with him inside and me outside his house. William Frost, the Dryden scholar, was Head of the English Department. He had contributed an essay to my early *Dryden's Mind and Art*. He was a Santa Barbara native and his brother was sheriff or some such authority. Santa Barbara was beautiful, but Garth said that the ocean was always too cold for swimming.

We flew from Los Angeles to Paris by way of Jamaica, Trinidad and Portugal. In Jamaica we stayed with Cheryl and Michael Dash who we knew from Nigeria. Kingston was suffering one of its periodic phases of politically inspired violence. Armed gangs broke into houses and raped women in front of their husband and children. The Dash's house had heavy iron bars on the window and entrance. The Kingston's art gallery was robbed the day before we visited and kept its doors locked until the taxi we telephoned actually arrived.

In Portugal we were ordinary tourists.

# Chapter 10: Paris Once More and After

*Paris, France 1977–78*
*Sterling and Dollar, Scotland 1978–79*

As during the year we would be covered by French medical insurance I had to be examined by health inspectors like any other immigrant. Except that I had an invitation to teach from the President which those examining my balls and arse found amusing.

My teaching duties were light, not lighter than most senior professors at French universities, except that I would not have theses and dissertations to supervise. I was supposed to give a weekly two hour doctoral seminar on African literature in English and for one semester I taught two ninety minute undergraduate classes on Shakespeare's *The Tempest* and Swift's *Gulliver's Travels*. As a visiting professor I could avoid departmental meetings and committees but I would be expected to share in some of the end of year oral examining. Robert Ellrodt, Head of the English Department at Paris 3, and an internationally known scholar of Sixteenth and Seventeenth century literature whom I admired, asked if I wanted to give lectures on the Seventeenth-Century to the doctoral candidates. Trying to impress him I suggested obscure topics and the idea was dropped.

The undergraduate classes were of a high level, they could have been at a good British university, but that may have been because my classes attracted native English speakers and the best French students. The oral examinations were disillusioning as they allowed favoritism and prejudice. After I gave an Irish woman 17 or 18 over 20 I was asked about the high grade and was stunned that the committee marked her down to an 11/20. I had not taught her but I looked at an essay which was as good as anything I might have written. She had a good degree and an MA in English from Queens University in Belfast. What was going on? I later learned that she had quarreled with her lecturer who wanted to prevent her from getting a 13/20 or better as that would allow her to register for an advanced degree in France which she needed for a French university position.

My doctoral seminar was supposed to be limited to 12 students, but everyone doing an advanced degree in any aspect of African studies heard of it, asked permission to attend, and soon there were over 40, most of who knew less English than I knew French. An excellent Cameroonian became translator and summarizer, but as the auditors began to participate I asked them to give reports and soon I had created a mess; we needed a four hour seminar to cover the material each week. The Africans were used to long political speeches, but a woman who already had a Ph.d. from Berkeley complained that a four hour seminar was unbearable, and she was possibly right. I would take the bus home across Paris with my head still spinning.

The Cameroonian woman came to me with a problem. Her dissertation supervisor was never available; she claimed that he insisted on sleeping with her. As a professorial bird of passage I could not supervise her dissertation and I asked Michel Fabre what to do, especially as I knew her supervisor. Michel arranged that Albert Gerard, Professor of Comparative Literature at the University of Liege, Belgium, would be supervisor. Gerard was one of the best African literature scholars in the world, who had taught at a university in the Congo to avoid Nazi Belgium, and who has published a book on four South African language literatures, a first of its kind, as well as books on English Romanticism and on Baroque European literature. When I read the student's dissertation I learned much about how to place literature within social, ideological and cultural contexts. I later helped edit a part of it for publication.

I learned during the year although the results were not always obvious. Michel explained that he had to give 12 two hour lectures for those entering the competition for the Aggregation and had to think up a variety of topics such as imagery, space, time, narrative techniques, textual genesis, structure, biography, contexts, ideology, aesthetics, and interpretative history. Although I was aware of, and at times made use of, such critical approaches I had not thought about them as part of a systematic analysis of writing.

Michel Fabre was remarkable. He was raised in a small provincial town that was socially, politically and culturally divided between Catholics, politically on the Right, led by the local priest, versus secularists, of whom his Socialist school teacher parents were prominent. It was a place where the faithful and laity did not speak to each other and where no one would expect to go to a university as there were no higher schools preparing the young for university entrance examinations. Such social division and lack of opportunity was common in provincial France before World War II. But a Jewish professor fleeing the new racial laws in Paris arrived, unexpectedly found refuge with the priest, Fabre's parents started talking to the priest, and Michel was taught Greek and Latin, which allowed him to go the Ecole Normal Superieure. His results were at the top of the national competition as they would be in future examinations. He made a surprising detour, which friends claimed would end his career, by working for the American military while in the French Navy (1959–62), then taught in the United States, and emerged as one of France's first post-War Americanists.

Seeking a topic for his doctoral dissertation, which then had to be on a dead author and only one person at a time was allowed to research the author, he read in a newspaper that Richard Wright had died in Paris. Fabre immediately visited Wright's widow, bought Wright's library and papers, and became France's first specialist in black American literature. His books on Wright brought him fame, invitations to Harvard, and support for conferences and other activities that the Americans found culturally useful during the Cold War. Fabre continued to work on black Americans in France, especially the novelist Chester Himes. After James Baldwin complained that Paris was for white tourists, Fabre wrote a now much imitated guide to Paris for black Americans. Along with his wife Geneviève, he founded the Centre for Afro-American Studies at the University of Paris 3. Geneviève also became a famous scholar, perhaps best known for her book on Black American Theatre.

Fabre was a committed Socialist, very much on the Left, who knew and helped the Black Panthers when they were in Paris, and was sympathetic to the decolonized. He became one of the first Professors of English in France attracted to the study of African, Caribbean and Indian writing in English, and wrote brilliantly about such authors as Derek Walcott when they first began to attract attention from academics. He became part of the American, Black American, African, and Commonwealth literary scenes. The Modern Language Association in the United States voted him one of its distinguished foreign members. He had an unusual range of interests, art and book collecting, political activity, music, cinema. After his death Harvard University held a conference in his honor at which literature had a minor place among the many pursuits that were mentioned.

The University of Paris 3, La Nouvelle Sorbonne, was a part of the Sorbonne which after the events of 1968 fragmented along political lines. Paris 4 was conservative, the English Department conducted interviews before accepting students, and did not require that lectures be given in English. Paris 3 was by French standards at that time middle of the road (traditional Communist and Socialist), and the Department required that lectures and classes be given in English. The University of Paris West, Nanterre, Paris 10, was "New Left" and attracted literary and cultural theorists, various militants, and Americans looking for the latest intellectual fashions. Other branches of the Sorbonne had their own personalities. The English Department of the immense Paris 7 was located on the right bank at rue Charles 5 where it had its building. It was formed after 1968 by those who objected to the traditional emphasis on centralization, a State Doctorate, research and publication. While it could not get away from the need for a State Doctorate for appointments (indeed with the absence of required publications competitive examination results became even more important), many of its professors had fewer publications than usually expected for appointment and promotion. Being located in Paris it became a powerful voice on national committees against the power of national committees

The standard of professorships at Paris 3 was high, at the time appointment was by a central national commission. Later when departments could hire their own staff there was more opportunity to hire friends. In the past it was common first to become a professor in the provinces and hope to be invited to a chair in Paris on the basis of publications and reputation. Now it would be foolish to leave Paris as you would be unlikely to return. Better to hang around waiting for a post to be created for you, and indeed it has become common to describe the necessary areas of interest to find a person that the department has decided in advance to hire. Several of the professors I knew during 1977–78 proudly told me that they were Huguenots. They were unusually friendly to me in comparison to most French. I gathered that although the term meant "protestant" instead of Catholic it also suggested that they were likely to have opposed the fascists, were sympathetic to Jews and Israel, and still thought of themselves as a minority with a moral seriousness. As a well-known French Roman Catholic theologian had recently died in the bed of his mistress, the reporting of which led to the closure of a Paris English language magazine, it was not difficult for Huguenots to feel superior.

Michel seemed to know everyone. We invited the American professor of French, Susan Suleiman, and her husband to dinner with the Fabres and of course they had met earlier at Harvard. Susan had edited a book of Paul Nizan's essays (1971), Adele had written her French Universary doctorate (1970) on Nizan and both knew Henriette Nizan. We had come to know Henriette through our old friend Madelaine Rousseau and I used to enjoy seeing her and her son-in-law Olivier Todd when we visited Paris.

Henriette, from a distinguished Jewish family—Claude Levi-Strauss was her cousin—had married Paul Nizan when he was a leading Communist intellectual. They had visited the Soviet Union as guests of the government and seen nothing wrong. Then when Germany invaded France Nizan broke with the French Communist Party for accepting the Soviet-German pact. He was killed fighting at Dunkirk and subsequently vilified by the Party. The New Left made him a hero and his reputation as an intellectual and writer has risen while his friend Jean Paul Sartre is now thought foolish for justifying Soviet and other leftist tyrannies.

I liked Henriette as having had a full life including several lovers and surviving without complaint. She told us of having discouraged a rapist by pretending that she had cancer. She had spent the war years in Hollywood as a teacher and writing French subtitles for films. She was a feminist without loudly proclaiming herself one unlike Simone de Beauvoir whom she said should learn how to dress and fix her hair. I think Henriette and Todd found us of interest because of our teaching in Africa. I thought everything about them interesting whether the traditional French oysters at Christmas or Todd's hand drawn postal card from Picasso. They were Left intellectuals who had minds and personalities of their own. Olivier had been a close member of the Sartre circle, broke with it and moved to the political center where he became famous as an editor and journalist and later became internationally famous for his biography of Albert Camus.

Adele was writing a book about *The Writings of Camara Laye* (1980), who was famous for *L'Enfant Noir* and *Le Regard du Roi*—among the first and best post-war French black African novels. Laye was exiled in Senegal for opposing the tyranny in his native Guinea and was in Paris for medical treatment. Adele began interviewing him, invited him home for a meal, and soon we were friends. I found him personable yet strangely naive for a writer, which I thought might be explained by his illness. It never occurred to Adele to ask him about rumors that he had not written his novels. She would have thought such questioning impolite although later when she tried to disprove such claims she ended up writing a book, *Rereading Camara Laye* (2003), showing that the second novel had indeed been written by someone else, a pro-fascist Belgian homosexual who had escaped to

Paris after being sentenced as a collaborator, and that the first novel required much help. (The actual author of *Le Regard du Roi* was one of those on the left bank who lusted for black men and at whose apartment Laye had lived for a time. The woman who helped with the first novel remains a mystery, her name known to an African government minister who was a Professor of Philosophy in Switzerland when she and Laye visited, but he insists her identity should remain a secret.)

We took Camara Laye to a large party at the Fabre's celebrating Geneviève's award of the State Doctorate. Geneviève remarked that when Michel was celebrating his Doctorate she prepared food and was responsible for the party; she was still preparing the food and making the preparations to celebrate her Doctorate. At the party we met the Guadeloupian writer Maryse Conde and her British husband. She had gone to Africa seeking roots, found that African men mistreated women and were more attached to tribe than their partners, and returned to the Caribbean to write novels that reflected her disillusioning experience. She and Richard came to dinner and we remained in contact over the years more acquaintances than close friends.

Donna Kuizinger had found us our nice large apartment during the summer when she was in Paris. It was owned by an American woman who married a Frenchman who preferred to live in the provinces with his family which was more significant to him than his wife. Such a situation is not uncommon in France where until recently, unless there was a marriage contract specifying otherwise, property was inherited by family and a wife could be evicted by the heirs. The apartment was near the Place Periere in the 17th. The quiet hotel next to us was a *maison de passe* where respectable businessmen took their mistresses. When I was a visiting Professor at Charles V (1990–91) Adele and myself were invited to dinner by the woman who was Head of the Department. When I mentioned the *maison de passe* her husband said that his father was a habitué and he had been taken there for his first experience of a woman.

Nicole attended the Ecole Bilingue, which had been highly recommended to us as where foreign diplomats sent their children. I was impressed by the teaching that was from the first orally in French with emphasis on pronunciation. Only after the children felt secure in speaking was there written work including grammar. Among her early lessons was the biology of a *moule* (mussel), which I thought advanced for such an early age and yet typically French in being about both sexual reproduction and something eaten. We used to color in the drawings as section markers for a cook book. At school Nicole made friends with a Swedish girl on whom she modeled herself and would only wear clothing we had to buy in a Swedish shop. She also had an American friend whose mother seemed amazed that we took Nicole with us to such restaurants as Le Duc and Le Bernardin (then small and run by the chef and his sister) when Nicole as a vegetarian could hardly eat anything on the menu and complained. But we had ever since she was a baby tried to take her with us wherever we were going, whether to a party in Deia or a midnight film in Paris.

While the teaching at her school was excellent her guitar instructor at home was terrible. Instead of the fun of playing melodies she was given chords to practice week after week. She decided not to play guitar any longer and would never try again with another teacher. This was upsetting as Adele and I both enjoy music and still thought of ourselves as amateur musicians. In Los Angeles Adele bought sheet music to play the piano in the house we rented and later when she was teaching in Indiana we bought an old upright piano. The King Trio would need become a piano-drum duo.

There are often casting calls in Paris for films and Nicole found herself with a choice between a famous French director and George Hill's *A Little Romance* (*1979*) with Lawrence Olivier, Diane Lane, Sally Kellerman and Broderick Crawford. Although she was a nameless extra she appears on some cinema posters. She had to wake up early to take the first metro to where the crew assembled for a bus to the lot, ate with the crew, returned late tired, and was paid union rates. She found waiting around for hours between takes and retakes boring and never wanted to be in films again. She also had an attack of appendicitis and was rushed to a hospital for an operation, where she spent several days. The expenses were covered by my university insurance.

Adele and I had been returning to Paris on holidays ever since we met in 1954 on our way to Europe. We once felt proud that we were accepted by regulars at a tiny steak, chop and cheese house in a small street near Les Halles; the others were administrators and business executives who lunched every day with a bottle or two of wine and some brandies and probably slept the remainder of the afternoon. They passionately discussed where to eat on Monday when the restaurant was closed. Our tastes had evolved with the Parisians and now was the time of the New Cuisine. We purchased books by Michel Guerard, the Troisgros brothers, and Paul Bocuse. Although over the decades Raymond Olivier's Le Grand Vefour was our favorite restaurant, we now had to try Alain Senderens and Alain Ducasse. The *Gault Millau* had replaced the *Guide Michelin* as the arbitrator of restaurant rankings until it became too sure of itself and started claiming to have judged new restaurants before they opened. One night we ran into a couple we knew from Deia in our favorite fish restaurant; they religiously followed the *Gault Millau* and had telephoned from Germany to reserve a table a week in advance.

Another night the three of us went to a new fashionable restaurant near Les Halles and I saw how unattractive and bad skinned famous television and movie stars could be close up. I asked the owner, who was low voiced, heavy and bearded like myself, some detail about the food and was invited to spend an evening in the kitchen. There was a mistake; he thought I was a professor of Cuisine, not a professor of English literature. As they say, something was lost in translation. I arrived the next afternoon at 5, was given white kitchen garb which, as I did not know how to wear it, kept falling down, was lectured in French by the young chef on the principles of the new cuisine and spent the next ten hours watching with my eyes wide open to what was for me a new world. First the chef inspected all the products in the refrigerator. Three mornings each week he would go to Rungis to purchase what they were serving but not today when there were deliveries. As everything had to be prepared from fresh ingredients he now began composing the dishes that would appear on the menu; he weighed the portions of meat, fish, fowl, teaspoons of caviar, to fix prices. He then had to explain to the kitchen staff what would be involved with each dish including the portions, the composition, the appearance, and responsibilities.

There were only six in the kitchen, all men, including an obviously homosexual pastry chef and an Algerian dishwasher. Adele later complained that she had not been invited, but it would have been awkward to have a woman in that tiny space where there was continual body contact as men shoved past each other, often wordlessly completing a task started by someone else, offered opinions by doing rather than speaking. I had never seen such co-operation in my life. I want to make comparisons to bees, ants, but even they seem too poetic for such intense work. I have since been in other restaurant kitchens and I am no longer surprised that even professionals make mistakes and have bandaged fingers, but I remain impressed by how well these men understood what they were doing and instinctively exchanged roles adjusting the appearance of a plate, moving to the towel draped on the side of the cooker some scallops so that they would stay warm without becoming dry, freshly chopping parsley into a moist mince. The young head chef made in advance, while he explained to me what he was doing different flavored *beurre blanc* sauces that were kept over warm water. I was a not so old dog learning new tricks. Use champagne vinegar for the reductions as it is more intensely flavored; prepare the court bouillon with lime peel and thyme.

We joined the waiters and others in a larger basement room for a meal of freshly made beef stroganoff, with lots of paprika, fresh cream, and pasta, much more filling than the fashionable cuisine that they would serve. About eight a hung over proprietor arrived in the kitchen and ate standing at the butcher's block fried eggs and pate while drinking brandy. He later showed me his apartment above the restaurant; he entertained until early morning. He had had moved to

Paris from the Caribbean where he had his last restaurant and was being observed by several American groups looking for chefs who could manage quality restaurants they owned. That would be big money. He eventually took over a restaurant in Williamstown, Virginia. I arrived at our apartment early in the morning with a recipe for flourless biscuits. Begin with the whites of seventeen eggs. Adele asked how she was to divide seventeen for a smaller amount of biscuits.

I had made progress with *New English Literatures*, had discovered "modernization theory" which offered insight why and how a modernizing movement such as nationalism could sour and become tyrannical, but no new job had turned up until Derry Jeffares asked me whether I would like to replace Alastair Niven at the University of Stirling for a semester as a Visiting Professor.

## II   Stirling

Although I seldom saw Derry Jeffares after I left Leeds he remained a patron whom I could depend upon in an emergency. When the 1977–78 year ended in Paris I had no place to go and Derry arranged for me to replace Alastair Niven who was leaving the University of Stirling to be Director of the African Centre in London. Less than a decade previously I had, while visiting Richard and Ada Coe in Leamington Spa, hurriedly read through Alastair's long doctoral thesis, before examining him at Derry Jeffares's house in Leeds. Alastair was now at the start of a distinguished career in the Arts Council, the British Council, and later as Principal of Cumberland Lodge among those who made cultural policies. He was especially a forceful voice in support of the multiracial, multicultural England that emerged, surprisingly, during the Margaret Thatcher years when many on the Left were appointed to the National Theatre and other places of influence where despite objections to the government's economic views there was a conscious attempt to promote a new, less white Britain.

I would be a non-renewable visiting Professor for the first semester 1978–79. Jeanne and Derry were kind in helping us find a house in Dollar, a lovely town in which the roses bloomed until Christmas when the streets would be covered in many feet of snow and snow shoes would be a normal means of transportation. Derry and Jeanne were surprised by our desire for heating and comforts, they lived in an old stone house that was, I thought, freezing and depended on a coal fire. Derry had built himself a study in which he was not to be disturbed until super when the wine and anecdotes would flow for hours.

I had not realized that Derry really was worried financially. I knew he would at times invest badly, losing money on stamps for example, and had feared the bailiff when moving to Stirling he had been for an unexpectedly long period unable to sell his Leeds house. Derry now told me that he bargained for everything, wine, fuel, clothing, and had done so since his years in Dublin when there was no money but he had to keep up family appearances. The Jeffares were Anglo-Irish protestants who lacked the means that went with their heritage and often there were problems that fell on Derry's shoulders, such as sending to Canada an uncle who had been arrested in Phoenix Park. He often was selling something, such as beautiful but much too heavy Irish woolens that would have been unbearable to wear in centrally heated North America. His ethics were pragmatic. When a colleague who was departmental social chairman neglected to post announcements for a visiting lecture, Derry said that no one cared about the missing bottles of wine provided that you did your job properly. Which was very different from Bristol where the unfinished bottle of sherry was tightly corked and guarded by Henry Gifford until the departmental party next year.

I can now see that Derry remained Irish and heavily involved in Irish university and cultural affairs long after he settled in England. Many of the writers, literary editors and agents I met through him were Irish or Irish by origin. Derry did not get drunk but there was an Irish conviviality in which everyone was invited and expected to share the on-going and often slanderous gossip. Alongside the amusing stories about friends, the academic and literary world, publishing, life overseas, one topic of course was the Irish. Jeanne was Scots, so they formed a kind of Celtic Union, and it was not surprising that Derry should become a power behind the scenes in the Scottish Arts Council and other cultural nationalist organizations. As an American I often felt I had been included in a community of Irish, Scots, and Welsh who depended on the British and the British Commonwealth for their employment. Even Niven was a Scots. But then I often felt that without the Celts and Americans running the show the British would never survive as they were caught up in their social differences defending imperceptible snobberies and minor privileges. Others had to do the work.

My teaching duties at Stirling were light, and the students well disciplined. Tommy Dunne, who had previously chaired English departments in Africa, and who I replaced at Lagos, had instituted a system according to which essays and terms papers had their deadlines set well in advance and they had to be deposited into a central box by the set date or they automatically lost a fixed percentage depending on lateness. It was one of the few places I taught where you did not need to announce that a student was only allowed one dead grandmother a year. The campus was also beautifully landscaped with swans on the artificial lake that you first saw when entering the campus. I seldom saw Dunne who was reputed to keep his house at African temperatures that allowed him to be nude.

We had several dinners for visiting lecturers including Ruth Prawer Jhabvala, who reminisced as she often did, about the herrings of her youth, which she was rediscovering in New York, and P. Lal, who pretended shock when the diner conversation turned to bull fighting: pretended because when I visited him in Calcutta his wife said that he was only a vegetarian abroad. At home he was a meat eater. We drove to Edinburg to buy fresh fish, listen to jazz and see art films. Nicole studied at Dollar Academy, wore a uniform, and before we left we met her first boyfriend, about whom we had known nothing. Adele decided it was soon time we put her on the pill.

The main event of the year was meeting Sarah Mahaffy who had newly become an editor at Macmillan and who came to Derry hoping to publish books in Drama. Derry passed her on to me as I had surprising picked up a record in Drama ranging from my early book about Dryden's plays to having started a theatre department in Nigeria. At Stirling besides giving seminars on the New English literatures I tutored in Modern Drama. I gave Sarah names and contacts and they seemed to go no place until Adele told me that I was expected to propose something. I remember telling Sarah to see Tony Harrison, but Nick Hern or some other publisher was already involved with the National Theatre and Sarah thought she only needed open her mouth and utter her upper class vowels and Tony tuned out.

I liked Sarah. Although not an intellectual, indeed somewhat out of the literary circles I knew, she was ambitious and knew how to use her contacts. She enjoyed food, wines, and cooking, and we discussed recipes. She looked down on those with bourgeois values including sexual fidelity and I had a feeling that the friendships that helped her advancement through Macmillan and then in setting up her own publishing house, Boxtree, were not always Platonic. Sarah's family knew Robert Graves and there may have been an affair between Robert and her mother. Adele liked to conjecture that Sarah was one of Robert's children. There was a candor about Sarah, an appetite for life that invited such speculation. I remember her excitement one night when we went to the theatre with Eva Figes. When I found a temporary American home for the Modern Dramatists with Barney Rosset and Grove Press, Sarah took the series away from American Macmillan, which I thought and proved to be a mistake. Barney was more tough and erratic than Sarah; finding the first books did not sell as much as he hoped he drastically reduced the print order, started American offset printing, and soon the series was back with American Macmillan.

Sarah understood that any book with a connection to television or the cinema or celebrity would sell, and had no hesitation at Boxtree earning big money that way, but she also enjoyed researching illustrations for books and paying for them even for the academic studies with which I stuck her. She enjoyed the excellent reviews and praise although the Modern Dramatists were hardly likely to earn her much as an editor. Sarah would eventually sell the publishing house she founded, Boxtree, for a small fortune, while being enticed to join the board of directors at Palgrave/Macmillan where I gather that she was out maneuvered by another woman, and withdrew. Luckily she had married an American who worked in the city and who had inherited considerable money; they lived comfortably after in London and France.

I thought it absurd that I should be regarded as a Theatre authority but I like to fantasize and drew up a proposal for thirty introductory books on 19$^{th}$ and 20$^{th}$ century dramatists. My readings of Eric Bentley's anthologies and books about European theatre became handy. Sarah's advisors claimed that it was the best proposal that they had ever seen and the Modern Dramatist series eventually reached fifty volumes before it petered out. Once or twice it would briefly flare up again after a newly appointed editor at Palgrave/Macmillan perceived a demand for a still unfinished title or a few that might be updated.

# Chapter 11: Trying to Begin Anew

*Christchurch, New Zealand 1979–83*
*Florence, Alabama 1983–86*

I was very unhappily one of the three professors of English at the University of Canterbury, Christchurch, New Zealand where I spent four of the most miserable years of my life (1979–83). You should never get what you want. In Nigeria, during the civil war, emergencies, oil, water and electricity shortages, I thought how nice it would be to live some place quiet, settled, in a little house with a small garden and not worry about soldiers accidently discharging their gun over your head or faculty radicals trying to take over the university. I had, however, become addicted to such a life: I found New Zealand intolerably petty minded, a place where it was assumed that if there was not a law permitting it then it was not allowed and should not be thought of. It was a more sophisticated civilized version of the small town I had fled in my youth.

Christchurch was a small pretty city, which mostly consisted of small houses with small fenced gardens, even the downtown center appeared as it were constructed for a picture on a postal card. In the distance were some mountains. But it was cold, windy, isolated, somehow part of the past. There used to be a common joke about foreigners arriving in New Zealand and being told to set their watches back twenty five years. Another joke, an Australian visits New Zealand and returning home is asked how was it and replies "I don't know. It was closed". I remember driving from Christchurch north to Wellington and thinking if the car breaks down there are only sheep to help. Further away from the city there was nothing except signs suggesting someone had in the past tried to live there and I started wishing I still had sheep for company. Christchurch had more than its share of all the social snobberies that the British and their off-shoot seem to delight in. If you were not a family that came with one of the first four ships there were private schools and even an Agricultural College for the socially superior. As the city was flat to live on a street with a slight rise was to be socially elevated as I was delighted and amused to find about where we lived. If we had rented around the corner, where there was no rise, we would have gone down in the world.

The truth is that I had not wanted to go to New Zealand. When earlier I had commissioned essays for *Literatures of the World in English* (1974) I had confused the large Australian island of Tasmania with New Zealand. When friends assumed that whoever was supposed to have written the essay on New Zealand missed the deadline I kept my mouth shut. I had applied for the Christchurch professorship because nothing else was available and until late I was expecting, an offer from the University d'Abidjan, Cote d'Ivoire, for Adele and myself. By the time it came, several years later, I felt defeated and was. After three decades abroad, I was now thinking of returning to the United States. I had turned down professorships in Fiji, Northern Ireland, and even Nigeria, but no one in the USA seemed to want me. I am still not certain what spun me into a depression in New Zealand, but the continual rejection of my applications for a professorship in the USA did not help my view of life.

There is no sense in complaining about the English Department at the University of Canterbury, New Zealand. It was quarreling, unhappy, and petty minded before I got there and remained so long after I left. Where I had either been head of department or a visiting distinguished professor I was now expected to take orders from an undistinguished professor who, I understand, came to New Zealand from a large British school and chaired accordingly. Trips to conferences abroad, even to conferences within New Zealand, had to have approval far in advance, with explanations of replacements during that period, and, of course, money had to be budgeted. It was assumed that non-teaching time would be spent socializing in the senior common room and Friday nights drinking at the Staff Club bar with the Vice-Chancellor. There was continual resentment and petty mindedness. If a member of the department had nervous problems and asked me to help teach his graduate seminar I was later told that I should have had official permission. When the lecturer on New Zealand literature went on sabbatical no one would volunteer to do the lectures on the decades before 1940, so I thought it my duty as a professor to take on the task which I enjoyed and I gather so did my students who appreciated the international outlook as the first forty years of their literature was often influenced by American writers (such as Walt Whitman), the international feminist movement, and by modernism. Within the department I heard bitching about my being an American lecturing on Kiwi lit, a literature that they seemed not to know. Whenever I offered an idea I would be told that no, that is American and not our way, when I actually had picked it up in the UK, Nigeria or France.

The students, however, were often good, and the recently retired Head of Department had created a good degree structure starting with an introduction to criticism, close study of some set texts, and moving on in the second year through the study of poetry, Shakespeare, and various periods and special subjects. The university was based on the system, common in Scotland where the first three years begin with three subject areas, then two in the second year and specialization in the third. The fourth, or MA year, is similar to the British B A Honours.

I had students whom I still remember. Bill Direen was already known as New Zealand's best art-rock guitarist, and directed an excellent private production in the basement of an abandoned factory of Strindberg's *Miss Julie*. My daughter Nicole sung and recorded with his rock group and when she left to do a B.A. at Bryn Mawr in the USA I eventually began playing percussion with Bill's group, creating odd sounds and rhythms around his almost military solid drummer. Bill's girlfriend played piano. As I had no drum set in New Zealand and we seemed always short financially (the university did not pay me well) I kept borrowing and trying to make do with odd equipment such as a three drum tree I found in a local music school and a contraption I put together with coat hangers, a broken cymbal, cow bell and wood blocks. I sometimes listened to a cassette I had recorded playing solo in which I sounded like a cross of my early hero, the New Orleans drummer Baby Dodds, and the Yoruba drumming I had learned while teaching at Ibadan. This episode of my always disappearing musical career concluded at rehearsal one night when Bill announced that because of a sore throat he was cancelling our opening for Nick Cave and Bad Seed. I felt it predictable, Bill wanted to remain an underground legend, far from the spotlight of success.

This was not the end of Direen in my life. After we had left New Zealand and Adele was teaching in Indiana I received a late night telephone call from Bill who had a gig playing guitar for a night or two at a famous Lower East Side punk rock club and wanted to take a taxi from the airport to spend a couple of nights with us. I had to explain that we did not live in New Jersey and that Indiana was probably a thousand miles from New York and not a taxi drive. Even if it were I would have been reluctant to give a bed or even space on the floor to a New Zealander as they were notorious for travelling around the world, long outstaying their welcome by many months with once innocent hosts.

Decades latter while retired in Paris I came across Bill, who had married a small, pale, quiet French woman, who tried to support the two of them as a part time librarian. Bill had a few gigs around Europe with other very neglected art rock musicians, some which took place in underground rooms rather than basement clubs. He had published several volumes of his poetry in extremely limited editions. His email address and telephone seldom worked but I would every year or two get a message that he was dying of disease for his earlier sins, then he seemed cured but still inaccessibly underground. One evening during 2014 I found in my mail box a beautifully crafted thin volume of his versions of poems from various languages and a note saying that he had inherited money and was returning to New Zealand to remodel a house he had purchased there. Which would keep him out of trouble for a few years. As the book was the 52$^{nd}$ of 53 copies I was puzzled to know what to do with it as I sell off or otherwise get rid of books in Paris to make space for more. Luckily the New Zealand poet and book reviewer Janet Kemp (Jan until she moved to Germany where Jan seems always to be a male name) came through Paris for a conference and was willing to take it off my hands, review it, and give Bill's volume a space on her shelves.

I do not remember whether or not Stephanie Johnson was a student of mine. She was part of our social world in which brighter students, some faculty, and creative people, would come together in various ways ranging from parties, conferences, to alternative theatre and rock. She was beautiful, bright, and stood out. Ian Richards, one of my better students, attracted women easily, but told me that Stephanie was the one he felt beyond him. (He would later teach at a Japanese university while waiting for a post in New Zealand, marry a Japanese woman, settle in Osaka, and continue to write creatively and about New Zealand literature, mostly on-line.)

The first time I saw Stephanie was at the New Zealand Writers conference in Wellington soon after I came to Christchurch. I had to travel and stay in Wellington on my own money as, of course, I had not requested financing at the start of the academic year, before my arrival. The grayness, steep hills, and rain of Wellington seemed a fitting setting for the depression that I was starting to feel. Many of the writers I met there acknowledged being or once having suffered from depression and I thought perhaps this is the national illness. Then a beautiful young couple appeared at the conference, a blond princess and her dark haired prince, Stephanie Johnson and her then poet mate Michael Morrissey. I would later meet Michael who had a poet-in-residency at Christchurch (he was part of the literary establishment, his mother was a good well-known poet) and who would, after they broke up, write a nasty satire on Stephanie and her problems with her leg which, she told me, had been poorly reconstructed when she was younger. Stephanie would go to Australia, become a successful dramatist, and I am told married an Australian novelist with whom she returned home where she eventually founded the Auckland Writers Festival. She became a successful New Zealand novelist, writing amusing, middle brow, comedy about local society. Her novels sometimes come my way for review and I enjoy reading them for their wit, intelligence and awareness of the intellectual world outside, alluded to more in passing and narrative events than through opinions.

The New Zealand cultural world was small and I felt it was heavily guarded against outsiders and the outside world. Michael Harlow, a poet from America who then lived in Christchurch seemed excluded from lists and anthologies of New Zealand literature. During that trip to Wellington to the Writers' Congress, I went to see whoever was then nationally responsible for the Arts (I seem to remember it was a minister for Sports) and was sent on to the head of the Arts Council where I explained that New Zealand was not part of the Commonwealth Arts scene that was developing rapidly across the former British Empire. There was a need to fund travel by writers, artists, and scholars to international readings, festivals, and conferences, and create residencies. I was asked to write a paper with my ideas, which was published but I thought ignored until one of Karl Stead's post-graduate students came to me for further suggestions. In Nigeria or England I could be Nigerian or British, but in New Zealand I would always be an American. I remember going to see someone at the Canadian high commission and explaining that Canadian studies could piggy-back on the Commonwealth Arts scene but as a start there would need to be Canadian books in libraries. Again the project was thought not for an American and a Canadian was found who did not know what to do next.

I had known the poetry of Ian Wedde through Silkin and *Stand*. In New Zealand he was a postman and I thought well okay Wallace Stevens walked to his insurance office while making up verses he would dictate to his secretary. The smallness of the literary world kept surprising me. *Landfall* had sent me a volume to review by a prize winning Otago poet. It seemed full of rain, depression, and being at the edge of the world, which I found common to many local poets and which corresponded to my own feelings about life in Canterbury. Some time after the review was published the poet showed up at our small house, objected to the review and then invited me to a pub for a drink. To get rid of him I innocently accepted and soon found myself alone at a table while being threateningly glared at by his unkempt fiends. I was not beaten up, but I did have an authentic taste of New Zealand literary life where everyone was encouraged by their friends and where the usual critical commentaries I knew were thought offensive. He was not a bad poet, but a minor talent by the standards of England, Nigeria, Canada, or the United States and I made the mistake of not calling him a big fish.

The planning and creative abilities I had developed in Nigeria were useless at the University of Canterbury where I soon retreated into writing my *History of Seventeenth-Century Literature*, editing the *Modern Dramatists* series of books, and applying for posts abroad. I was undergoing a mid-life crisis when after early earning a doctorate at 27 and being appointed a full professor at 34, and after having expanded an English department into the projected Arts Faculty at Amhadu Bello, I found myself alienated, gossiped about, frustrated, depressed, and doing all the wrong things to get out of my situation in New Zealand where parties seem to consist of uninteresting people who had gone to school and university together talking to each other about their school and university years. Excitement appeared to consist of exchanging partners, finding new partners, affairs with students,—not my cup of tea although faculty and administrative wives were often on offer. Our social life was far less interesting. A member of the English invited Adele and myself to lunch and immediately afterwards without any transition asked me to commission him as author of a book for the *Modern Dramatists*. What style!

I arrived in Christchurch tired and worn out having travelled from Scotland by way of I forget where except it included a week driving around Fiji in a rented car. The university put us in an unheated empty house were we tried to survive until we found something to rent. I early attended a concert of Renaissance and Baroque guitar and was impressed by the ambition of the program and very much unimpressed by the guitarist who several times stopped mid-composition to retune, very unprofessional after the excellent amateur guitarists I regularly, almost nightly, listened to during the summers in Deia. I mentioned the concert to the head of the school of music and was told that I had heard that best classical guitarist in New Zealand. And that just about summarizes my experience of Christchurch. Decades later, living in Paris, I would read about the terrible earthquake that destroyed Christchurch and, inadvertently, I laughed. I did not realize that I still hated my years there so much.

    One reason was that my marriage seemed to have become a boxing match with Adele and myself continually arguing. I suppose it was normal that if I were going through a mid-life crisis she would as well. I thought, wrongly, that my publications would eventually come to my rescue and wanted to devote my time to them instead of adjusting to the small world to which I had brought us. I do not think I have ever consciously adjusted to the demands of others and certainly had no desire to do so in Christchurch. Adele, learned that there was no chance of a lectureship in French in the already overstaffed university department, took a part-time position teaching French in a private school, and developed an intense interest in the feminist movement. Adele knew nothing about cooking when I met her, had become an excellent cook through studying French cookbooks, and now a feminist cooked no more. We used to share cooking, I now did it all. Having a chance to edit a series of books on women writers she felt insecure to do so on her own and turned to our old acquaintance Eva Figes as co-editor. Several letters from Eva to Adele should have been warning that Eva was going to be obstruct commissioning books about her contemporaries or by critics she did not know or disapproved of. Another theme for Adele and myself to quarrel about.

    When Nicole, who was graduating high school, said that she could not live at home while attending university because of our quarrels I fully understood and gave her what help I could in rapidly obtaining applications to some American universities, arranging though my cousin Mimi, a high official at the Educational Testing Services, to take their SAT examinations. Soon after Nicole left for Bryn Mawr College Adele and I separated for some weeks. Although we reunited our relationship remained quarrelsome for a few years as we gradually became once more a couple. I suppose if either of us had found someone really attractive that would not have happened, but I kept finding other women not as beautiful, intelligent, desirable, active, interesting, well read, well dressed and of course they did not share the same history. Adele and I seemed fated for each other. A blond attractive secretary only needed to order fried food that she ate with her fingers to turn me off, and that was before she explained that she, her boyfriend, and everyone she knew had herpes. I once read about a woman's international sex experiences; she claimed that Americans ask for your telephone number at 1:45 AM. Have I really become that kind of American?

Conferences are a normal part of academic life and can be addictive. A French physicist I know attended so many foreign conferences that I asked whether he had a girl in every airport as I knew that while he was away his wife would go out to play. They soon divorced. Another friend seems to be part of at least one conference a week and has complained that if universities expect you to go to conferences they should not expect you to teach. I seldom went to conferences unless they were in exotic places, such as Budapest (August 1976) or Malta (March 1978) where I became friends with Francis Ebejer, the bilingual novelist-dramatist whom I would write about. He claimed that his father had helped standardize the Maltese language and he was showing its wide-spread European and Mediterranean origins through the various models of his writings.

In Christchurch I felt locked in and wanted to get out. I several times went to conferences in Australia, perhaps hoping someone would offer me a professorship there. I had long admired A. D. Hope for his poetry and literary criticism. I had republished an essay of his about Dryden's *All for Love* in the first book that I had edited and I had commissioned another essay from him for another book about John Dryden. I would write about him in *Sewanee* Review. Michael Wilding and I shared a scholarly interest in the English Seventeenth Century, we had both written about Andrew Marvell, were friends with Jon Silkin (he was Australian editor of *Stand*), and Michael was one of a group of new Australian fiction writers that I was reading including Peter Carey and Frank Moorhouse.

A conference on American Studies (remember I was looking for excuses to visit Sydney) during 1982 began with a rather too thorough examination by customs at the airport. (I later learned that some American with a yacht was sending drugs to Australia.) At the conference I discovered that Richard Gibson, a friend of Walter Bowe whom I had been introduced to in Philadelphia, and whose writing I had discussed with Lionel Trilling, was now thought by scholars to be a double agent, a black radical active internationally while also working for the American government. I would sometimes see his name in the news and even letters to newspapers in which he defended his liberationist adventures.

During 1981 I managed to go to a Commonwealth literature conference in Frankfurt, Germany, that proved fruitful. I met (George) Kofi Awoonor (Williams) whose writings I had been sent to judge for a professorship in Ghana while I was professor at Amhadu Bello. Awoonor had recently returned from the USA, where he was already a professor and famous writer, to Ghana where he had been imprisoned for supposed involvement in an attempted coup against the government. I was pleased that Ghanaian universities remained independent from politicians and replied that if Awoonor, whose writings I admired, were not qualified for a professorship I did not know who else was. Now in Frankfurt he was ranting about the need for political revolutions and mocking Wole Soyinka as a romantic individualist for, during the Western Region crisis, taking over a radio station at gun point and broadcasting a tape calling for an uprising. Awoonor claimed that Wole had used a toy wooden gun. I knew Awoonor was lying probably from jealousy, but I could not stop his assertions. Awoonor eventually became an ambassador for Ghana; he died, along with his son, when Islamic militants attacked a shopping center in Nairobi Kenya, September 2013.

That same night in Frankfurt I met Alamgir Hashmi, a Pakistani poet and critic, who remains a friend. After I returned to Canterbury he asked me to contribute an essay to what was the first book about the Muslim literary imagination. My essay included a discussion of Salman Rushdie's recently published *Midnights Children* and in reply to an enquiring letter Rushdie wrote it was amusing that Hashmi and I should consider him a Muslim. This we before *Satanic Verses* and I have kept the letter. I would write about Hashmi several times as I thought him the best of the Pakistani poets and was interested in his attempts to avoid the politics of decolonization and continue to see literature as a criticism of life. This was the start of my involvement with Pakistani literature which I continued to write about as a new generation of novelists appeared and gained international attention. After Hashmi returned to Pakistan he twice had me invited to give lectures there, but as at the time I was a visiting professor in Israel and the invitations could only be sent to me through Switzerland that was a bit foolish. I never questioned Alamgir about his life in Lahore, but I did hear that there had been an attempt to kidnap him in public at gunpoint which was foiled by a crowd of bystanders.

Official New Zealand thought of itself as the land of sport. That was the masculine culture it promoted while ignoring the alternative and minority views that were becoming stronger, perhaps a new majority. Nicole at school was told to get a backpack and prepare for a weekend of hiking as part of her curriculum. She was much more at home disco dancing and the weight on her back literally floored her. We had to telephone the school to say that she was too ill for the excursion that weekend. The major public event of 1981, indeed of my years in Christchurch, was the wide-spread protest caused by the long Springbok rugby tour, the police violence against the demonstrators, and the Conservative government's pretense that sport is apolitical and should be supported by every means possible although this angered those who objected to treating apartheid South Africa as normal. The government's protection of the South Africans and the violence against protesters soon radicalized many women, the young, Maoris, the Left, everyone who was against racism. Soon the aim became to disrupt matches and this meant occupying the playing field.

I skeptically listened to heated discussions, I was a foreigner in what was becoming a revolution against the sports culture that dominated New Zealand, and then realized that I could not be a bystander. My students, even the most frail female students, were building wooden shields to protect themselves against the long police batons, and Adele and Nicole intended to join them. Someone loaned me a first world war German helmet, one of those absurdities with a spike on top, and there I was in the front line trying to break through the police cordon as batons violently crashed down. The government had formed special riot squads to defend the tour and they were prepared to do damage. Somehow, perhaps because of my professorial manner, I was the first to break through and did not know what I should do next. I was not even certain I was protesting the South African tour, I might have just been reverting to a juvenile delinquent against law, order, and the police. As I could not politely tap one of the enraged riot squad on the shoulder and ask to return to my friends, I waited in the stands until Nicole and Adele showed up, which they did, and the police ejected us. Adele claims that as she was being mishandled I threatened one of the riot squad that I would call the police but she may have that mixed up with when I was sunbathing nude on a Goa beach and a plain clothes detective tried to arrest me and I threatened to call the police. If I agreed, would he have taken naked me to the police station while seated on the back of his bicycle?

Those against the tour were excited when pictures of the protest in Auckland, where bags of white flour were dropped from an airplane, appeared on the front page of the *New York Times*. New Zealand was news. Afterwards we were invited to some alternative culture parties by the young, but to me this was ho-hum after the many deaths in Nigeria during the Western Regional crisis, the civil war, the several coup d'états. Still it was better than being bored with the party talk about which schools in Christchurch people had attended. Adele likes to point out that until then she assumed that a rugby ball was round, like a soccer ball, not like an American football. For all I cared it could be a triangle.

Although Christchurch felt provincial after the excitements of Paris or Nigeria, New Zealand was part of the wider world and we had visitors such as Martin Esslin who gave lectures at the university and with whom we discussed British friends in common, such as Eva Figes,. The Malayalam scholar and poet Ayyappa Paniker came through, we became friends, and I would later be attached to his department at the University of Kerala in Trivandrum while researching my first book on Indian poetry.

C. K. Stead, who I admire as a literary critic and novelist, came to dinner while visiting Christchurch and we ate with him when we visited Auckland. I thought his book *The New Poetic* the finest study of how T S Eliot had become a poet who similar to the cubists had eliminated narrative continuity and other kinds of conscious authorial meaning from his work. I followed his history of New Zealand poetry that set local within the international fashions and I agreed that New Zealand was making itself absurdly politically correct in over-estimating Maori writing and treating writing by whites as inherently guilty.

I was not surprised to learn that some of his colleagues at the University of Auckland resented his international reputation, his time for his own writing, and his claims that New Zealand should continue its relationship to Europe rather than retreating into the self-preening enclosure that many demanded. Karl had known, indeed was often close friends with, those who had made New Zealand and Australian culture—Frank Sargeson, Janet Frame, Barry Humphries. He was also one of the regional treasures but that stirred up those who could only publish locally and whose narrow vision was actually a copy of the retreat from modernism and internationalism that was current abroad.

When I left the University of Canterbury the going away party was only attended by about half the English faculty and was mostly filled with students which I thought summed up my relationship with the department. The students gave me a jade tie clip that I still have. As at least two members of the department were later appointed professor at other New Zealand universities I suppose they were not as awful as I thought although the two included the one who felt that a lunch should to be repaid by a book contract.

## II

I had no idea at the time my return to the United States and taking a supposed distinguished professorship at the University of North Alabama would end the first half of my life and begin a second, on hindsight more fruitful, phase. I had removed myself from the British Commonwealth universities educational system and would never find my way back, yet my move to a very minor American university, which I had hoped would make me available for job interviews, was distinctly lowering in the eyes of those doing the hiring in American university English departments where Commonwealth or the New English Literatures in which I now specialized remained unknown until later presented as Post colonialism, a Marxist influenced version of the post-Vietnam anti-Western views that became required in American Humanities and Social Science faculties. I had no interest in such anti-Americanism, which I regarded as simply another version of American intellectual culture; my concern was with the immense rapidly developing body of writing that had emerged following decolonization and the previous histories of such literatures. I was interested in how writers criticized the governments and society that had developed in their own country since independence, not in continuing a diatribe against Western imperialism that I thought now dated, and while presented as radical, was a reactionary idealization of the past.

I began as a specialist on older British literature who had written books on John Dryden, Andrew Marvell, a *History of Seventeenth Century literature*, and except for editing books about African and West Indian writing, and my wide ranging *New English Literatures*, I treated the new national literatures as an interest rather than a professional field. My more specialized books about V S Naipaul, Derek Walcott, Indian writers, and black British literature were in the future as I moved on from introductory surveys to trying to understand uniqueness.

I was a white man who now specialized in and expected to teach what Americans crudely and often mistakenly assumed was "black" literature, although many of their "blacks" were light skinned Indians, Sri Lankans, Pakistanis, West Indians, South Africans and the movement for cultural and political decolonization included such mostly white British Dominions as Canada, Australia and New Zealand. Somehow in American thinking white South Africans seemed to be regarded as honorary blacks, engaged in the struggle against apartheid although many of the white South African expatriates were apolitical at home. Most American "blacks" are of course part, sometimes almost, white, and when real African and West Indian blacks began being hired "black" was changed to the exclusionary "African-American".

On the way from New Zealand to Alabama I stopped in Canada for a Commonwealth literature conference at Guelph University, August 1983, where the English Department was chaired by Douglas Killam, who had been a friend and Reader in my department at the University of Lagos. Doug was part of a significant and wealthy Canadian family—I was a few years later to hold a Killam fellowship for visitors at the University of Calgary—but his branch of the family had split off, gone its own way, and missed out on the influence and wealth of the family name. He several times told me of his family's history, clearly felt cheated and seemed intent on establishing his place in Canadian society, applying for university presidencies and sending his son to the most prestigious school in Toronto. His attractive wife was treated with radiotherapy for cancer which resulted in her losing her hair, and he had taken up with his pretty married secretary. Authoritarian rule and too much international travel led to a revolt by his department, especially after the university president with whom he was a buddy died, but that was in the future. He would retire to Oxford England with his secretary for years before returning to Canada.

At the Guelph conference I once more saw such old Commonwealth literature friends as Anna Rutherford and Kirsten Petersen, Hena Maes-Jelinek, Alamgir Hashmi (whose marriage to a Swiss woman was becoming shaky), and met the Canadian-Australian novelist Janette Turner Hospital whom I would meet again several times at conferences over the decades and regard as a friend. I introduced her reading in Paris of one of her novels translated for the then prestigious Serpent a Plume publishing house; and I interviewed her when I was writing a biography of Derek Walcott. We and others went dancing at a local pub in Guelph and I was immediately struck by the energy, cohesion and swing of what was really no more than a local rock group. I was back in North America with the richness of its jazz influenced popular music.

## III  Florence, Alabama 1983–86

From now on my life may seem confusing as it was for over a decade shaped more by opportunities and failures than choices. I was awarded a ten month research fellowship to India but the University of North Alabama required me first to take up my distinguished professorship for a semester. After we left Alabama Adele began teaching at Ball State University in Muncie Indiana for seventeen years, which remained a base as I moved from being a visiting research professor in Calgary, Canada, to a visiting professorship at Ben Gurion University, Israel, six months research once more in India, another semester at Ben Gurion, followed by a summer in Paris, an important conference at the University of Kent, Canterbury, England, where I met Derek Walcott, who became the subject of two major books that I would write. At the same conference I met several of the younger "black British" writers who would be the start of another of my books. I was also appointed to another visiting professorship in Paris. I travelled between continents, sometimes with, sometimes without Adele, until after the death of our daughter Nicole in Paris, I accepted that my career was less important than my marriage, and the transcontinental wandering, to France, the Caribbean, and England, was usually as a couple until Adele retired from her university position and we moved to Paris.

American popular music was why I had decided to take up an otherwise unattractive distinguished (soon to be extinguished when the stock market dropped and the money ran out) professorship at the University of North Alabama, a former teachers college that had hardly changed except for its title. The English Department was a dreadful place where conference attendance and publications were discouraged as taking energy from teaching fifteen hours a week. Being a distinguished professor meant I taught three, instead of five courses. I understand that the university and Florence have now changed for the better. There was no one at the time of interest there except me and students rightly wondered how I got there.

The reason was Muscle Shoals recording studios in a small town across river. During the 1960s and 70s Fame in Muscle Shoals along with Stax of Memphis were in the forefront of southern soul music. The local radio stations mixed white and black music and the two crossed over racial boundaries. Wilson Pickett, Bob Seegar, the Staples Singers, came to Muscle Shoals to record with the famous rhythm section. There had been hit after hit that I danced to in Lagos. I assumed that no matter how uninteresting the university there would be nightly jam sessions, great music, famous musicians, partying, and maybe I would even take up drums once more. Ha! None of that happened. Southern Soul was no longer top of the pops, the nights were past when such musicians as the Rolling Stones would record and party at Muscle Shoals, and the recording studio now bragged at being at the forefront of those recording sweet sentimentality, folk and Christian rock; its famous rhythm section was replaced by a dulcimer. During the 1990s when Bob Dylan wanted to record some Christian songs he went to Muscle Shoals to prove his authenticity; it was in the Bible belt. I was not the only one mistaken. An American bar singer in Germany with the suggestive name of Joy Ryder talked her company into a trip to Muscle Shoals to record some low down dirty funk and not only found that those days were now over, but she was quarantined in the rooms for visiting musicians as the studio executives wanted to keep her from disturbing the natives with her late hippie ways.

Not that drugs and too much booze were foreign to that part of Alabama. Florence itself was one of those 'dry' American cities where the sales of alcoholic beverages was prohibited; the boot legers across the Tennessee River financed the preachers to keep Florence dry. The area had a long established history of making hooch which the police found as profitable as the church leaders in their varied ways. In modern times the area was also part of the route for the distribution of drugs. After a private airplane crashed a curious bear died a happy death by overdosing on a parcel hanging from a tree. More amusing were the letters that were published in the local press from a female bootlegger to a friend about her life in prison and all the favors that sex could earn. Indeed she appeared to be enjoying herself and had a crush on a guard or two. As might be expected prostitution flourished at a local hotel for businessmen where some wives earned money more on their backs than on the side.

There were many excellent musicians who lived near Muscle Shoals, including some studio musicians who were flown to Nashville, Los Angeles or Paris, for recording sessions; and as famous musicians and their agents recorded locally, there was money for entertainment and at unsophisticated steak and lobster houses with long lists of incredibly expensive bottles of wine that would make three star Paris Michelin restaurants seem moderate. I avoided such places where I would overhear someone who looked like an uncultivated dirt farmer talking about his show in Nashville or London.

If you enjoyed good musicianship the Florence-Muscle Shoals area had its pleasures although they were more often country style fiddle, guitar, drum groups in smoke filled taverns where the men seemed always to have a large badly bleached blonde on one arm while holding a bottle of beer. Nicole came one weekend from Bryn Mawr, was blinded by heavy tobacco smoke on her contact lenses and would never return to Alabama. I enjoyed clubs in which tough rock musicians defiantly wore designer jeans taunting the audience with hints of possible bi-sexuality and risking a punch up or worse as the listeners all arrived in pickup trucks with guns. I still had some Nigerian embroidered shirts that I naively wore to such clubs and they were perhaps my best protection against the violence in the parking lots. I was so far out of it that I was not worth the effort.

The university had a Little Big Band that rehearsed weekly but I thought it uninteresting. Within it there was a good modern jazz group that sounded like the latest Blue Note recordings and in which the rhythm would shift to uneven beats similar to Philly Jo Jones and Elvin Jones with Miles Davis. We sometimes danced at clubs outside the city but the demand for jazz was slight. There was more in Birmingham where the first time, knowing nothing about Southern tornados, we drove through the increasing rain and wind until suddenly there was a dead spot and we raced for the nearest lit building which turned out to be the jazz club I was seeking. A seemingly immense black bartender, probably the owner, wrapped his arms around Adele and myself, offered drinks, and explained that we were in the center of a tornado. I felt I was re-enacting a cliché, being comforted by a wise black "uncle". We returned to Birmingham another time and the same black trumpet player was playing in public with two white musicians. My mind filled with racial clichés I thought that is to make the combo more acceptable to whites, but I was told that was because local black musicians would not play for such a low salary.

Being from the North I was often wrong footed by race, class, and culture in Alabama. The head of the English department played banjo in the departmental office. He belonged, as did many, to a, to me previously unknown, Christian sect. He led a touring folk music group and explained that they shared a similar repertoire to black groups. I was told that the county did not have strong racial antagonisms and had succeeded from the Succession during the American civil war. Yet on our first arrival in Florence I was told by a colleague's husband that someone lived near the "nigger" part of town (at which Adele wanted to leave Florence immediately). He proved to be an opinionated fool who argued that the region had been settled by those of Scots descent and consisted of brave, strong, rational, individualists. We went to a dance at a large third world looking run down shack outside of Florence to listen to Little Milton (a Muscle Shoals recording artist) and both the audience and the dancing couples were racially mixed. I assumed that was because they were farmers and field hands who lived near each other. Invited to a party of writers and academics some distance from Florence I listened to our hosts mention the social changes and that their son that weekend invited a black friend from school to stay over. I understood this to be a first. But I also gathered that they were worried that in future it might be a black girlfriend. I asked a black musician I knew in Florence about a black blues club I had heard about. He said he had never been, would not go, and advised me to avoid such a rough place. I thought this nonsense. I had played drums in Harlem during the racially charged early 50s, danced in black clubs where my partner and myself were the only whites, learned to park my car near street lamps, what could scare me? Adele and I drove to the club and saw big black guys wearing Stetson hats and obviously packing guns. My middle class black musician friend was right, I chickened out and drove on.

The secretary of the English department accused me of laughing at "us" but that was not correct; I felt I was observing a world more strange than I had known in Nigeria. One of my better students had taught locally in high school and I overheard her telling another student that you were expected to use tenth grade text books for twelfth grade. No wonder the academic level was low. My one black colleague invited us to dinner and expected me to say "grace". I had never heard of such a thing and would be reminded of it when later teaching in Israel where an invitation to a Friday dinner meant I should recite a blessing in Hebrew, an equally alien concept. She explained that during the period of segregation it was difficult for black men to get decent jobs as the unions would not accept them, so families invested their money in educating their daughters who could be hired as nurses and teachers. This seems a better sociological explanation of the pathology of black families, where men are too often absent, than claims about surviving African culture.

Most of my colleagues should not have been teaching at a university; they bragged of lecturing from notes that they had taken decades earlier in Graduate School. The idea of actually reading a work of literature and explaining it was foreign to them and to their students, one of whom showed me her weekly study plan and argued that she could not spend more than 20 minutes a day on my Shakespeare course. There were three forty minute classes a week and Shakespeare had been previously and would after I left be taught with a lecture on a play, followed by the showing of a film adaptation of the play, a class of questions and discussion, and the fourth class would be a test. Play covered. Next four classes another play. I tried to introduce the close reading of a few plays (about six a semester) by requiring the students to purchase inexpensive paperbacks of the texts and supplying them with cassettes of the plays in performance to which they could listen while reading. Classes would be spent on discussion and close reading of scenes.

It was mostly a failure. If I saw a student bouncing up and down while teaching at Bristol I would assume he had taken too many amphetamines or some other stimulant. In Florence he would suddenly erupt with "I believe in Jesus, I am saved". The only discussion I can remember was a rambling semi-confession about the dangers of losing your virginity to a preacher who would "cheat" on you. I had become a Shakespeare expert because I wanted the job and to prove to myself I was qualified I contracted to write a book about *Coriolanus*. Although no one said anything about it I think the administration was hoping that I imitate the University of Alabama—Tuscaloosa's Master of Fine Arts which at that time was central the staging of an annual Shakespeare festival. As I was afraid to correct the pronunciation of Dido and Aeneas from "Dildo and Anus" without causing embarrassment there was little likelihood we would stage a play.

Florence was a pretty provincial small city where the church you attended was your social life and students would argue that Christian meant the sect to which they belonged. Shortly after arriving we took a colleague and her husband (a bank official) to a locally fashionable restaurant in Muscle Shoals overlooking the river and they sat stiffly, uncomfortable, drinking some fruit juice. One night I telephoned the husband to say that because of a delay in being paid I might need a three dollar overdraft on my bank account for the weekend and was told that I would need apply for a loan and mortgage our car or house (which we did not have). I thought him unmentionable but later was impressed when he resigned from his position over some moral disagreement about policy and they needed live on his wife's poor university salary. Not everyone was naïve although their experiences and values were unlike any I had known. A few students had parents working in the middle eastern oil fields, others had spent time in Europe although at biblical hostels. Education seemed to consist of learning how to clinch a business deal or expected restatement of traditional morals and beliefs.

There were glimmers of something more. Our first landlady was a tough, hard drinking, chain smoking, old bird who insisted even in hospital after an operation on continuing her ways. W C Handy, the supposed father of the blues—he was the first to write and sell sheet music of the music and composed such classics as St Louis Blues -- came from Florence. A Jewish entrepreneur had opened the first cinemas in the south, grown rich, and settled in Florence. To entice a son and daughter-in-law to stay he contracted Frank Lloyd Wright to build them a house complete with furnishings. It is now a tourist site. I was always surprised when Little Milton or Percy Sledge ("When a Man Loves a Women") would give a local performance. Once our old friend Jon Silkin came to Florence as part of one of his American university tours. Besides being a good poet and reader Jon was a tremendous performer. He sold out all the books and copies of *Stand* he had brought with him. The students had seen nothing like him previously. Alas I had. He stayed in our spare bed room, and left all the guest towels untouched. He was still trying to convince the world that he really was working class, a 1930s proletariat, unwashed in body and clothing.

The most memorable event of my eighteen months at the University of North Alabama was the month long disappearance of a manuscript and notes of a book I was writing. Before leaving the University of Canterbury I had applied for an American Institute of Indian Studies fellowship for ten months to write a book on Indian poetry in English, a subject about which I and hardly anyone knew much, but which I hoped which take me to India where everyone else seems to have spent time discovering, enjoying, meditating, taking drugs, and dropping out. I was awarded the fellowship and had to choose between staying on at Christchurch and asking for a sabbatical or asking the University of North Alabama for a delay in my appointment. I was asked to come to Florence for a semester and then take up the fellowship. While in India (the subject of another chapter) I had acquired the materials and started writing a major book which I tried to finish during the coming eighteen months in Alabama. I was also given a student to type my manuscript; she was paid by an American government student job program. Everything was fine for a while, then she, my manuscript, my drafts, my notes and everything I had showed her disappeared and no one could find her for a month until she was eventually located at her mother's house recovering from a long deep period on drugs. No one had warned me that she was an addict. Happily she returned, although without any explanation to me, typed my manuscript that was eventually published, by which time I had resigned, was once more seeking another professorship, and Adele had what I thought bad offers from two universities and tired of staying at home returned to full time teaching. We attended Nicole's graduation from Bryn Mawr, afterwards she left for Paris where she would work in art galleries.

Adele spent the next seventeen years at Ball State University, Muncie, Indiana, ascending the academic ladder from Assistant (despite having already published several books) to full Professor, retiring at the age of 70 when we moved to Paris. I never was offered that distinguished professorship at a major American university and spent more than a decade commuting between Muncie and visiting professorships abroad or supplementing our income with grant funded travel, research and the writing of books. At first I was depressed at having ruined a once high flying career, then realized that this was what I always wanted, the chance to write research and write books without worrying about demanding students and awful colleagues.

# Chapter 12: Poetical India

*India 1984*

While I was teaching at the University of Canterbury Adele, Nicole, and myself used the 1981–82 Christmas break (the equivalent of a summer holiday in the Northern hemisphere) for a trip through Bali, Nepal, India, and Thailand. In Bali we stayed at an inexpensive beach resort apparently favored by Australian and French tourists, many of whom were nudists. There were two unattractive French women seeking sex and drugs, a man with what may have been the longest penis in the world (in my imagination it reaches the ground but that is clearly impossible), and a couple of Americans who leave their usual employment at opposite sides of the world every year for hot times in Bali. We did the usual touristy things. Adele bought a dragon design silver ring that many decades later she still often wears. I was massaged by an old hag on the beach who knew precisely how to make me erect; I was asked to pay twice the price for an ear cleaning by a drunken madman who waited until he had a sharpened heated hair pin in my ear and I of course agreed. How do the tourist agencies find a dead body for a Hindu fire burial every day? Adele and Nicole visited a Hindu temple while menstruating and for several months afterwards Adele missed her period and Nicole suffered intense menstrual cramps.

Bangkok meant a cheap hotel notorious for junkie dealers and developing a love for Thai food that I still often cook. Even then the city was noisy, dirty, smoke filled, with transportation a nightmare. We did the usual tourist sights, a visit to a market on a canal, saw many brilliant Buddha's, watched Thai dancers while eating a banquet at a hotel.

Katmandu consisted of more drug dealers, a recent hippie history including American apple pies and representations of a stoned Ganesh that I now wish I had bought. Nicole and I listened to temple chanting and drummers while taking turns on an unhygienic communal toke. A meal at a clean modern western style restaurant among beautiful young people in smart clothes, I thought the most sophisticated New Year's Eve I would ever have. We still make the yoghurt dessert (dried dates and raisins, spices, no hash).

Our first trip through India was brief—New Delhi, Bombay, Madras. Adele and I both gave talks at the Institute of Technology in Madras and the three of us were fed omelets, eggs being an exception made for non-vegetarian foreigners. I bought some of the recently published Oxford University Press volumes of poetry in English, and my involvement with Indian literature began. Actually it began in Stirling; when researching India for *New English Literatures* I found many thin volumes by Pritish Nandy which I mentioned. Then after I published an essay in *Sewanee Review* about A. D. Hope and Australian poetry I was sent two books by Jayanta Mahapatra, an Indian whom I had never heard of and should have as his writing was excellent. *Rain of Rites* was published by Georgia University Press and some poems had a distinguished award from *Poetry* magazine. I was impressed and began corresponding with Jayanta who, amazingly, edited and published a high quality journal of Indian literature, *Chandrabhaga*, from Cuttack, a, to me, obscure town in Orissa, an area known for its tigers and deadly mosquitoes.

Similar to several other poets I was to learn about he had recently had a year at the International Writing Program at the University of Iowa. Jayanta claimed that his grandfather had converted to Protestantism to avoid starvation and that this meant he was raised alien to the rich traditional culture of Orissa, symbolized by its famous temples. Although a physicist by training he developed an interest in modern poetry and his own lyrics were often ambiguously about his waiting for the muse to arrive while he sat alone, his distance from the traditions among which he lived; others were about sexual desire. He felt isolated in Cuttack from the English language poets in major cities, especially Bombay. His poetry could be repetitive in its quiet at times depressive mood, but it had a similar awareness of the unconscious process of where poems start that was then common to such Americans as Robert Bly and James Wright.

Jayanta's poetry gave me the idea for a year in India, mostly on the beaches, reading Indian poetry in English when not greeting the many from Deia who wintered in Goa and Bali. As no one had written a book about Indian poetry in English, the American Institute of Indian Studies awarded me a ten month fellowship and I was on my way, except that the University of North Alabama demanded that I first take up my appointment there for a semester in 1983. While in the USA I travelled to Austin Texas to meet R Parthasarthy, one of several Indian writers attached to the University. "Partha" had been involved with starting the Oxford University Press Indian poets series, edited the influential, cannon forming *Ten Twentieth-Century Indian Poets*, published a volume of linked short lyrics about his life—*Homecoming*—and I hoped to get from him the basic background information and contact addresses I would need. We wasted most of a day trying to find a cassette recorder, microphone, and cassette tapes for what Partha insisted would be a recorded interview during which he repeated, sometimes word for word, what he had said in other interviews and written in essays. And of course afterwards we found that the cassette machine had not functioned and the cassette tapes were blank. A good start.

Partha's poetry also concerned alienation although in his case it was the supposed disruptions by education, travel, and language from the Tamil society and culture in which he was born and raised before being sent to Bombay for further schooling. A year as a British Council Scholar at Leeds University (where he wrote an extended paper on the Indian poet A K Ramanujan for Geoffrey Hill), a year at the International Writing Program, and Partha was supposedly ready for a life returning to his roots in India. Instead after a few years he moved permanently to the United States where he taught and translated Tamil literature. Some Indians mocked that he could only be a Tamil nationalist outside of India. Partha admired and wrote about the great Indian poet, translator and scholar A K Ramanujan, but whereas "Raman" saw tradition as evolving, treated it ironically, and regarded himself as the hyphen between Indian-American, Partha's lament for his roots appeared reactionary to others such the excellent poet, translator, editor and essayist Arvind Krishna Mehrotra who mercilessly tore apart Partha's nostalgia in an often republished essay, "The Emperor has no Clothes", that originally appeared in *Chandrabhaga*. I was rapidly entering the then small world of modern Indian poetry in English.

## II

Shortly after arriving in New Delhi, December 1983, I was invited to dinner with K N Daruwalla at his home with his wife, family, and a very attractive and impressively dressed, woman. "Keki" was clearly in ministerial circles and part of the governing elite although I did not know how or why at the time. Later as we talked I learned that his father had studied in England, returned to a series of ill paid jobs tutoring Indian royal children and teaching English at universities. Keki regarded his father as a failure, someone who never transformed his qualifications into a comfortable settled home for his family, who were always on the move because of the partition of India. Keki's childhood consisted of changing schools and the language in which he was taught as his father moved on to the next job. As soon as possible Keki left home by taking the first civil service examination available and that was how he became a police office. Bright, educated, multilingual he soon moved up the ranks and by the time I met him he was secretly part of a security team that advised the prime minister. I would meet him often over the decades including in London where he pretended to be in charge of visas.

I used to wonder how a senior police officer could write such tough minded skeptical poetry about both the government and the governed, but that was part of a world view he developed early in which life had no purpose and naïve idealism was as dangerous as those who repressed it. Civilization destroyed natural wild instincts including the will to survive. Although as far as I could tell he was what the West would consider an atheist he kept close ties to his Parsi community and followed some of its rites. His early poetry made use of his experience patrolling in rioting towns in which the police feared being attacked by acid, villages where vaccinations and reasons for them were a foreign European invasion, and the rituals of Shia Muslims and other religions that were alien to him.

I knew little about the background and origins of modern Indian poetry in English and was stunned when Daruwalla showed me magazines and anthologies in which he had published. He even had some letters from Robert Graves who had read his poems in an obscure Indian literary journal and wrote appreciatively. I did not know what I was doing or why but I started taking notes on the titles of the Indian and foreign literary journals and photocopying what I could. I would continue to do this as I met poets and was showed copies of what they had published. Besides the volumes of poetry and literary magazines there were political journals, vaguely modelled upon such British publications as the *New Statesman*, in which besides a section of book reviews there were often poems. The poets I was interested in were also friends with or affiliated with a democratic left opposing the pro-soviet socialist tendencies of the Indian government since independence. Something was of interest here but I knew too little to trust my instincts.

As I sought poets to interview we toured, from Delhi to Jaipur, where at a restaurant I had an amazing Jain meal, flavorful because the taste of the vegetables and spices was not covered by onions, garlic, or other roots with human-like limbs that that religion forbid eating. We stayed at what was once a palace that had become a place for tourists. It was stuffed with stuffed animals and I think was later used in a movie. We joined a bus tour to the Taj Mahal and some famous now ruined forts. I was not spiritual enough for the holy city of Varanasi; where others found the Indian soul I saw only filth, mud, cow dung, and smelly scantily clad priests trying to earn a living. How could anyone bathe in such stinking water, let along drink it? The head of the university English department was welcoming and for years I would receive invitations to come to conferences, but I was told that English was a foreign language (which was true about how he spoke it) and the real India could only be written in an Indian language. There were many who would tell me such nonsense. History was moving on and they were being left behind although their species would never become extinct.

In Allahabad we met a thin young bearded Arvind Krishna Mehrotra, who was riding a motorcycle and would have been at home on the cover of an issue of *Evergreen Review*. He had brought the alternative culture to India where he recreated it Indian-style. He showed me mimeographed limited edition publications he had edited such as *damn* you (a polite Indianisation of the New York literary journal *Fuck You*), *Ezra*, *Fakir*. Although they appeared inconsequential they had a surprising influence on modern Indian poetry. Arvind also showed me many obscure American literary journals of the 1960s and 70s in which he and his friends had published, as had the Bengali Hungryalists and other aspiring Indian writers of the period. This was part of a history of Indian literature I had not expected nor thought about, the international connections during the beatnik and hippie periods. I had gone to India in search of the drop out joys I had missed earlier, now I was finding that Indians had been dropping in with the dropouts. I photocopied everything.

I had enjoyed what little of Arvind's poetry I had read with its avant-garde mixture of autobiography, the factual, the surreal, amusing irony, and probably some faux autobiography. He was one of the main reasons I wanted to learn about Indian poetry. We got along well, walked along the banks of Ganges and Yamus which is supposed to form a confluence with a third, invisible, mythic river, and where every twelve years there is an immense spiritual celebration that in the past had resulted in murderous battles between groups of Hindu priests wanting to control the offerings. We ate with his wife, Wandana, an artist, and an excellent cook. We still have her recipes, which are subtle and complicated. (Their child, Palash Krishna Mehrotra, has since become a one of the better young India writers and journalists whose fiction I enjoy reviewing). We went to a restaurant whose proprietor formerly had a business in Nigeria. He told horrific stories about bureaucratic incompetence, delays, corruption, bribes (all of which I knew) and concluded with the punch line that it was ten times worse in India.

Arvind appeared a misplaced hippie in India, but he had made an important decision to remain in Allahabad and concentrate on reading, writing, and translating poetry while seeing himself as part of the modern world. Although the university's standard's slid downwards from a former prestigious British colonial educational institution to a dangerous, at times murderous, battlefield between warring Indian castes seeking a paper qualification that might and often did not open the way to a government job, and the teaching of English literature now took place in Hindi, Arvind imagined himself in New York, had such visitors as Wallace Shawn whom he had met in his travels, and lived among books, colleagues, and friends. One was Madhusudan Prasad, who invited us to dinner at his home, where his wife sometimes peeked out from the kitchen. Prasad asked me for an essay about Jayanta Mahapatra for a book he was editing; "The Shapes of Solitude" was the basis of much that I later published about Jayanta. Prasad, however, had little knowledge of the realities of the publishing world. Having decided to collect stories by Anita Desai into a book he had not realized that he needed her permission and to make matters worse he assumed that he had the right to give her work to others to republish, a costly error that her agent made him correct.

Almost silently over the decades Arvind became one of the major names of Indian writing and scholarship. He was one of the four editors, along with Arun Kolatkar, Gieve Patel and Adil Jussawalla, of Clearing House, a publishing co-operative that for a time brought out now classic volumes of Indian Writing in English and which along with Oxford University Press created the cannon. He edited the highly influential *Twelve Modern Indian Poets* (1992)—which replaced Parthasarthy's ten poet anthology although he excluded himself, edited the *History of Indian Literature in English*, and his *Absent Traveller* transformed ancient two thousand year old Prakrit love poems into modern English, his *Songs of Kabir* made the fifteenth-century mystic into a great poet of our time, and, perhaps his most important contribution to Indian culture, edited *Arun Kolaktar Collected Poems in English* for Bloodaxe in England, which made that great, recently deceased poet available to the world in contrast to the limited editions by a small publisher that caused his work to be neglected in India. A selection of Arvind's essays is republished as *Partial Recall*, his *Collected Poems* are now Penguin Modern Classics, and during 2009 he was nominated for the Chair of Poetry at Oxford University.

He felt from the first that the India of Nehru-Gandhi and the traditionalists was alien to himself, but except in some early ranting poems he did not waste energy in polemics opposing the politicians and gurus. He looked up to Arun Kolatkar who lived a modern bohemian life in Bombay and wrote without expressing opinions concerning cultural issues. Arvind and Arun however, as did other modern Indian poets in English, shared a vision of India that contradicted the Sanskritic Brahmanism and idealization of the peasantry that had become the official national culture. They identified with and translated the bhakti writers of the past who came from non-Brahman castes and who wrote in the vernacular languages rather than Sanskrit. Many, such as Kabir, were uneducated and their songs part of an oral tradition that was much latter written and published. The care for the ordinary, the contemporary, the class-less, found modern expression in acknowledging the presence of the Beatles, the Beats, the dirt, the incongruities, and the realities in the India they knew. At first regarded as iconoclasts they introduced what became the aesthetic of modern Indian literature as writers began to focus on real lives in real locations instead of the mythological India claimed by the nationalists and traditionalists.

By the time I left Allahabad I had a list and addresses of poets to meet and interview. In Cuttack we stayed with Jayanta and his wife, enjoyed her cooking, and slept on cots under heavy mosquito nets, Adele created and typed a bibliography of Jayanta's writings from his piles of magazines and books. I remember someone else publishing it. We met Bibhu Padhi, who then taught at Ravenscroft College, and besides being a poet had a scholarly interest in D H Lawrence. Although Cuttuck seemed like it should be the centre of nowhere it was on a major highway, rare in India at that time, and there was almost as much danger from the speeding trucks as from the deadly mosquitoes supposedly resistant to anything science had devised.

Adele still tells a story about our visit to see the Bengali tigers in a zoo where we watched young cubs being exercised. At the time there were still many wild tigers and one jumped over the zoo walls to mate with a female in the tiger enclosure and could not get out. He unhappily joined the exhibit. I thought of it as an example of much of the world's literature, the basic myth of the heroic independent warrior male trapped into domesticity and society through desiring a female mate. Marriage.

I asked Jayanta whether it was true as I had been told that he had a love marriage and he said no one in India is alone with a woman long enough for a love marriage. I had the impression that he began writing poetry in his forties to woo a woman, to show that he could do it, and while he continued to teach physics, poetry became the main interest of his life. He kept up with the latest American poetry through journals in his college library. When he began to write he often spent time in Calcutta where Pritish Nandy, who had become a self-promoting leader of the city's youth culture, published several series of underground, often surreal, chapbooks, which included Jayanta's early exercises at modernist poetry. A few of Nandy's poems were actually good, although he would later as he became editor of the *Illustrated Weekly of Bombay*, and then a film producer disdain his verse as "merde".

We stayed at an apartment owned by the American Institute of Indian Studies in Calcutta. Everyone except me was a genuine Indianologist, studying some deep subject and at ease with Sanskrit, Hindu, and or Bengali. P Lal was one of the founders of modern Indian poetry in English; a university teacher, polemicist, publisher, anthologist, translator from Sanskrit who pioneered the term "transcreation", and who was a minor lyric poet. He had, or once had, a writers workshop that included foreigners based in Calcutta along with some locals at their weekly meetings. He travelled abroad, especially to the United States, doing readings and selling his Writer Workshop volumes to libraries. A Bengali Brahmo, raised in a modern version of Brahmanism influenced by the West, he had a love, rather than arranged, marriage, and appeared the perfect middleman between India and the West. Alas India was moving on. His educated daughter wanted an arranged marriage and some of the English language poets he had published rebelled against the way he was starting to accept anyone who would pay for copies of their book to be Writers Workshop authors; they wanted standards and quality whereas Lal claimed let the future decide who was a real poet and aimed at quantity. Those who wanted to safeguard the gate of standards were led by Nissim Ezekiel, a Bombay poet, and eventually the Bombay school of poetry would win and Lal would feel betrayed by those he initially helped.

After I had published my *Modern Indian Poetry in English*, which recounted this history, and was on the side of Ezekiel and the Bombay poets I contacted Lal when I was preparing a second edition for his views about newer poets and received an angry reply that as I was the expert on Indian poetry why was I bothering to ask him. In a way he was right to be annoyed, I had his help when I began, had followed his advice about which of his Writers Workshop volumes to purchase, and had treated him in my book as a minor writer, one of the late fey aestheticists found throughout the former British Empire whose slight lyrics had challenged the antiquated notions of poetry still common after the triumphs of Modernism. Ezekiel, Daruwalla, Mehrotra, Mahapatra were the heirs of Modernism, Lal was not. But he did know good poetry from bad. He had published important books by Ezekiel, Daruwalla and Adil Jussawalla, he told me about Agha Shahid Ali long before anyone else had mentioned him. Among the seeming unending flow of Writer Workshop volumes there were still some interesting new poets and I had the impression that Lal did not expect them to purchase advance copies of their books as required by others. The editing however was a scandal. Manuscripts appeared to have gone from the post to the printer and then beautifully bound copies sent to the author without anyone noticing the many absurd printing errors, the misspellings, the weird and missing punctuation, the crazy lineation. Lal seemed to have stopped caring and the Writers Workshop, still one of the few opportunities for the publication of English language poetry in India, had the reputation of a vanity press. He did continue to translate the classical Sanskrit texts and I understand that his versions are much better than competent although I admit to not having read them. He once showed me his file of letters from famous writers around the world. The ones I saw seemed thank you notes for copies of his publications, for hospitality in Calcutta, but I assume that the American university library that purchased them found more of interest.

We went to Hyderabad so I could interview Shiv Kumar, a well-known academic who had retired as Vice-Chancellor of Hyderabad University and then taught part of the year in the United States. He was one of Parthasarathy's ten Indian poets and continues to publish good poetry, fiction, and literary criticism although he dropped from the evolving cannon of modern Indian poets. Charming, amusing, even sly, he was obviously an excellent teacher who I thought would give his American students high grades and receive favorable student evaluations. He seemed uninvolved in the Indian literary scene but was obsessive about a lawsuit over some property that would, in the Indian manner, slowly worm its way through the courts for several generations until no one remembered it.

It was perhaps from Shiv Kumar I discovered Srinivas Rayaprol, an engineer in Secunderbad, a sister city of Hyderabad (we were staying at the University of Osmania guest house in Secunderbad), another pioneering modern Indian poet in English who had dropped from sight. Rayaprol studied engineering in California and came to know such American poets as Kenneth Rexford, Kenneth Patchen, e e cummings, and William Carlos Williams with whom he had often corresponded after his return to India. He had also published a now obscure international literary journal, *East-West*, for several years. Henry Miller was among the contributors. Here in Secunderbad was an unexpected and seemingly neglected branch of New Directions, a decade before the start of modern Indian Poetry in English, and from transplanted American rather than British seeds. Although Rayaprol had stopped writing poetry so he could devote himself to his career he did translate verse from Telegu into English. I was amazed to discover Rayaprol and mentioned him in *Modern Indian Poetry in English*, but he seemed a surprising precursor of what happened soon in Bombay, a dead end unfortunately. Then later I was amazed to be asked to contribute an Introduction to the first book republishing some of his poems, a Writers Workshop volume. Jeet Thayil was to include some of Rayaprol in his anthologies and after his death his children wrote asking me if I still had any of the ten or dozen issues of *East-West* as they had a publisher interested in republishing the set. I no longer had.

While in Hyderabad-Secunderbad I began research in the American library to see if I could find more journals in which Indians had published. As a result I published two preliminary essays, mostly bibliographical, about the influence of American literature on Indian writing. I had not thought much about them until I started getting requests for copies and more information from Indians who either thought I had opened a previously unexplored field or who hoped this might lead to a year or fellowship at an American university. I had forgotten about this episode until recently someone wrote asking for a copy and how I had without major university resources found so many now hopelessly exotic underground publications which were then what the poets had in their own files. I really did not take my publications seriously beyond feeling I should do something with my Fellowship so I did not keep track of my scholarly articles; somehow others would later discover them and once even pay me to republish a talk I had written for a publication. I am not being modest; I thought

I was destined for something more than what I was doing, a feeling that I have had since childhood.

Calcutta seemed perpetually sick with strikes, manifestations, demonstrations, electricity failures, and other results of its Communist government while industry, finance, the arts and jobs fled. This period of its history can be found in the early novels of Amit Chaudhuri. The capital of Indian finance and culture was now obviously Bombay where the company that employed Chaudhuri's father moved. Bombay was noisy, filthy, overcrowded and alive all hours of the day and night. It seemed a great city squeezed into a tiny space overflowing with skyscrapers, modern restaurants, traditional restaurants, every kind of restaurant, coffee bars, milk bars, clothing shops, shoe stores, streets filled with beggars, stands that sold perfumes, clothing, food, electronics, and inexpensive books that had been so badly printed half the pages were blank. To have an idea of the variety, excitement, and vitality of Bombay at this time read Arun Kolatkar's great *Kala Ghoda* sequence of poems with its delight in the ordinary, the living, the absurd, the incongruous, including the city's history, its dogs, its street lamps, its food vendors, its traffic, monuments, languages, improvisations.

There were English and German backpackers, Americans and Australians seeking or selling drugs, people on their way to enlightenment by gurus on mountain tops, pretty men and women seeking fame in Bollywood films. The world was coming to Bombay, including many rich Arabs who had come for its entertainments and hospitals, I would gawk at in the terrible heat Muslim women covered from head to toe barely able to see out of their Niqib. The newspapers carried stories about young Indian college girls being killed by being thrown out of windows by Arabs who presumably regretted their own sexual behavior or felt that their wealth made them beyond local law. Money was flowing in from the Gulf and while some of it went into business and Bollywood other parts funded gangsters and the bombing of Hindu sites in revenge for attacks on Muslims.

At first we stayed at the YWCA, which disliked having males but as Adele and I were a respectable married couple who had long ago booked, the Y made an exception. On later visits to Bombay we stayed at Bentley's Hotel, a long established refuge for foreign beatniks, hippies, druggies, and semi-impoverished artists on the way to or from Goa. Adele thought it unwashed and there were indeed many legged creatures at times scuttling about that more fastidious hotel owners might have tried to eliminate, but the rooms were large, the hotel moderate priced for Bombay, and where else could a poet/musician friend wake you up in the middle of the night, spread a white cloth on the floor, and sing Indian classical music until he tired and it was time to go to an all-night fish restaurant for fresh lobster or crab cooked in much sauce. Watching him eat the dripping crustacean using his dirty fingers was an advanced introduction to Indian culture that I would have preferred to avoid.

Bombay meant visiting regularly Nissim Ezekiel whose lair was the office of PEN near Marine Drive. The Bombay PEN All Indian Centre was a branch of International PEN and it was fitting that Nissim should be in charge—Secretary-treasurer, editor of its newsletter—as, besides being a poet, editor, publisher, essayist, art critic, dramatist, university professor, he seemed to have contacts at all the newspapers, magazines, and businesses likely to publish writers or make a contribution to the artistic and modern cultural life of the city.

Nissim was the son of secular educated Jewish educators. Himself early an intellectual he was helped to study in England by an influential wealthy friend who latter established the Indian national theatre. In London Nissim was an impoverished student, attending lectures, readings in libraries, living in a flat shared with a relative, published a few poems, and had a still obscure affair with some woman which contributed to his decision to return to India. He worked his way home on a steamer, his first volume of poetry, was sent to him from England while his ship was docked in France. Back in Bombay he asked his mother to arrange for him to be married, sought well-paying jobs, and tried to reform his ways. Much of his poetry would be concerned with the conflicts between his desires and his wish for an ethical life grounded in social conventions. He left his self-published second and third volumes of poetry in Bombay bookshops to be sold. He was also poetry editor by the Bombay *Illustrated Weekly*, a publication edited by an Englishman who wanted to encourage the development of an Indian literary scene. I spent time searching the files of the *Weekly* and was struck by its coverage of P. Lal's Writers Workshop publications and other literary news, and its treatment of the local literary scene as if it were a race track with up and coming favorites. Nissim, an already known poet, was to find and encourage others including Dom Moraes before Dom left for England.

Eventually Nissim became leader of the Bombay poets and someone that other poets in India, such as A K Ramanujan, would visit. He did not like Jayanta's poetry and antagonism was mutual. He wrote stunningly perceptive reviews of the first two volumes of Daruwalla's poetry. He highly praised the poetry of Adil Jussawalla. He normally was in the front line defending freedom, even during the censorships imposed by Mrs. Gandhi during the Emergency, but he defended the banning of Salman Rushdie's *Satanic Verses,* as the novel was likely to cause riots. That was a mistake as India is now troubled by religious censorship from any group that claims hurt feelings. Besides local newspapers Nissim had influence on *Opinion* (an opposition political journal edited by a friend), *Quest* (a Congress for Cultural Freedom supported intellectual review), edited *Indian Poetry* (a significant journal for translation from Indian languages), published a book of poetry by Gieve Patel, and much else. He was part of the wave of Indian university teachers who first taught American literature; an ex-radical dissent Marxist he became a leading Indian intellectual on the side of the West during the Cold War. As he advanced through the university system to a full professor teaching post-graduate classes his students themselves became university teachers, literary critics, editors, and writers. There was a Bombay school of poetry that derived from his writing, in which short, well crafted, poems concerned problems of contemporary urban life. He adapted the tone of a secular man of the world talking to others like himself in a society plagued with lack of knowledge, dishonesty, incompetence, the morally, culturally, spiritually malaprop. The poetry of Saleem Peeradina is an example, but Nissim's influence can be seen on others, such as Eunice de Souza, who developed an even more stripped down, imagistic, quietly ironic presentation.

Eunice, who taught at St Xavier's College, and became a significant editor of Indian literature as well as a poet, was one of his women as was Linda Hess, an American who brought him the delights of the alternative or underground culture of the 1960s that Nissim explored as part of his openness towards and wariness of the world. He tried LSD, published poster poems and experimented with other characteristics of the counter culture, while increasing his stature as a public intellectual. When I got to know him, and I would see him often for a few years, he was aware that he had become internationally known, but he felt he was a small fish compared with Derek Walcott, A K Ramanujan, Wole Soyinka, and the other giants that were starting to emerge from the former colonies and dominions of the British Empire. He mentioned that his poems were taught in South Africa but otherwise his books seldom sold abroad. He kept no bibliography of his writings (a friend was trying to establish one), gave away copies of his publications without asking for their return, and would eventually entrust his biography to a former student, a homosexual who disliked the "womanizer" that was his subject.

It is part of the paradox that drove his life between desire and conformity that he remained married to a wife who never forgave him for infidelities. He gave her most of his salary, shopped and brought her food, took her to receptions and openings as his wife, yet he lived alone, an impoverished poet-intellectual although he enjoyed company and good food. Adele and I several times ate with him at a small café near the PEN office before he one day said he would enjoy a big meal and suggested a fancier, more expensive restaurant. The café was part of his self-discipline; the restaurant meal was like one of his women. I was saddened many years later to learn that he died alone, among filth, after suffering from Alzheimer disease. He was a great man. Now of course almost everything publishers can find by him is in print.

Nissim's rival, if that is the correct word, was Dom Moraes. They had known each other much earlier after Nissim returned from and before Dom left for London. They did not like each other or each other's poetry. They had rival establishments, Dom and his wife Leila entertaining visitors nightly at home in their apartment, to food and drink that they, as I was later to learn, seldom could afford, just as I imagine he could not afford the meal he treated us to at the Taj Mahal hotel, one of his favorite places to meet and drink. He had been raised around the famous, powerful, and moneyed, had earned well during his years abroad, and was said to the best-paid journalist in India. He felt part of an elite and continued in the role quietly mumbling some gossip about his years in England while a drunken Leila hysterically screamed that the bailiff is coming. He was generous. I still have a lovely antique foot scraper, with two elephants entwining their trunks, which he offered when he noticed me eyeing it. Such foot scrapers are an Indian art form that I doubt exists elsewhere; I remember visiting a foot scraper museum in some city. I even purchased a more modern one to keep my antique company.

Dom always had promising young writers he was promoting. Dom's family was part Goan and his mother was obsessively Roman Catholic, so his apartment became a meeting place for Goans and Catholic artists, such as cartoonists, journalists, editors and poets. I first met Jeet Thayil and David Davidar through him. Jeet then had a large notebook of Baudelairean sonnets, David edited a Bombay publication and was writing short fiction. Manohar Shetty, who was to marry a Catholic Goan, had worked with Leila on a magazine she edited.

Don had once been a famous English poet and now was a famous Indian journalist. His father was a newspaper editor who knew all the leaders of the national independence movement. His wife however became mentally disturbed and when not violent withdrew into religion.

Dom sometimes travelled internationally with his father, but perhaps as a scapegoat was sacrificed to his mother's company. He withdrew into literature, wrote a book on cricket when he was 13 years old, and fled to England where he became part of the then famous Soho bohemia, which included Francis Bacon and Lucian Freud, when he was not at Oxford. He was at 18 seduced and then married to the older, already twice married, foul mouthed, often drunk, drug using, Henrietta, an artist model, who had been painted by and slept with many of her employers; he one day went to buy some cigarettes and did not return. He would write several prize winning volumes of poetry, become famous, edit well-paying British magazines, and make the mistake of moving to the USA as an investigative reporter for UNESCO, which eventually returned him to India at the request of Mrs. Gandhi who wanted him to write her autobiography. He never escaped as his UK reputation was now in the past and at first there was well paying writing in India. He wrote three enjoyable if premature autobiographies, was hired to write books about touring, and so on until he became a high quality hack. Meanwhile his muse of poetry fled, perhaps because he had married several times without bothering to get divorced. He kept fantasizing a return to England, a Poetry Professorship at Oxford, and the kind of homage that W. H. Auden had when Dom as a student knew him.

When I was writing *Modern Indian Poetry* I did not know what to do about Dom. He was the most famous poet in India, and I was indebted to him for friendship, gossip, company, meals and a foot scraper, but he had not published any books of verse since his return and was dismissive of Indian poets. I decided to talk about his British career and list his publications, and hope that another opportunity would present itself to return to his poetry.

The third Bombay poet important to my research and social life was Adil Jussawalla who after decades I still regard as a close friend. His French wife's family has a small apartment in Paris where I see him perhaps every two years, if his health permits. Adil, although not as old as Nissim or Dom, was one of important modern Indian poets in English who early published a major book and who continued to have a major role in the creation of a new literature. He was one of the four editors of the Clearing House cooperative which published some of the best books of Indian poetry, he edited the *Penguin New Indian Literature* anthology and he kept poetry readings going in Bombay over the decades as they went in and out of fashion. He knew everyone or appeared to, in most Indian vernacular literatures, and stored every publication, letter, and carbon copy of a poem, which came his way. It would be impossible to write a history of modern Indian literature without his aid. I borrowed, photocopied, returned, borrowed, and photocopied some more, after nights of talk, eating tandoori chicken and drinking with him and his wife in their lovely apartment overlooking the harbor.

Adil's father had been sent to England on a Parsi scholarship to study the then German practice of natural healing which he brought back to India and that the leaders of the nationalist movement regarded as a return to native practices. He became famous, wealthy, wrote books and was so involved with his work that his family seldom saw him. Adil went to England for his further education, stayed on some years as a teacher of English, became part of the intellectual scene, and returned a disillusioned emigrant who knew he was unlikely to fit in among ordinary Indians. His first two volumes of poetry, his last for decades, reflected such awareness, the second being a highly experimental attempt to give voice to the *Missing Person*, the alienated highly educated, highly conscious, bourgeois Indian. It was probably influenced by reading Frantz Fanon, familiarity with Marxist analysis of class and false consciousness and even the non-narrative modernist epic laments of T. S. Eliot and Aimé Césaire.

When I met them Adil and Veronik were already living in a highly desirable flat on fashionable Cuffe Parade. Adil, who earned money as a journalist and editor, was amusing, witty, ironic, highly civilized, soft-spoken, and usually seemed diffident, tolerant, tentative, although he occasionally displayed anger if he felt he or others had been misrepresented or slighted. He was also beginning to drink too much, a characteristic of the Bombay literary scene at the time He appeared to me an intermediary between Nissim and Dom, friends with both, but also close to the various "street people", other poets, painters, journalists, actors, teachers, hanger-ons, who were part of the cultural scene at the time. He introduced me to many young writers and their often ephemeral publications; but I would notice that over the decades he knew and was visited by all the Indians, especially former Bombayites, who lived abroad—Farrukh Dhondy, Salman Rushdie, Homi Bhabha. In his immediate circle he was especially close to Arun Kolatkar, Gieve Patel, Arvind Krishna Mehrotra and Eunice de Souza. Manohar Shetty was another friend. Although I would ignore him in my own writings, Kersey Katrak, a well off Parsi advertising executive was also a highly productive poet and a friend of Adil and Nissim. The small world of Indian poets was actually fairly large in Bombay at the time and it was necessary to follow chance, instincts, availability, and social networks, as well as your own critical judgments. The Katraks always seemed to be away, and although I read his poetry I was not impressed.

Bombay was rapidly sprouting new magazines for women, film goers, men wanting an Indian equivalent of *Playboy* or *Esquire*, that give expression to the opportunities of what was becoming a modern city in a mostly still stagnant India. Many of the new journals were edited, or had sections edited by those with literary ambitions or who were friends with writers. Some published, even paid well, for poetry, unlike a decade or so previously when poetry was lucky to find a place in the back pages of a political journal. Many of the poets were becoming known socially as part of the admired cultural world. Gieve Patel, besides being a published poet and part of the Clearing House cooperative, was a medical doctor, a highly regarded painter, and, along with his wife, prominent in the English language theatre. He, as did Nissim, published a volume of plays. We had dinner at his house, wandered around the sculptured hanging gardens, with their animal shaped bushes, on Malabar Hills where he lived, and I went to his medical clinic for advice about an ear infection. Gieve was always affable, but over the years I felt that his interest had shifted from poetry to his paintings that were shown internationally and we more likely to discuss transportation and insurance charges and gallery percentages than literary projects.

I cannot now remember when and how I met many of the Bombay poets, such as Santan Rodrigues and Eunice de Souza. Wherever it was we talked about their publications, New Ground (the new publishing co-operative they formed), student literary magazines. There was Arvind's brother-in-law, who briefly published a strange little magazine and who admired Adolph Hitler although as a Jain he literally could not hurt a mouse. Having caught one in a trap he, although an impoverished student, hired a taxi to take it to a wooded area outside Bombay and set it free.

Adele and myself several times ate with Manohar Shetty in the excellent restaurant he was unwillingly managing for his family, as he no longer had a job as a journalist or magazine editor. He was lonely, drinking beer, welcomed company and a chance to talk about his poetry. Copies of his first book were piled up in the kitchen, seemingly under or between every pot and pan. The family traditionally had restaurants, and was wealthy as a result, but he had been sent to an English language school and was unwilling to return to the tradition or the language the family used which he claimed had no creative literature of its own. Unlike many of the younger poets I met who had turned against the British literature that they read in school and that continued to be an influence on Moraes, Manohar was more interested in recent British than American poetry.

Arun Kolatkar was more difficult to find and I was told to seek him at a restaurant where he held court at lunch once a week with his Marathi publisher and Marathi speaking friends. At first Adele and I sat at another table until we were introduced and allowed to join. A few years later, on a second research trip to India, he was more available; I had been told—by Adil or Arvind?—to bring him a newly published book he wanted to read by an American poet. He took us to visit Arun Shejval, who played a pakheva, (a large tension drum), who showed Adele how to count rhythmic cycles, and we went to dinner at Arun's small apartment which, as it was stuffed with books of literature, meant eating at a table outside. As the small grants we were living on meant mostly travelling by three wheelers, I was surprised when we went to his home by taxi, which he paid, from an impressively large wad of bills. He was by profession a well-paid commercial designer, so well-known that he kept no portfolio on the assumption that if you did not know his work you could not afford him. This side of Arun was mostly unknown except to his close friends until the brilliant French scholar Laetitia Zecchini began writing about him and examined his life. It became obvious that his love of surreal contrasts, visualizing scenes, amusing juxtapositions, and allusions to modern life, in his poetry, and his concern of the visual appearance of his poems and the books' covers, had origins his work as an artist and in advertising. But when I first met him and for long after he had only published one real volume of poetry in English, the classic *Jejuri*, about a day trip with some friend to a religious site, and while traditionalists were wrong to feel that the sequence of poems was satiric and offensive, it was difficult to understand the precise nature of his ironies, seeming distance, and the values behind the descriptions. Arun was, however, instinctively an artist and bohemian. Many of his poems are constructed with shifting perspectives and will temporarily change focus as the visual takes over the meaning. He respected those who had challenged society and was excited when I told him that Malay Roy Choudhury, whom I thought hardly poet, had been moved by his bank to Bombay. I had to get the address for him. Malay was the founder of *The Hungry Generation*, the Bengali Hungryalists, and had been imprisoned for his writings. Arun soon visited Malay; here was someone he respected. I am told by Bengali writers that Malay's later Bengali prose is, unlike his English verse, great.

 Although I no longer took Pritish Nandy seriously as a poet and accepted his word that the days of his fame were a time of seeking celebrity, as when he published a book of poems with Kamala Das strongly implying that they were lovers, I interviewed him and needed his permission for access to the *Illustrated Weekly* files. The night we dined with him and his beautiful wife at his large modern flat furnished with a combination of Indian art and an Indian version of modern Danish furniture, Mrs. Gandhi had been assassinated and he was continually watching television and on the telephone. Political thugs forced vendors, shops, car drivers, taxis, everyone to close, go home stay in, and we were unhappily making the long long journey home by foot when a car surprisingly offered us a lift.

III

Goa at last! So also wrote some of those whose names I recognized from Deia in the guest book at the restaurant on immense pillars above the incredibly beautiful sandy beach between Baga and Calangute. After a few days in spacious rooms at a guest house we rented a small two room house on a dirt, muddy when it rained, lane that led directly to the beach where we swam, sunbathed, and read. There were mud huts and houses, even a small hotel or two, near us, and I would hear programmes broadcast in a local language where the routines with the alterations of music and drama sounded as if they were being translated from 1940s American radio. Indian gangsters supposedly lived in some the buildings on the lane, and at the top there was a small, enclosed field where water buffaloes were kept and I fell in love with these beautiful creatures with their shiny black skins and large glistening eyes. I was angry when I saw them herded by being hit with long bamboo sticks, and I decided never to eat four-legged animals again. Near us was a communal well from which I sometimes pulled buckets of water when our supply stopped. Soon we had made friends and I became an extra in two probably Maharastra or Telegu films. In one Adele and I were seated on the beach at night with other "hippies" waiting for a boat of marihuana to arrive, at which point we and the others had to get up, shout with joy, and dance when handed something to smoke. Many of the extras expected real pot and arguments started when they were given tobacco; we left. In another film

I was dressed in a 1930's style suit with broad stripes and wide thickly padded shoulders; I had to stand with my arms folded as one of maybe eight body guards around a table where the young star keep misreading his cues for seeming hours. Was he really so illiterate? I got bored and left during a break to eat. I wondered whether my absence would ruin the previous scenes but was told that Indian audiences would never notice the lack of continuity.

During the early 1980s south Goa was still romantic and not the wildly over-packed international carnival it now is. During a visit two decades later I was disgusted by the crowds, the way tightly jammed together buildings had replaced open spaces, and by the general filth as if no one had bothered to pick up their mess after a drunken party. During the 1980s you could still walk at night along the road from Baga to Calangute and meet only a few others. The drug culture had however taken root earlier with the hippies, and if asked whether you wanted to join a trip tomorrow it was not about an automobile ride. Some then famous visiting rock group had left powerful amplifiers and other electronic equipment at Anjuna and its Full Moon parties were wild psychedelic affairs with dancing in open air around a fire. It must be the only time I can recall someone running through a crowd with a box of white powder offering a handful to everyone in sight until his friends caught up with him and stopped it. The police looked on from outside the arena, where there were stalls selling food, drinks, clothing, hippie apparatus. Looking back on it the police were unusually tolerant with foreigners despite the calls by some Goans (mostly those who did not directly depend on the tourist trade) to bring a rapid end to this local Sodom and Gomorra, a call still being made. Decades later when visiting Bangalore the chant had now become "ban the low backed dresses" that were popular at night among sophisticated local young women. An example of police tolerance in 1980's Goa: I watch a small slightly bearded, drunken or drugged homosexual South American dancer go crazy and start to pull down the poles supporting a Calangute restaurant-bar. He beat up the first policeman, and wildly fought the others who eventually took him away to prison. We speculated that he would be deported, or perhaps beaten or raped to death in prison. Instead he showed up at the café the next evening with a few bruises and small bandages.

Everyone had a story about him. He sometimes was in drag and Fernando claimed to have been drunk or drugged one night and only discovered he was not a woman when they were having sex. Fernando himself liked to brag that the mother of his German wife was one of Hitler's secretaries. He used to taunt us as "straights" but he bought drugs from and bedded a very fat seemingly always sweating and greased cheerful prostitute whom I cannot imagine making a significant part of my anatomy erect.

Drugs and prostitution were part of the scene. I used to have coffee at a café whose owner left Kerala to earn a living elsewhere and who was planning to marry a former prostitute, a surprising to me occurrence among uprooted Indians. It was at his café I met the two drug addicts who were trying to recover from their heroin addiction by running for hours in the morning. I would jog with them under the sun along the water's edge, some days the soles of my feet peeling, increasingly wondering if I would make it home yet learning the unexpected high that comes from long distance running.

To get to Anjuna from Baga meant walking a tricky path over a small wooded hill or at night risking your life as an unhelmeted paying passenger on the back of a motorcycle. We were not hippies or druggies, but others were. A writer/translator I looked up in Anjuna could hardly sit still as he waited for his local dealer. The daughters of a friend from Deia were already part of an annual migration of partying in which drugs were brought from Ibiza to Java and Goa and the proceeds exchanged for clothes, stones, and silver taken on the return journey. Eventually their father joined them as he could not afford to retire in Paris and he had his children for support in his old age while in Goa, Bali, and Ibiza.

Goa was of course one of the Portuguese colonies "liberated" by and incorporated into India, an event then still resented by many Goans who regarded the Indians as hut (in contrast to house) builders and unhygienic especially for their culture of defecating along the beaches so the sea will remove their feces. Looking on Face Book and You Tube, where the 1980s hippie parties are now a paradisiacal legend that pre-dates the Goan "trance" parties of later decades, there are similar complaints by Goans and foreign tourists who want to ban Indians from Goa. We went to a ball at an old Goan club in Panjim where the furniture was made from heavy elaborately carved local wood and the dances were a variety of waltzes from the first half of the twentieth century, perhaps older. When was the last time you danced the "hesitation" waltz and its sound a-likes? There was a distinctive Goan culture of literature, music, food, architecture, furniture, clothing, and manners being challenged by India as well as by modernity. In looking for older Indian poets who wrote in English I found a once well-known Goan, Joseph Furtado, who wrote in English, Portuguese, and Latin and who was published and praised in London. Apparently even at the turn of the twentieth century Goans went to Bombay seeking jobs and wealth. Eunice de Souza had praised one of his poems for its vernacular English.

## IV

Our travels from Bombay and Goa south to Trivandrum, where I was supposedly to be based with Ayyappa Paniker, in his English Department, took us through Aurangabad (to visit the famous rock caves at Ellora and Ajanta) and Mysore (to visit the Palace and the statue of the Nani (bull) on Chamuli Hills and at both there were welcoming academics, especially in Mysore where Professor Gowda took us to see the sights in his car which ran out of petrol on the Hills. Abandoning us to seek petrol he warned us to keep the windows shut to keep out robbers. Gowda was in a fight for supremacy with a rival English professor who also had a research institute, also had his own sources of money, and who was backed by the then powers-to-be in the study and promotion of Commonwealth literature; so Gowda was happy to have me as an ally and I was willing to publish a few articles in his

*Literary Half-Yearly* and even guest edited two special issues one on Chinua Achebe (dated 1980), the other on T S Eliot (1988). His English Department included the brilliant Kannada writer U. R. Anantha Murthy, best known for his novel *Samaskara* (1965), translated into English by A K Ramanujan. Murthy did not believe that English was a fit language for the psychological intricacies of the Brahman mind that this fiction explores; he represents a questioning liberal modernist stream of Indian literature that often shocks the conventionally minded Hindu.

I thought I had fallen out with Gowda after being sent without warning a semi-literate, badly informed Ph.d. dissertation by one of his students that his university asked me to examine and return. I replied that I never agreed to read it, that it should be failed, and, as I was then unemployed, I could not afford the postage to return it nor had I any university which would. Gowda sent me an angry telegram saying that I could not stop the award of the doctorate. I was happy to bring our acquaintance to an end, but then I would receive requests to write for this or that and even eventually agreed to contribute an essay to a volume that someone I knew was editing to honor Gowda.

In Cochin we were the usual tourists (the old Synagogue, waterways, spice shops, a beach hotel where we rented a room over the water that one of the Kennedy's had used for his honeymoon) and then in Trivandrum Ayyappa Panikar found us a large modern house at Kovalam beach where we spent our time except the one afternoon each week when I was expected to be with his post graduate students. They seemed young, lively, articulate and smart if not always well read. We were invited to some of their social occasions including a long bus trip to a place in the mountains of Kerala and to a party where we ate jackfruit, tapioca, and other local foods. Unlike the students the university library was depressing with piles and piles of donated dusty books and magazines in languages that no one could catalogue, including major European languages.

Ayyappa was lively, respected, witty. When the owner of our house learned we would not pay the rent she first requested she named a much lower sum. Ayyappa compared this to a famous telegram during the colonial period that offered a low salary to an Indian and included an offer of a much higher salary if the first was deemed insufficient. Besides being a professor in charge of his department, he was an internationally famous scholar of medieval Indian writing, a translator, and was said to be the first modern poet in Kerala who discarded the archaic rhetoric, diction, and stanzaic forms of Malayalam to create a contemporary manner. When I interviewed Kamala Das, to whom he introduced me, she told me that one reason she wrote in English was that she would not have been able to master the intricate Malayalam forms in which her mother wrote.

Ayyappa took us to or pointed out and explained traditional Kerala cultural events, such as the Kathakali and other dances originally performed in temples lit by oil lamps at nights before the elite, and a three week cycle of nine-hour nightly shadow puppets (tholpavakootu) performing their medieval version of the Ramayana (kamba Ramayana) which, as no one could understand the mixture of ancient Tamil, Malayalam and Sanskrit, the patrons often slept though missing the at times obscene jokes and ironies of the voices impersonating the puppets whose shadows are cast upon the back lit screen, each performance lasting from sunset until dawn.

Kamala Das was, and even after her death probably remains, the best known of the modern Indian poets who wrote in English. She had a natural facility for rhythm, cadence, words, concise argument, tone, striking phrases. Her poetry sounds, and probably was, natural, as if written without much thought or revision. She told me that she wrote it every day and had many unpublished poems in notebooks in her closet. She also claimed that her supposed affairs were only products of her imagination. It was seldom clear what was fantasy and what was fact. She liked to shock others, and seemed both to need to be mentioned in newspapers and to dismiss talk about her as false. Typically before she died she married a Muslim and claimed to have converted to his faith. That once more made headlines, reminding people that she still existed. She had an active, at times depressive, imagination, and long after I had last seen her I would at times get highly poetic yet depressively suicidal letters in which she wrote of clouds in her mind and wanting to throw herself in the water.

The implied stories of her adulteries that made her famous often were with nameless men, perhaps mere longings. Her husband spoke blandly of his need to accept her reputation as part of living with a famous writer, but then in her autobiography she implied that her husband had a homosexual affair with her cousin, an Irish-Indian author. Kamala like Ayyappa, was from an upper caste family and had a natural sense of herself as in charge and being in a spotlight. Perhaps it is because of arranged marriages but some upper caste Indian men seemed surprisingly tolerant of, even encouraged, their wife having a lover. The women claimed that their husband wanted to get on with his own life. Although Kamala could seem moody and vague, she was also a conventional mother, talking about her son, servants, property, retiring, the success of the cultural gatherings she held in Bombay; even the food she ate was bland.

Among many Malayalams inheritance was a right of women, which meant that a woman could say to a man this is my house and my bed, pack your bags and go. Inheritance had not changed, but the notion of women as head of the family had. Still I would like to think that Kamala's apparent impudence and fullness about herself belonged to this tradition. Meena Alexander, a Syrian Christian from a Malayalam family, taught in the USA where she at times in her poetry imagined two grandmothers, one traditional and unthinkingly rooted to her society, the other a fierce feminist troublemaker.

Kerala was then one of the two Indian states governed by Marxists and unlike Bengal was well run and clean. There was a constant water and electricity supply. It was impossible not to contrast mentally the washed and painted houses, the tidy attractive gardens, and well-landscaped trees and bushes, with the filth and disorder of the north. Humans seemed to live here not their animals. Although the state was poor the percentages for literacy and education were high.

Tapioca, jackfruit, an immense variety of bananas, yoghurt, and strange vegetables, like drum sticks, were central to the cuisine. The rich curries and tandoori of the north were scarce. The food seemed mostly vegetarian, served on long thin rolled breads. Even the yoghurts served for dessert had small hot peppers. There were Hindus living not always amicably alongside Muslim communities and there were many Syrian Christians who traced their origins to St Thomas visiting Kerala and who were now divided into three groups after the Portuguese invasion attempted to impose Latin liturgical ritual on the Syrians and a third branch attempted to compromise between Syrian and Portuguese religious practices. The ancient Jewish community of Cochin and the coast had mostly migrated to Israel. I later found that Susan Visvanathan's sociological writings and some of her fiction offered a detailed portrait of the variety of Syrian Christians and their history.

Trivandrum was a large open attractive modern city with boulevards lined with palm trees, the opposite of the crowded labyrinths of Bombay, Delhi, Benares, Calcutta, Hyderabad. There were inviting hotels at which to have vegetarian thalis and smart shops. I still, after more than three decades, have stylish neckties from Trivandrum. The sandy beaches of Kerala were lined with coconut trees, as if the state were intended for picture postcards of tropical skies and sunsets.

We lived in a large modern house in Kovalam near "Eve" beach named for the topless hippies of the 1970s. The house was designed by a modern architect and was unique. The owner had hoped to rent the rooms on a short term basis to tourists but it had not worked out that way and had rapidly negotiated what we assumed was a rental of the house for several months. After our daughter Nicole arrived for the summer from Bryn Mawr College, she had one of the rooms. When the monsoons forced us from Kerala we discovered that the owner wanted to charge for Nicole as well and on the basis of a daily rental, which led to a tempestuous quarrel that Adele had to calm and settle.

Eve beach then was a smaller, kinder, version of Goa with few hotels and many fewer tourists. Papayas, coconuts and pineapples were sold cut up, and eaten on the beach where there were a few vegetarian restaurants. I was not, as I was in Goa, continually aware of traders, drugs and prostitutes, but I was aware of similar relationships I knew from other beaches. There was the naïve young woman who was certain that once her Indian friend had a British visa he would continue their romantic relationship; there was the older British man who kept inviting young attractive Indian males for meals at his house and was said to offer them money for their favors. The very sympathetic young Indian girl who sold us papayas was beaten by her father to hand over the money she earned. Hearing her sad story we handed over still more money so she could buy clothes for school.

Swimming was, as so often for me, a daily but at times perilous delight. One day I was caught in a rip tide that I fought unsuccessfully and thought I was going to drown. I remember thinking what will Adele and Nicole do as the traveler cheques are in my name. Then an Indian swimmer shouted to stop fighting the tide that was carrying me out to sea; try to swim out the side of the tide and he was right; something I should have been taught decades ago. I swam for a while naked at a small cove near our house just after the hotel on a promontory. (I think that there is still a hotel there but with a different name.) I must have been watched swimming as one day I returned to the beach to find my prescription sunglasses, bathing trunks, towel, small change and flip-flops missing. I had to walk home in my birthday suit.

When not swimming I was starting to work seriously on sections of my book although I was still unclear of its shape. I was still mostly reading and had been sent a ten year run, about a hundred issues, of a provincial poetry journal that included mostly second choice lyrics by the better poets I was following, minor poetry by minor poets, and to my amazement an excellent poem by someone who had never been mentioned to me. About a week later a thin bearded handsome Christ-like young man knocked on our door and said that Ayyappa Paniker told him I was interested in meeting poets. He was the son of a local filmmaker, had appeared in some of his father's films, and was vaguely studying in the English department in Trivandrum. It turned out that he had written that poem while on LSD in Goa and knew he had not written anything as good since. I read through some of his poems, which had some brilliant lines but did not hang together, and told him to send some of the better ones to Jayanta who published them in *Chandrabhagha*.

We decided to leave monsoon soaked Kovalam and return to Goa, but how was I to send my books and photocopies there. I was then mostly publishing with Macmillan UK and approached the Englishman in charge of Macmillan in Trivandrum about my problem which he solved by offering to send a trunk to Macmillan in Goa. It took us some time to reach Goa and I was certain that my many months collecting materials and research notes had been lost in the heavy rains, but they were indeed awaiting me and I only needed to arrange to bring them home.

Before returning to Goa we decided to do some touring. There was a quick day trip by bus to the southern tip of India, a visit to the astonishingly decorated temples of Madurai where I had a fantasy of someday retiring as a beggar-drummer. Then a disastrous trip, at first by taxi, to other temples in Tamil Nadu. It was, I think, in Chidambaram where the taxi driver felt he did not want to go further and abandoned us at a hotel that could provide one small room for the three of us and a few cold chapattis to eat. The next day we were told that the room was booked as part of a temple tour but that there was no transportation available leaving that day. If we agreed to be happy tourists in some photo-shoot however we could have our room and more cold chapattis. This went on for several days until I decided I had enough and persuaded the driver of a three wheeler (there seemed to be no taxis available for days) to take us to the nearest train station where after much talk and some bribes the agent discovered he could somehow find us space on a slow train to Madras. Someone from the hotel followed us to the station and claimed we owed money for days we would have stayed if they had their way or something equally absurd, but after the now usual loud argument and the offer of some money we were not so merrily on our way to Madras while the nonair-conditioned train seemingly stopped at every village, where we purchased coca-cola bottles filled with strange, unrefreshing, dangerous orange drinks in which were dead flies. The bottles must have been ancient as at that time coca-cola was prohibited in India.

Madras was civilization, pomfrets eaten to Indian music in hotels, a former British colonial hotel had its faded charms as a place to stay, we made our first trip to the temples and caves along the sea front at Mallapuram, and tried to return by air to Panjim. Halfway there India Air went on strike and we were stranded for days in rainy, expensive Bangalore. Every morning I and others would turn up at the Indian Air office to be given seat numbers, told to remain, and much later be told the airline was still on strike. After about a week I gave up and hired a taxi to take us to Goa. During the freezing overnight drive through the rain when I was certain that the driver could see nothing and his sleeping relief driver would be of no use, I cursed myself for having subjected Adele and Nicole to such a dangerous journey, but the next morning the sun came up and we were in Goa and soon at what I thought our house in the muddy lane between Calangute and Baga where our return was welcomed as if we were part of the local community.

It was time to make sense of what I had been doing which meant starting by making a chronology of poetry and other publications and events. I needed a basis of establishing a canon of significant poems and anthologies and determining what if any sociological basis there was to the poets who wrote in English, their education, religion, even their birthdates. I had not understood when I started to make chronologies, tables, finding social friendships and social networks, I was already creating the foundations for my book. At first I did not know what I was going to say and made several false starts, for example writing a chapter about Surrealism in the avant-garde poets before I saw that I was on the wrong track. One of my poets Gopal Honnelgere had mostly published in extremely limited hand-colored editions. The surrealism of other poets often was secondary to other modernist influences. I needed an historical introduction, a chapter on Nissim and his influence, and a way to discuss such poets as Parthasarthy, Mehrotra, Jussawalla, Mahaparta, Kamala Das, and so on.

By the time I ended my ten months and left India a book was rapidly taking shape, Nicole would leave from Bombay to return to America and Adele and I would go to Germany and then Italy where I had short fellowships to help fill in the year. We left during the annual Ganesh festival when immense statues of the popular elephant headed god were taken to the harbor to be set afloat. I sent my books to Alabama with a metal statue of Ganesh with a spiritual third eye that I had bought in Goa and a paper mache Ganesh reading while standing that I bought in Bombay. The later never arrived and I can remember his large ears seemingly flapping. No one has found me a replacement to join my growing collection of Ganapati. I also long for the stoned hippie Ganesh poster that earlier I did not buy in Nepal. Before I even started kindergarden I had decided that I wanted nothing to do with gods or religions, but as Ganesh, like the Yoruba trickster god Egba, is sometimes said to be the god of writers I am attracted to him. The elephant head, floppy ears, and big belly have nothing to do with it I say.

I see now that the book I was working on, which has been republished almost annually, was the culmination of several previous publications, such as my survey of the history of English Seventeenth Century Literature, my seeking a social and cultural network to explain the symbolism of Andrew Marvell's poetry, my attempts to explain why Nigerian literature was distinctive from other African literature, my closeness of some Nigerian writers and their distrust of decolonizing nationalist politicians, and my experience at Ahmadu Bello University. I was interested in details, connections, aesthetic choices, and tried to avoid large totalizing generalizations.

Reinhard Sander, whom I knew from Nigeria, had arranged for my two weeks as a visiting professor at the University of Bayreuth, a university supposedly created for the study of Africa. There I became friends with Eckhart Breitinger who published a series of scholarly studies of African literature and culture. In honor of the Austrian scholar Ulli Beier of Nigerian literature there was a Mbari Haus, originally created for Ulli before he decided to move to Australia. There I learned that I was the Bruce King who, unknown previously to myself, had influenced several African musicologists, especially Professor Wolfgang Bender, by the porridge of notes I had thrown together soon after I went to Ibadan, published in *Jazz Monthly* as "Introducing the High Life". I did not have a copy and had to be sent a photocopy.

I would see Reinhard several times again. He published some of the early scholarly books on Caribbean literature and married the West Indian literary critic Rhonda Cobham, but while American universities were competing to hire her, an articulate black female, he, a white male, was unable to find a tenured position on the American mainland. Tired of being asked whether he perhaps had some black blood from an American soldier stationed in Germany he had to settle for a professorship in Puerto Rico and estrangement and separation from Rhonda.

During the month at the Rockefeller Bellagio centre I wrote several more chapters of my book.

# Chapter 13: In Adele's Footsteps to Middle America

*Muncie, Indiana 1986–2003*

The move from Alabama to Indiana started badly when the redneck sounding owner of the moving company claimed we owned him several hundred dollars more than the estimate and if I did not pay immediately he would unload our load (African art, books, computer, clothing) in Nashville, Tennessee. We angrily wired him the extra money. I remain annoyed. What did the estimate mean?

Muncie was drab, flat, uninteresting. In a famous sociological survey it had been barely disguised as Middletown, a dead end working class town in mid-America. I thought those were probably the best days. The working class areas that still existed appeared weird, eccentric, fanciful, lower bohemian, lively, in contrast to the standardized grayness of other parts of the city. Monotony thy name is Muncie. This was not the typical mid-west American city of Hollywood movies, no space creature or space ship would bother to invade here. A hollowed out city center, identical red brick buildings the same height, size and color, small unattractive houses close to each other, uninteresting suburban houses with uninteresting gardens, garishly lit highways lined with ugly shops that looked like a badly designed pinball machine, small shopping malls where it was necessary to go for fruit and vegetables. I would walk along White River watching the birds until they migrated leaving me behind.

During the seventeen years Adele taught at Ball State University we never found a good restaurant in Muncie or a place to dance in the city. The one Japanese restaurant had Mexican waiters dressed as Orientals probably on the assumption that no one would notice, which also describes the food. I am told there are now three Thai restaurants in Muncie, but during the decades we were there a very working class restaurant served some Thai food once a week for lunch; the owner while a soldier had married a Vietnamese. As the woman was either away or the specials had usually disappeared before Adele's teaching ended, what was left were the starches and pork and beef dishes that made up the Muncie diet. There was a club that had live jazz but the owner, himself a musician as well as an ex-hippie medical doctor, insisted on the sound being unbearably loud. It was better to listen or try to dance outside. One of my first memories of Muncie was of a dry cleaner badly ripping the front of a silky designer jacket I had bought on sale in Paris. How can one claim replacement value of a thousand dollars from an impoverished looking shop that probably took half a month to earn that much? I was in the wrong place.

I thought of Muncie as a temporary move, another job on the way to another job, and emotionally never unpacked my baggage, indeed I was away as a visiting professor, or on grants, perhaps half the time, but Adele, who wanted to return to university teaching, stuck it out and moved up the ladder from temporary visiting assistant professor to tenured then emeritus full professor. For five years she was head of her department replacing the reincarnation of Hitler who retired. No matter what, Adele was underpaid and over worked in a bad department that the university, which had abolished the language requirement, would have liked to close but kept because required by the state. Departments such as History that needed French specialists hired their own. She was in a combined department which taught mostly introductory language courses in French, German, and Spanish (the largest section of the department) and included a Latin section that taught vocabulary and other remedial courses as Classical Civilization. Dismal.

Her publications, which then included books on *Camus*, *Proust*, Paul *Nizan* and *Camara Laye* and her co-editorship of the Women Writers and the Modern Dramatists, meant little in a department where a conference presentation counted for annual research, and she was thought superior for reviewing books. One year the University honored her as the professor who had done the best research that year, but the person who presented the award must have thought her books an unmentionable topic and only commented upon her reviewing. While at Ball State she would write a book on *French Women Novelists* and a second book on Camara Laye, but they were like messages in a bottle, sent to the world, unnoticed. When we were in Alabama she had just missed a professorship in California because the university President decided to appoint a female scientist. Once she had moved to Ball State the only offer was administrative and she decided against it. I knew the score. My tenure at North Alabama diminished my standing in the American academic world and that was unlikely to change.

There were good departments in areas such as Architecture, Journalism, History, and no doubt others of which I was not aware, but we hardly met people from them. Unlike in Africa, people only knew their colleagues, neighbors, and fellow churchgoers. I tried to introduce myself to the English Department but there was little interest except from someone in Irish studies who thought of herself as 'postcolonial' and who would leave for a Rock and Roll museum (her real interest), and someone in Linguistics who kept seeing us dancing in Redkey and New Orleans. We had a few friends, including a Trinidadian who we introduced to the German that became her husband. We would often eat with our friends, especially when they brought Trinidadian roti from their visits to her relatives in Brooklyn. I enjoyed dancing with her.

Their wedding brought his divorced parents from Germany, a good-looking couple who danced well in an older European style, more waltz than foxtrot. Although I can waltz competently I have never mastered the art of moving forward on the dance floor while turning non-stop in circles. Until I can do that I will never feel a European, just as I am reminded that I am not Oriental by that stiff unsmiling old fashioned ball room style that some Chinese still learn.

Our Trinidadian friend was later moved to administration, which was thought a step up, whereas in the British system it was to be avoided. If you wanted to be an administrator you would not have written a doctoral dissertation, scholarly articles, books, or become a lecturer. Business would have paid more. University teaching was for the cultured, the scholarly, those who did not seek power over others. It was a different, more subtle, glorification. In an American university, however, the administrators are the ruling class, with their bosses often brought in from the outside, and the teaching faculty the diminishing in numbers worker-bees, whose many tasks include filling in the never ending arrival of forms that the ever growing administrative class create.

The Provost for the Humanities and Social Sciences, originally an Historian, seemed intrigued at having a well published independent scholar in the community, and besides being friendly, arranged over the years for student assistants to help me with my books. They would be assigned to Adele who had to account for their hours and arranged for their being paid. The Humanities had a few excellent students who had been wooed to Ball State by promises of student aid and working with researchers and it was handy to have Adele and myself to give them something to do. Adele increasing was involved with a broad Honors program intended for such students and her better students became our research assistants. One, out of the blue, emailed Adele recently to thank her for introducing her to postcolonial literatures, a topic the former student now herself taught and researched.

Adele was known for her appearance and French clothing. Female students would gather after her classes to look at her rings; when she retired someone said it was the end of a fashion show. When a bright attractive student, who was one of my research assistants, said that she wanted to be known for her brains Adele said that I claimed beauty went with intelligence; the student replied that I could say that because I married Adele. I will not pretend that such remarks do not please me. I know that I have a beautiful, intelligent, energetic, well dressed wife and I feel lucky to be married to her. There was a period of about five years when we often bitterly fought but otherwise it is a happy marriage and life would be much less interesting without her.

Although she seldom pushed herself forward, perhaps because people liked her and she had a record of accomplishments, Adele was asked to become editor of *Women* in *French Studies*, which was then in its early stages, and during her five years, 1996–2001, became an established academic journal. As this happened quietly members of her department were not aware until they heard from friends at other universities. She also kept bringing in writers and academics for readings and lectures by getting other departments to share in the costs. This began as a continuation of my Indian and other connections. One of the first was a reading by A K Ramanujan and Aghad Shahid Ali. We followed it with a many coursed meal at a Chinese restaurant that agreed to provide a banquet at a fixed cost. Muncie possibly had not previously had such meals as people stuffed themselves so much on the early courses that the latter ones were left untouched. Raman obviously expected to be the star and I do not think he even knew of Shahid at the time, and was puzzled that Shahid was getting so much attention. He would not have known that Shahid had attended school in Muncie when his father was studying for a Ph.d in Education, the first such doctorate earned by a Kashmiri. Shahid would return to Muncie a few years later for another reading; this time he was invited as an honored guest. We picked him up at the airport and as we entered the highway our car went into a skid on the snow and I remember thinking what if we have an accident and kill the honored guest.

Other visiting lecturers included the Indian poet Saleem Peeradina who had moved to a teaching position in the USA, Jeet Thayil who was studying creative writing at Sarah Lawrence and arrived with a beautiful equally stoned Indian girlfriend who worked in a bar in California, Jon Silkin on one of his American tours, Jackie Kay who seemed still high from her experience at a California Lesbian theatre and tried to seduce Adele in the car while I was driving to Muncie, Bernadine Evaristo who like Jackie was included in a book I was writing on the changing nature of British literature, and the Canadian postcolonial critic Diana Brydon who being invited by a women's group keep addressing black "sistren" when her audience consisted of bored middle class married white female academics.

The Provost spent summers at Oxford and was more conscious than others at Ball State of status outside the American academy. He was happy to invite Michel Fabre from Paris 3 who was in the USA for a time. The Provost was also happy to sponsor David Dabydeen, who of course we knew as a writer, but who the Provost respected as a professor from prestigious Warwick University. Money was made available to entertain David and among the guests we invited a friend who taught Spanish and always complained about the lack of available men. One day she told me that she had a date that weekend and I joked that he would be gay. When I next saw her she said I was right. David was the opposite of gay but liked to drink and we had to persuade him that our friend would see him home. The next day he complemented her bed skills, and congratulated me on my forethought; she used Spanish to indicate her approval of him.

The Provost recognized that Adele was one of the few on campus of interest to visiting lecturers in the Humanities. When Christopher Ricks came from Boston University and although asking about me—I left Bristol at the time he replaced Frank Kermode—told Adele to stop pretending she was just my wife. He had read and admired her book *Camus*. Tony Judt gave a brilliant lecture on France immediately after the Second World War that was followed by an embarrassing silence when he asked if there were any questions. I dislike asking questions but this time I had one which set off a mini-lecture in reply and which oddly resulted in the Provost and members of the History Department congratulating me as if I had saved them from humiliation. Judt came to us for a meal and we talked about some of the difficulties when a couple are often separated.

One of Adele's students invited us to stay with her family in Istanbul during the summer. They turned out to be financially well off and, for us, surprising in other ways. The husband was usually away (making them well off) in Saudi Arabia. Besides his wife there were several female relatives living in the apartment. They all were highly clothes conscious and devoted much time to reading fashion magazines and having clothes and shoes made. They mocked religious Muslims and thought it a scandal that one of their relatives who had studied at an American university had married a very religious American. They assumed that Adele's student would also marry in the USA as having been abroad Turkish men would assume that she was no longer innocent. (And yes she did marry an American who turned out to be a religious freak.) When Nicole and her boyfriend joined us our Turkish hosts were surprised when we said that they were not engaged.

I enjoyed their company, even the fashion talk. When asked whether some shoes I had bought on sale in Paris were originals or a copy and I said a copy of a copy, my remark was repeated as sophisticated. We were taken to a large modern hotel where we drank outrageously colored cocktails and watched the dancing at a wedding. The music and dancing could have been Jewish. I enjoyed such street food as mussels cooked with rice and spices, and I bought a small Turkish cookbook in English that I soon mastered except for the pastries. Turkish food lacked the hot peppers and many spices of most middle-eastern cooking, but in compensation there were clearly distinguished flavors such as in lentils and tomatoes garnished with a sprinkling of parsley. Iranian food, by comparison, is much more complex and with flavors competing for attention.

We visited friends of the family who were in shipbuilding and the lingua franca was German. This was the Westernized secular, urban elite that had developed and profited during the decades of Army rule. They sent their children to prestigious English, French and German language schools, were tolerant of Christians and Jews, vacationed in Paris and Rome, supported the "classical" Turkish music of the past, and felt superior to the small town religious Muslims who would in the near future gain political control of Turkey and begin to disassemble the tolerant liberal sophisticated culture of the recent past.

You could see the cultural and spiritual contrast clearly when we moved to an apartment down the coast owed by our hosts where they went for weekends. Those in the modern apartment blocks wore bikinis, read European literature, listened to the latest pop music, watched European TV shows, and dressed like Europeans. But in nearby villages the women were covered in black, regarded us with suspicion, and clearly the local people did not share in the boom that Turkey's incorporation in a global economy had brought many. They looked like the past but in the Muslim world they would be the future.

Trips abroad made Muncie more tolerable. There were four or five trips to the Caribbean when I was researching my two books on Derek Walcott, several summers were spent in Paris especially when I had a visiting professorship or when Adele had summer grants from BSU. We spent a few summers in Muncie exploring the pleasures of Indianapolis and Redkey.

Redkey Indiana was a surprise, a small town, about a twenty mile, forty minutes, drive north of Muncie, through smaller towns where the only life on Saturday night appeared to be a dimly lit coca-machine in a closed gas station and the liveliest action would be a car or pickup truck suddenly emerging from a dirt road. There was Albany, a town of about 2,000, where the main road branched and on the side I could see a tavern with parked cars in front, but I never was tempted to wonder whether anything was happening here besides getting drunk while watching television. I recently looked it up on Google and recognized the Duck Inn which serves chicken fingers, fish tails, steaks and pork chops but no duck.

As you approached Redkey there were signs of civilization, a well-lit gasoline station, a 7/11 dough nut and coffee restaurant, and something else that I cannot remember beyond it being odd, maybe for reindeer testing It was a small well laid out town of slightly more than 1,300 population with a railway track which ran through the center. Redkey was originally built around the railway when the tracks were laid. Its big period was (if I remember correctly) just before the Great Depression when some Republican president regularly ate in a Redkey restaurant, which still opened on Sundays to serve the same bland meal. There were streets laid out on the usual American geometric grid with no one moving until you got to South Meridian Street when suddenly there were lights, parked cars, people, and the Key Palace Theatre, a restored 1950s movie house that had become one of the ten best places to listen and dance to live blues in America.

Charlie Noble was born in Muncie and educated in another small town before leaving in hope of a career as an actor. He was in major theatre companies in Los Angeles and Chicago while supporting himself as a construction worker before returning to Indiana where he sometimes acted in the Muncie Civic Theatre. He bought a 220 seat movie theatre in Redkey with the idea of converting it for plays and for a few years remodeled it himself, building large stone columns and a wooden plank sidewalk outside, making the doors inside the theatre into something you might find on Shoot Out Saloon in a Western, covering the toilet rooms with varnished photographic covers from old film magazines, and other eclectic decorations, before in 1986 opening the Key Palace Theatre, which after a few weeks of indecision started having blues bands on Saturday nights, and attracted blues fans and people seeking something to do on weekends. Charlie hired big name acts, introduced every show himself, told the same or similar jokes every week, and had a warm open personality. The Key Palace felt intimate and was so off the beaten track, it soon became a legend. Musicians loved Charlie and continue to speak of him with respect although he died in 2010 and the theatre was eventually sold by his heirs.

We enjoyed dancing to such well-known musicians as John Prime, Marcia Ball, John Hammond, Maurice John Vaughan, Larry Garner, Leon Redbone, Anthony Gomez, Stanley Mitchart, Terrance Simien, Big Bill Morganfield, Eddie House Rockin Shaw, and such groups as Mississippi Heat, Saffire, Howard and the White Boys and Big Al and the Heavyweights. As we often arrived late and immediately began dancing in the space in front of the stage it was rumored that we delayed our entrance until the music started to make an impression by rushing down the central aisle directly on to the dance floor. After such rumors how could we resist? Sometimes two generations of blues musicians played, first together, then the young replaced their famed parent. Johnny Copeland, before he retired, introduced his daughter Shemekia who would be on the annual program. Although most of the musicians drove down from Chicago, and some got lost in the small towns on the way, others were on tour from places like New Orleans and Baton Rouge. Mitchart came directly from Nashville. Charlie would go to the annual Memphis Blues Festival to size up new talents and make contacts.

As the Key Palace became popular a second room was opened on the side of the theater for talk, food, and drink, Besides the popcorn, soft drinks and a limited selection of beers there were now immense hamburgers and French fries There were shows also on Fridays and on holidays, especially on New Year's Eve with, I remember, in 2000 a marriage on stage at midnight. Regulars got to know each other and some musicians got to know those in the audience. Years after we had left Muncie we were dancing in a Paris hotel to an American blues band when one of the musicians remembered us from Redkey. He then introduced us to the band's leader: "You were dancing in Redkey Indiana and are now dancing in Paris. Introduce me to your agent."

Although the Key Palace was Charlie Noble's personal success there was interest throughout the mid-west in the blues. Older jazz was thought Dixieland played for white Republican graduates of fraternity houses; modern jazz was promoted by black radio stations. The blues revival of the 1960s resulted in many of the working and middle class playing guitar and harmonica, The blues were a step up from pop music as one became older. Many cities had blues clubs and their own blues magazines. Some of the people we met at Key Palace were middle executives at local factories, small town professionals, a secretary in the administration of Anderson (another Indiana city), a former nurse who had for a time been a stripper in Vegas and was now probably the mistress of the car dealer she accompanied,. One night we arrived with two other Ball State couples and the six PhDs began the dancing. There were others from Ball State University, and a full length documentary was made by a professor. Despite the closeness to Muncie, the Key Palace was not a hangout for academics, who were a tiny part of the audience, but had a mixed crowd—except racially. Redkey is about 98% white and the few blacks in the audience, unlike on stage, were noticeable, not unwelcome, but unlikely to return.

The Slippery Noodle in Indianapolis was the oldest bar in Indiana and also regularly featured blues. We often went and danced but it was less intimate and warm than the Key Palace. It had music regularly as it also was a restaurant but there was not much space to dance except between tables. I remember hearing the Tower of Power, Gatemouth Brown, and other name acts including Pinetop Perkins, who was then in his 90s and kept interrupting his piano playing to dash for the men's room. I had heard Perkins previously in Chicago after a MLA convention. It was New Year's Eve, and one of our party, was asleep long before midnight, as she was a well brought up woman who never stayed up late. One night we were talking to Sugar Blue after the show at the Slippery Noodle and were harassed by a car full of drunken white idiots who were certain, because of Sugar's leather French beret and performing clothes, that he was a homosexual needing a beating. We however were discussing Paris where we had seen him perform and where he impressed the Rolling Stones who had used him on some albums. Unlike many harmonica players he followed modern jazz and played a bop saxophone influenced style of long many noted flowing runs through altered chords without the staccato accents, obvious syncopations, and heavy beats common to blues harmonica.

The many blues musicians in Indianapolis were usually white and played in the same clubs that also had folk and local pop groups. Some were excellent, such as Andra Fay, who became part of nationally known Saffire: the Uppity Blues Women. I heard her first in a duo in Indianapolis, then with Saffire at Redkey, and then with Saffire at a women's group sponsored event at Doc's in Muncie (where I first met and danced with Charlene, our Trinidadian friend.) Yank Rachell, an older country-blues musician originally from Tennessee, who played mandolin and had moved to Indianapolis in 1958 often performed with a backup group of white blues musicians. I cannot name the many small clubs that sometimes had blues groups, the bands and clubs have long disappeared, but I do remember Boogie Blues that opened on 86th street at the upscale Keystone Shopping Mall; it featured name blues and world music groups, and served southern, especially New Orleans, food, until the owner went bankrupt. She had inherited the money that allowed her to start the club, and, according to rumor, a well-known blues musician had sweet talked her into supporting his drug habit until she could no longer afford him or the club.

Besides dancing at blues clubs we also went weekly to a famous jazz club that was once called The Place to Start and which after several changes in ownership became and still is The Jazz Kitchen. Touring acts would often play on Mondays on their way to or from better paying clubs or concerts in larger cities. One night a week there was the Mid-Coast Swing Band modeled perhaps on the smaller Count Basie groups, to which we danced, another night was for live Latin music, usually very up tempo, to which we sometimes danced. Dancing on other nights, as when electric violinist Cathy Morris had her Brazilian Lounge Band, was a problem as the owner tried to fit in as many tables and chairs as possible, but we were a fixture and most musicians like to have dancers. Indianapolis had a jazz history, a few locals when younger had been in bands led by Count Basie or Duke Ellington, and could still play well while bringing along promising younger musicians. There were also jam sessions, usually when visiting stars, such as Nicholas Payton, sat in, even bringing their own groups, after playing at concerts and theatres. We first met Sylvia Howard, a singer who became our entry to the black Americans in Paris jazz scene, when she returned to Indianapolis one summer to visit with her family.

We often danced at the Madame Walker Ballroom where some of the black singers and musicians played as well as the Jazz Kitchen. The Madam Walker was the center for the black community in Indianapolis in an area that had half a decade earlier been lively with black night clubs, jazz clubs and restaurants, until the Great Depression and especially the Second World War brought an end to what now seemed a golden age of African-American social life. Madam Walker herself was famous as both the first American and the first black American self-made female millionaire. Her business was in black cosmetics and you can still buy such products as hair strengtheners and skin lighteners.

The Madam Walker was a large building with a theatre, ball room, library, and many other rooms used for rehearsals, meetings, business, and social events. The theatre in the past had such artists at Louis Armstrong and Duke Ellington and restoration had brought out its decorations—early twentieth-century Ethiopian modernism. It had black theatre groups, gospel shows, and we attended performances by visiting jazz musicians, sometimes flawed by a bad sound system. The Ballroom on alternate Fridays had a calendar of jazz, blues, funk and soul musicians and a menu of southern soul food. The ballroom was great place to dance and we sometimes had the floor to ourselves until intermission when it came alive with black women sexily grooving to the Electric, Latin, and other syncopated glides, which, as I do not like line dancing, it was our turn to watch.

We had been dancing together for decades, were swing dancers, and had spent decade in Nigeria often dancing. One of the better black jazz singers in Indianapolis used to joke about our showing the women at the Madam Walker how to dance. Then one night we were doing our thing when the musicians accompanying a singer started beating time while watching us; the vocalist lost her cool and said over the microphone that she did "not rain on our parade", which I thought fair enough and, embarrassed, sat down. We danced discreetly in a corner the remainder of the evening. This however was also an embarrassment for the Madame Walker and for weeks we kept getting telephone apologizing and asking us to return. Even the singer looked shame faced when we next saw her. I should have realized that our dancing was annoying her as she had some weeks previously said that she did not mind, which if I had more social sense I should have understood as meaning the opposite.

The dancers at the Jazz Kitchen, Fountain Square, Mickey's Irish Pub, and other mainly white places mostly wore tee shirts and jeans.

The self-conscious Swing Dancers at Fountain Square were in retro 1940s clothing, out of old films, which the women found in vintage shops; some men even wore Zoot suits made by custom tailors in Los Angles. Blacks however dressed up for going out, the men wore suits and the women often had sequins and gowns. Those who made it as a businessman and in the professions did not want to be mistaken for hippies and beats, did not dress down to reject bourgeois origins. Large cars, even chauffer driven limousines, would be parked outside black clubs.

We were almost groupies of Dog Talk, a popular Indianapolis band, and danced to them at the Jazz Kitchen, the Indiana Roof Ballroom (usually for big bands and ball room dancers—some men in tuxedos), the Rathkeller Biergarten, and Mickey's Irish Pub (very working class, billiards, drunks, and independent rock groups on Fridays and Saturdays). Dog Talk was started by its drummer to play his quirky whacky songs requiring intense precision and musical sophistication. I got a kick from the unexpected percussion breaks performed on weird instruments. The group soon had a following including us, and then transformed itself into a "cover" band playing a wide range of popular, Latin, jazz, and soul hits. It had some of the better musicians in Indianapolis, incorporated itself, paid its members an annual salary, and charged large sums for appearances especially for the private parties, that, as Indianapolis became a centre for businesses moving to the mid-West, were lucrative for musicians. Charlie at Key Palace hired them once or twice but thought them too expensive. It was cheaper to get nationally known blues bands.

Cliff White, the Dog Talk vocalist, started in light opera and musical comedy but was often hired by black churches for gospel singing. Besides his excellent voice and range he was a sensuous performer; women often rubbed themselves against and propositioned him, but he kept himself from temptation perhaps because his wife often waited for him after the show. Bill Lancton is one of the better guitarists in Indianapolis and proud of his professionalism and range. Although he is a jazz musician I sometimes wondered about the "country" licks I would hear in the midst of a solo; I later learned that he had spent time in the house band of the Grand Ole Opry. Lancton and Cathy Morris often teamed up in "Les Chat du Swing", a modern electronic version of the Django Rinehart and Stephane Grappelli Hot Club of France. The first time they played one of the summer Jazz at the Zoo dances the banner unfortunately proclaimed "Les Shat du Swing". At another Zoo concert I was fascinated by Cathy athletically jumping up and down "bang, bang, bang" louder than any rocker heating an audience. I later learned that the drummer had brain cancer and she was keeping the beat.

There were good dancers in Indianapolis and the best became dance teachers. As the dancers were mostly Lindy hoppers, whose foundations were in the late 1930s Shag rather than the 1940s jitterbug I knew, I never felt the need to compete if only because I did not have the same stamina. I felt competitive, however, with a trio I kept seeing at Latin events with the man, an oriental, and his two female partners dancing many of the same steps I used except he kept the two women active as they moved around him, changing his partner and often spinning off the other behind him. I had tried jitterbugging with two women but how did he manage the coordination and get the women to dance separately when not being led by him? Envious I had to ask and found out—it should have been obvious—that they had learned together at a Latin dance studio and that they were themselves dance teachers.

There were also surprisingly good restaurants in Indianapolis. Not the often mediocre and worse upper scale French, Italian and fusion restaurants but small inexpensive ethnic places such as the Queen of Sheba (better than any Ethiopian restaurant I know in Paris), the Coyote with its immense Snapper Veracruzana (provided you do not mind the fixtures hanging unfixed from the walls like some small village Mexican dive, or the police trying to find drug dealers), or the excellent snapper served on weekends at a small Caribbean restaurant with a Trinidadian owner and recorded music. There were Korean restaurants where the shrimps or chicken you grilled at the table were accompanied by a dozen small highly spiced dishes. The best chicken kebabs and some of the better humus I had was served at an Egyptian restaurant. My favorite restaurant by far was Yummy, which had dim sum on Sundays and the owner claimed that his imported chefs were bored by the small demands of the local Chinese community and used us to show what they could do if challenged. The food was great.

As Chinese New Year was approaching and I felt that no oriental would eat the bland Chinese food usually served in Indianapolis I telephoned several restaurants to ask where they would go for Chinese New Year and was told Yummy, which proved to be small, unprepossessing, typical in appearance, but inside had plaques and photographs commemorating recent banquets and a menu for a New Year's diner. Our first meal there was very good but the six course New Year menu, mostly duck, was unlike the ten course, mostly sea food, banquet I expected so we talked to the owner who offered to create a feast if I paid for a table of ten. In future years we had two tables of ten, and increasingly sophisticated dishes using dried scallops, razor clams, duck breasts, sea cucumber, egg white omelet topping, each course a work of art.

Bored, intelligent, ambitious, and energetic, he had left the Hong Kong administrative bureaucracy for the USA and before long had decided to create a Hong Kong restaurant in Indianapolis, the first in the city to offer dim sum from carts on Sunday. His name sounded like Yummy, the restaurant's name. The restaurant not only sold to take away the usual soy chicken and duck, there were several flavor variations on the roasted fowls. He was dissatisfied with the fish available locally and began to raise his own in tanks that lined three walls of a small room. They became a series of breeding tanks containing fish of different sizes. He knew that Chinese would pay more for live fish, especially for his white carp that he claimed had a more delicate texture.

He was always discovering solutions. The electricity was expensive to light and oxygenate the fish tanks so he hooked up a series of batteries. He knew a Chinese scientist at Johns Hopkins University who advised him on genetically improving the fish he raised. He then decided that Indiana farmers did little all winter when instead they could be hired to raise his fish in ponds and tanks. At first he would travel in his van to Toronto to bring back sausages, clams, and oriental spices and vegetables he could not get from the many impressive large oriental markets in Indianapolis. Then he started travelling throughout the mid-west selling his fish to Chinese and fish restaurants and most days Yummy was run by a friend. Next he decided to sell Yummy to the friend and devote himself to raising and selling fish. This was around the time we were selling our furniture and other goods in preparation of moving to Paris. He waited several weeks until we were down to what seemed left over when he arrived around midnight with a large truck and a non-English speaking assistant and we rapidly traded this desk that his sister might use for two lobster meals, the chairs his friend might want for three courses of peking duck, why don't we let him have these cassette tapes that his cooks (who seemed to work seven days a week and live in dormitory conditions) would like? Okay a course of dim sums. I do not know what happened to Mr Yummy, but among the many authentic oriental restaurants in east Indianapolis his was by far the best, and needless to say I miss talking to him. I understand that after he left the restaurant changed owners and names and closed.

Another of our favorites was Yats, near the Jazz Kitchen, a fast food restaurant serving immense plates of New Orleans creole and cajun food. The menu was the only aspect of Yats that was exclusive as you stood at the counter to order and briefly waited for plates piled high with various peppery, strongly spiced, red beans, rice, versions of etouffee, gumbos, jambalayas, maque choux, and some exotic Mexican and Thai influenced dishes that were ladled out of large steaming pots in the kitchen. There was a changing menu of seven to ten dishes chalked on a blackboard, and from the time you came in to the time you ordered the menu changed as one dish ran out and was replaced by another. This was definitely not fine dining. There were no wait staff. You carried your plates back to whatever table was free and on the way paused to help yourself to water, cold tea or lemonade from one of those dispensers that American cafeterias use. The food however was tasty, warm, filling, some of the best in town, served over rice or buttered toast, and at six or seven dollars before tax about a tenth of the price for the bad imitation Louisiana meals at other places in Indianapolis. At first there were no desserts, then New Orleans butternut cream pie became the choice. Tins of beer were sold, chirizo and Thai paste joined chicken and crawfish.

Joe Vuskovich, the owner, was from New Orleans where he had a restaurant, still had family, and visited on holidays. On his travels he had learned the new technique of flash freezing which he took to Yats and the branches run by his wife and former employees. I think that there are now a dozen in Indiana and Ohio. Immense pots of food were cooked, frozen in large bags, and then opened into steaming pots to be served at the restaurants. It is natural to dismiss this as mass food cooking, which it is, except that increasingly many expensive restaurants in the USA and France depend on flash freezing. The original location alongside the Jazz Kitchen meant that Yats was favored by musicians and record producers and there were on some holidays, erratically, parties with dancing to live bands, to which I looked forward. The restaurant would close but there would be free food and you brought your own drinks and began dancing like in New Orleans where everyone was a partner. Before we left Muncie for Paris we gave Joe some prints and art objects that seemed suited for Yats. I am told that a large print is still there.

Indianapolis had a small but active Renaissance and Baroque music scene and if you had the energy there were good films to be found, although this could mean driving to the southern part of the city to Indian films that played before packed Indian audiences. A really popular Indian film might play all weekend almost non-stop before perhaps five thousand. The city had a large immigrant population and there was often something to do. Butler University regularly had visiting modern dance companies and visiting writers. When a now Noble laureate Wole Soyinka was invited there was a local version of a Yoruba apala band to sing his praises. Although apala is normally played behind the beat with tension between the delay and felt pulse, which is contrasted by talking drum solos that erupt over the regular rhythmic pattern, this band was dragging like a zombie, and as I did not want to take over the talking drum I borrowed the sekere (the beaded gourd) to make the sound more lively. The drummer might have been surprised but adjusted as I led the small band from behind. Wole recognized Adele and myself, gave us embraces, which he usually does when we meet, and that I cannot help but contrast to Derek Walcott's air of distant superiority. But they are different people and my relationship with each has a different history.

## II

When Adele started teaching at Ball State University I thought of it as temporary and I was often away with visiting professorships in Israel and Paris, research fellowships in Canada and India, and assumed that I would through my publications and experience be hired by a major American university. I was not black, a woman, gay, or a literary theorist, but I was internationally known and was one of the founders of a field variously called Commonwealth, New, International or World Literature in English. Offers still came from foreign universities but none from America. Multiculturalism meant the new American minorities not the lands they had left. Post colonialism meant a Marxist, New Left, criticism of America and Western colonialism, a perspective influenced by the War in Vietnam, rather than studying the literatures and realities of the non-Western world. I remember an absurd interview when I was told that the university wanted to replace its required first year Great Books course with non-Western texts. I remember thinking that they probably have semi-lettered students who cannot now read their required book a week, and they hope to have these replaced by a survey of such immense epics as *Mahabharata* and *Ramayana*. They are being driven by ideas not any familiarity with the literature.

Two events necessitated choosing between teaching abroad and settling in Muncie. The death of our daughter Nicole in 1990 made me feel I wanted to spend my time with Adele, my career in teaching was far less important than our marriage. Then Adele in 1995 developed lung cancer. The first surgeon she consulted told her she had only months to live, but we telephoned around America for ideas and were told of a team in Muncie specializing in cancer. They advised an emergency operation and she survived, but with one lung. After that her life took precedence over all else. Amazingly she was soon back to dancing, teaching, and normality. She was a survivor. When I had originally met Adele I thought her healthy unlike so many of the young women I knew, and this remained true. She was optimistic and cheerfully bounced back. After her lung operation we purchased our first house; until then we regarded ourselves as wandering scholars and had lived in rented housing. I was not embarrassed that she had the regular job as I was still writing and would be awarded grants for my research. Adele seemed in my family tradition; my mother and her two sisters were energetic women and my mother and her younger sister had taken over and expanded their family business after the death of a husband.

In these memoirs I have tried to avoid precise chronology in the hope of more narrative continuity, but also because facts can be misleading. The period of a fellowship or grant might require research, writing, and submission of the application a year or two in advance. The resulting book might take another five or seven years. A joke among experienced researchers is that the fellowship year is used to research and write the application for the next fellowship. During the first three years after Nicole's death Adele was head of her department, which meant that she had a reduced teaching load while administering the department. The extra salary went to others who took over while we went away, usually for my research on Derek Walcott, during the summers.

Feeling that I should earn some money I kept agreeing to contribute short essays on authors to various encyclopedia and surveys of contemporary literature. If I sometimes felt like a hack writer, I also was proud of the range I had to cover and what I was mastering. In the process I was learning about Ghanaian and South African poets, early Indian poets, and much else. Often there were no previous critical comments I could use and I had to read the author rapidly from scratch. It was a good general education in areas of my field. I now had a London agent, Charlotte Howard of Fox & Howard, introduced to me by Derry Jeffares when Charlotte and I happened to be visiting him in Stirling.

I was an out of work white male English professor who survived on his wife's salary while writing books and essays for an area of literary studies that hardly existed in the USA, but reader's reports and reviews sometimes claimed that I was famous. I did not feel that way. I wanted to explode with violence when a well-paid white foundation executive, or a second rate Indian literary critic teaching black American literature, would ask me whether I felt guilty as a white man because blacks had been discriminated against in the academic world. I had spent my life in jazz circles, teaching for a decade in Nigeria, writing about authors from the new nations, and now I was being told by those with large salaries that I was a guilty white man who was paying a racial debt by being unemployed. I knew enough about the world to understand that if I did complain publicly, and strengthen the political Right, I would lose friendships and any chance of re-entering teaching. Black friends asked me what I had done wrong to have so many publications and not have a distinguished professorship and I could not reply it was because I am a white male. I met Henry Louis Gates Jr at a MLA conference and he told me that he knew my books and he volunteered to write references for me. He also suggested that I apply to black universities, but when I did there would be private telephone calls asking others whether I was a "soul brother".

As Adele became part of the Ball State scene she was awarded summer research grants almost every five years which sometimes enabled us to go to Paris which if she had a paid half year sabbatical we stretched out over a twelve or fifteen month period. Sometimes we would be in Paris one summer and the next in Trinidad and St Lucia for my Derek Walcott research. We visited Prague. In 1991 she was made a full professor. During 1994 she had a half year sabbatical which enabled us to live in Paris for the year. As she was on the BSU European Studies committee she was invited to a meeting in Denmark of European academics who were hoping to institute a year abroad plan for their students.

The trip to Denmark would be important to me. In 1992 Derek Walcott was awarded the Nobel Prize in Literature After the meeting in Denmark we travelled to Stockholm where, as someone writing a book about Walcott's Trinidad Theatre, I was welcomed to the National Theatre where I conducted interviews about their production of one of Derek's plays, learned of a Swedish actor who was once part of Derek's circle in Port of Spain, watched rehearsals, and eventually saw a production of Shakespeare's *The Winter Tale* that Ingmar Bergman directed and which was as somber and concluded as sadly as a Bergman film. Although the Nobel Prize committees have a rule against discussing their choices the chairman of the literature award asked the Royal Library to give me assistance which meant that a baroness in their employment was assigned to photocopy from their files hundreds of articles from newspapers, journals, and other publications in many languages about Walcott.

While at the Library I was shown their copies of my books about African, New National and Commonwealth literatures, all heavily marked, and I had the impression that they were actively consulted by the Literature committee. In *New English Literatures* I had written at length about Soyinka, Walcott and Gordimer. I was asked what I thought of several French Caribbean writers who were probably under consideration. Later as I interviewed Swedish publishers, translators and literary agents, I was told of how they heard from secretaries at the Academy what books were being ordered to be consulted by the Committee. Once someone was discussed it would take several years before progressing to a candidate for the award. Meantime European publishers would purchase rights and commission translations in the hope of having a winner ready. A Canadian author who I had written about in *New English Literatures* was then being read, and was thought a possibility, but he never received the award.

During 1999 Adele was at Westminster College, Oxford, for a term on an exchange. We went to Tübingen, Germany, for a conference and on a bus met Janet Wilson, a friendly New Zealander who had parlayed her degree in Medieval English into teaching Linguistics and New Zealand literature at Northampton University College outside of Oxford. Northampton University College had recently been formed from several teachers colleges, but Janet had previously taught at universities in Australia and New Zealand and she seemed to know and be liked by everyone at the conference. On our way back to Oxford we ran into her again at Heathrow airport. Her expected ride had not appeared, could she borrow some money to take a bus to Oxford. This began a long, mutually profitable, and continuing friendship. Janet earlier when working on her doctorate had an affair with Jack Pole, the well-known historian of American Studies. So had many women. Jack was now aging, suffering from Parkinson disease, but Janet owned a house near him, remained friends, and often looked after him and his house. We were invited to a dinner she cooked at which one of his former girlfriends—intellectual, exotic, a former journalist for Left publications—was a scratching, scathing cat all evening. The house was filled with Jack's paintings, signs of his continuing interest in cricket, his books and a large library, and we talked briefly about the book on which he was still at work. We were also invited to an excellent meal one night at High Table at St Catherine's College where we followed rituals about which way to pass the port and sherry, and listened to a pretentious American who assumed that this was our first trip abroad

Kirpal Singh was then editor of *World Literature Written in English* and needed to unload it after financial support vanished. *WLWE* was one of the first journals of international English literature and had been started and sent freely by the University of Texas. When Texas withdrew funding it had moved to the University of Guelph under the editorship of my friend Doug Killam. When Killam was deposed as head of his department by younger academics they took over the journal, had difficulty bringing out a few issues, and it passed on to Kirpal Singh, a poet who taught in Singapore, who I had known over the years and who visited us in Muncie and was one of Adele's visiting speakers. As I was not part of the Ball State faculty, did not have financial means for an academic journal I started thinking about who might.

England, unlike the USA, Australia and Canada, was then expanding support for research, especially publications, and I thought Janet Wilson needed something to advance her career which otherwise seemed a bit wobbly as her College, which was being upgraded to a university, was getting rid of staff in English. Janet reluctantly agreed to take over the editorship of *WLWE* thinking it would only be for a few issues before she handed it to others but with her usual luck and sociability found herself talking to a Routledge editor at an Oxford party who said we are looking for academic journals and we would like one in postcolonial studies. *WLWE* after a few catch up issues during 2004 became the *Journal of Postcolonial Writing* and Janet was now on her way to being the one member of her department secure from losing her job. Promotion followed to professor of English, Director for Research in the School of Arts, and eventually head, chair or vice chair of the European Association of Commonwealth Literature, the British Postcolonial Studies Association and the New Zealand Studies Network. Most of her time was now spent on administration, conferences, international meetings, representing Northampton University abroad.

I began writing for Janet during the transition from *WLWE* to *JPW* and have continued mostly in the form of review articles long after I stopped writing for other academic and literary journals. Although I receive a small payment for my reviews what matters is the freedom to decide what I consider worth attention. Janet also edited an international festschrift in my honor which pleases Adele and to me is evidence that my books and articles attracted some notice. If we go to Oxford Janet arranges our accommodations and we sometimes see her in Paris. Janet, an attractive woman, had decades previously had a love affair with the New Zealand poet Kevin Ireland who was then married. After Kevin's wife died in 2007 they restarted an intercontinental relationship travelling between England and New Zealand with many conferences in between. They married during 2012, remain intercontinental travelers, and Kevin became someone I think of as a friend although we do not see him that often. (As he is almost my age, about a decade and a half older than Janet, I once asked him if he also needed Viagra and he claimed that during the flight to England he already started to become erect.)

## III

It was while working on my Walcott books I unexpectedly found myself once more involved with Indian literature. *Modern Indian Poetry in English* had become a classic and was being reprinted almost yearly and Oxford University Press India asked me to write a preface for a new printing. I began trying to update the chronology and reading what I could find of what had been published during the past decade and that led to writing to friends in India about new authors who I might mention. It became clear that I needed to write another book and not a preface, but the way around that was to add five new chapters along with the updated chronology to the existing book. I had recently published a long essay on Agha Shahid Ali being nostalgic for a life he imagined and never actually had. I could update that, and I wanted to have chapters on new, younger, Indian poets, and specifically on new female poets. I would have a chapter surveying what my established poets had published since the first version of *MIPE* and a chapter on changing social, cultural and publishing conditions. The new edition would be long, even a bit hump backed, but the five new chapters would make it better as well as bring it up to date.

The younger male poets presented few problems. Jeet Thayil shared *Gemini* with Vijay Nambisan a Penguin book introduced by Dom Moraes, both would go in. I was interested in Tabish Khair's poetry as it was sociologically and politically conscious and written in an area where English was seldom a spoken rather than a written language. G V Prasad was one of the Tamil Naidu Brahmins driven out by the radical changes and violence in their state and who then were rejected by north Indian Brahmins. C P Surendran was a novelist and journalist who brought violence and realism into his verse. Bibhu Padhi wrote about the society of Cuttack and his family. I got Ranjit Hoskote wrong in assuming he was surrealist influenced but I was aware that his difficult poetry was related to his art criticism and knowledge of theory. The chapter came together easily, but the chapter on women writers was more difficult as, except for Imtiaz Dharker, there were fewer obvious poets and I was influenced by Eunice de Souza's recent anthology of *Nine Indian Women Poets* (2001), although I felt some of the poets did not deserve such recognition. Eventually I put together a chapter with good poets and a few others.

Soon afterwards I was asked to write a preface for a new printing of *Three Indian Poets*, which also once I thought about it meant an expanded second edition as Moraes and Ramanujan had published significant volumes since the first edition and, frankly, I wanted a new cover to replace the dreary looking existing edition. This meant expanding the introductory chapter and writing three more. Alas the Oxford editor who was responsible for the book apparently never knew what she was handling and the cover reads "edited" by Bruce King as if it were an anthology of poetry rather than a book of literary criticism.

I changed my usual publisher to Oxford University Press, England, after my editor at Macmillan would not advance payment on royalties for a book of essays on Samuel Beckett I wanted to edit. Macmillan had usually sold books of literary criticism to American publishers and that paid for production so English, European, and Commonwealth sales were pure profit. But that changed, there was suddenly less American demand for such books. Americans were preoccupied with politics, their minorities, cultural theory, racial and imperial guilt. Without an advance I could not promise to pay contributors; I let the project drop and began looking for another publisher. Routledge was still deeply involved in publishing books on literary theory and while the director I spoke with saw post colonialism as the likely future she thought of it as a continuation of critical theory which I thought was killing well written literary criticism.

A trip to Oxford was more productive if only because it was more conservative. Would I edit a collection of essays on New National Literatures? No, I had done versions of it several times already; instead I suggested a survey of critical topics and approaches which became *New National and Post-Colonial Literatures* (1996). Oxford University Press would also publish my two books about Derek Walcott. Then while still trying to think through how I could write a book about British minority writers I was asked to propose a volume about them for a new *Oxford History of English Literature*, which I jumped at, but which meant resubmitting proposals as the readers Oxford asked to evaluate the proposal could agree on nothing but that this is not the book each one wanted to see. Some felt that minority writers should be included with other contemporary British writers otherwise it was racial discrimination, although this would ignore the social and political context of such writing and was likely to reduce major authors to minor figures among what I thought of as inflated white British reputations. Then there were the theorists for whom anything less than abstract generalizations was nasty empiricism and then those who thought the racial political context was the only possible focus and the book I had in mind with its social history, notation of changing consciousness by immigrants and their children over the decades, and evaluations of authors and texts was much too traditional literary criticism, which it was. The proposal was eventually approved and I had to find money for travel to England, books, and other research. The

National Endowment for the Humanities again came through, this time with a one year grant as an Independent Scholar. It took me about five years to research and write the book.

My success with obtaining NEH grants astounded me as much as it did others, but I knew from the time I was heading English departments in Nigeria that I had skill at outlining projects and was good at writing letters of recommendation. I knew how to sell ideas and other people although I was told that my own job applications were confusing with my many and changing interests. The two Walcott grants had additional help that I did not know about at the time. The person who selected my applications and supported them knew my early work on John Dryden and was happy to have something in black and postcolonial studies to put forward which he knew would be solidly researched and not merely politically correct and journalistic. In the case of the British book I had earlier complained in a newsletter published for Independent scholars that the NEH treated us as second rate and was undemanding in what they supported, and I learned that my article had been read and noted within the NEH administration. So when I applied as an Independent Scholar I probably already had a leg up towards my fellowship. The *Internationalization of English Literature: 1948–2000* (2004) annoyed academic theorists who felt I had emphasized the aesthetic and changing social consciousness at the expense of the political, it was even felt to be too well written, but it had good reviews and there was a Chinese edition a few years later.

Usually I felt that there was an area of literature that I wanted to explore and would then write a vague proposal and hope the book would be contracted and that, after much pretense of knowing what I intended, apply for a grant. I never took book outlines seriously as I felt that if you could precisely say what you intended the book was not worth writing, it would be a mass of existing clichés. So my outlines were really blarney, the book would write itself, and usually, although not always, it did. Charlotte Howard, my agent, would come in at the contracting stage, after I found a publisher who wanted the book. She soon took a more active role. She convinced a senior editor at Continuum that they needed me to start a series of books and not only were the terms generous, but the editor was willing to fund Adele and myself to attend two literature conferences in Europe as her representatives, for which we did little beyond distribute some leaflets on tables. Apparently editors at publishing houses and some newspapers were expected and reluctant to travel to international cultural events. Being an ex-professor of English had become more interesting than complaining about the lack of a job. I had more than enough employment at home and it was of a kind I enjoyed. Even the people I was meeting were interesting. After writing about his books I met Mike Phillips, a British novelist originally from Guyana, whose black detective novels were a new direction and who knew Ronald Dathorne, also from Guyana, who was a colleague at Ibadan and who we had visited twice at the University of Kentucky when participating in conferences. We attended launches in London of books by David Dabydeen and Mike. Mike wrote his autobiographical *London Crossings: A Biography of Black Britain* for my short lived Literature, Culture, and Identity series (1999–2001). Nuruddin Farah wrote *Yesterday, Tomorrow—Somali Voices from the Diaspora*, and John Docker *1942 The Poetics of Diaspora*. The Continuum editor was an American who had followed her lover from Hong Kong to London. Charlotte's husband thought himself wealthy until a rich Russian paid in cash for the house next door on King's Road and installed an indoor swimming pool and gymnasium.

After teaching for seventeen years, 1986–2003, and at the age of seventy Adele retired from Ball State University and we would move to Paris where she had always wanted to live. She had an American pension and I even had a smaller one as a spouse and we both had pensions from a national teacher's association to which we had contributed. I was grateful to her for putting up with Ball State for so long; I would have resigned several times a year. I had, however, survived the years in Muncie because beyond my writing my life included Redkey, Indianapolis, Israel, India, the Caribbean, Prague, Stockholm, London and Paris. Even in America most of my friends were immigrants, while I remained emotionally a temporary visitor.

# Chapter 14: Unholy and Holy Lands

*Calgary Again, 1987*
*Beersheva, Israel 1988–89*
*India II 1989*

After my year in India I longed to return, but first I was for a ten week term during 1987 a visiting Killam scholar at the University of Calgary in Canada, followed by a semester as a visiting professor at Ben Gurion University of the Negev, Israel.

Calgary had changed since 1962 and was no longer the enlarged small town that tried to arrest washing machines at laundromats for working on Sunday. The University, now independent from and a rival to University of Alberta in Edmonton, was in a large building within walking distance to the small basement apartment I rented. The main floor of the house was taken for the year by a visiting American music professor who had lived overseas on army bases while doing his military service. Friendly, seemingly at ease when I first met him, he surprised me by complaining that Canadian tinned goods were bilingual and at times had names he could not recognize. His culture shock got worse, he had a breakdown and returned home to the USA. I could not believe this had happened. Maybe he was just lonely for his wife who he kept telephoning.

About once I month I would disappear for ten days to see Adele who was teaching French at Ball State University. It was a long trip through Seattle, Chicago, then a multi-stop local flight to Muncie. I would expect to be welcomed but was usually disappointed. Adele was busy with her classes and local problems. Then at Easter she came to Calgary and we saw old friends, including Joyce Doolittle, who was now a professor of Theatre, and her husband Quentin, a music professor. I went to concerts with Quentin, heard for the first time Mahler's great eccentric 9th symphony, and disagreed with him about a John Cage work based on a star chart which it took the performer several years to master. Quentin was skeptical, but I always found Cage's music good listening although I thought the ideas behind it foolish. Did the performers make the music interesting? There was a talent creating all these ideas that should not have worked in practice.

A partial film of the 1962 production of *Lysistrata* that Joyce had directed, Adele had starred in and I had joined, was an excuse for a party. Some people had changed little, indeed I had the same argument about the relationship of artistic form to society with the same person as I did during 1962, except that he was citing a more recent authority. Calgary, however, was now a city of skyscrapers and fancy restaurants and shops; women could even get a drink in the same bar as men, the rosewood and teak furniture we had sold a sociologist just before leaving for Ibadan was still being used. As I would do other times I wondered about the life I had fled and the comforts and security it gave. I was a rolling stone and had gathered no moss except experiences. Years later the same people who had stayed in one place advancing up the academic ladder towards retirement would listen my stories of experience and seem resentful that they had lost out on so much. If you live long enough you win even if, especially if, you made one bad decision after another.

No one was left from the English Department I had known except Ian Adam who had been both friend and whose scheduling me with six hours a week of the earliest classes had been the main reason I had resigned. Ian was instrumental in my temporary appointment. He was now a creative writer as everyone in Canada seemed to be, and a generalist in Commonwealth or postcolonial literatures. Everyone in department appeared to listen to him as the voice of wisdom. He spoke of the past almost with nostalgia, and had kept up with those who fled or left. He seemed interested in the personal lives of his colleagues and in postcolonial theory, two different entities. *ARIEL*, a journal started by Derry Jeffares at Leeds when Commonwealth or International English Literature, seemed a radical change in English studies, was now comfortably settled in Calgary although edited by those who seemed to think that literary essays were the past and scholarship needed post-colonial theory. Ian was raised on a farm and when I began kidding him about sex with animals he told me, no, that was considered perverted, but a turkey was okay. Was he pulling my leg?

I had met Victor Ramraj in Frankfurt during a conference. We shared an interest in V S Naipaul and I think he was surprised when I offered guest lectures on such Canadian authors at Mordechai Richler and the avant-garde dramatist George Walker. Through Ramraj I met Sam Selvon, also from Trinidad, who had left London and settled in Calgary as a writer attached to the English Department. One day Selvon, Victor and myself went to the races. It seemed like a comic story by Selvon who hid his choices, winnings, and losses from us although apparently betting on each race. We were supposed to think that he was winning but there was no evidence for it. Afterwards the three of us wandered through the streets of a lively section of Calgary where we saluted (to be polite) groups of young women, who correctly treated us as pathetic oldsters seeking a fling. I was reminded of Edward Horatio Jones, my African friend, insisting we try to chat up women around Piccadilly Circus in London during the 1950s, and wondered if Selvon was still repeating a custom of that post-war era. I felt we were in a scene from *The Lonely Londoners*. We ended in a sad bar, drinking too many sad beers, with a television set on the wall the only obvious life before Victor and Sam each returned home to his wife. The lonely Calgarians?

The main talk among male members of the English department Common Room was the possibilities of sex with female post-graduate students and female members of the department. The post-modern novelist and critic Aritha Van Herk attracted much speculation as she was shapely, good looking, and had a husband who was often away for long periods on research. She was also a good novelist with an international reputation unlike those who lusted after her. One of the research students came into my orbit as I needed transportation to movies, readings, and other cultural events and she was restless staying at home with a cheerful but seemingly maladroit bear of a husband and their two children. Even after she asked to see my room I once more felt I did not want complications, or maybe I afraid to go beyond a kiss, and I now marvel at my Puritanism or fears. She seemed actively to want to change her life and I did not want to be the one responsible. Karen, my young blond secretary from New Zealand, somehow turned up, less attractive than previously, now a mother of two, married to a man who from respect of her motherhood was unwilling to further touch her. They lived in Edmonton and I listened for hours on the telephone about her problems, but I was not going to be the one to solve them. Which reminds me of a story that Aritha, Ian, or someone at Calgary told me of an enclosed religious sect in Manitoba that to avoid the problems of inbreeding supposedly allowed one man a year for a night or two into the community as a stud.

Calgary now had a lively Asiatic scene with good restaurants, sea food and vegetable shops. Reading that the Chinese students were having a dinner for Chinese New Year I asked if I could join them. It was a great feast, a different seafood for each of the dozen or so courses, except for chicken feet in a three mushroom sauce which no one ate except me, the students being embarrassed by it. I enjoyed and was fascinated by the students. The males used their electronic devices to bet on the stock market against each other throughout the meal and were very competitive yet friendly in announcing results. The young women were flirtatious in an almost puritanical way, inviting yet seemingly sexually innocent. I felt an outsider observing the customs of another society yet we spoke the same language, dressed the same, and otherwise shared a similar North American culture.

II

While at Calgary I applied for a Fulbright Research Fellowship for a year in India and to Ben Gurion University in Israel that advertised senior positions in French and English. I went to Israel first to see what it was like and initially loved it although I was annoyed in being treated like a Jew returning to his homeland rather than part of an academic couple seeking employment at the same university. I accepted the routines of finding a rabbi who would recommend me and disclosing my Jewish parents as I was told that would help state support of my appointment, but after being in Israel I heard that financial support could be withdrawn after three years, that the position in French might be transformed into teaching English; none of which I suppose might put off someone with the desire to return to his spiritual home, but made Ben Gurion potentially unattractive, especially as Adele was reluctant to give up her American university position.

My first semester in Israel was lively. I was welcomed to family meals as everyone seemed to have married someone from or have a relative in Philadelphia. I had no idea what the Friday night prayers were about and was amused by the modern electronic devices used to avoid interfering with routine comforts on the Shabbat. I was interested in the heavily laden kumquat bushes in front of houses and that they were in three flavors, lemon, orange and lime.

Americans were regarded in Israel with confusing contradictions. Many Israelis had visited or studied in America and wanted to stay, others regarded it as a paradise of opportunities, good housing, high incomes, and peaceful neighbors; but American Jews were also resented for their wealth and for not leaving their comforts for a Jewish life in the Holy Land. The two countries were becoming surprisingly intertwined with Israelis scientists teaching part of each year at American universities, with America and Israel sharing military intelligence and with the two nations developing electronic and advanced weapons together, but their interests at times conflicted. Israelis had fought the British and Arabs for their country but increasingly depended on America for financial and military support; Israel felt under pressure to sacrifice its security to American foreign policy.

The religious monuments meant nothing to me, but the country was beautiful and I admired what the Israelis had achieved, especially when defending their country against Arab and Muslim attempts to murder them. I listened to broadcasts from Arab countries when I was in India and Israel; peace was absolutely impossible, the Arabs were even claiming that the Jews were not Jews, and, in a series I listened to twice, blamed all modern history as a Jewish conspiracy to create a Jewish state. Weren't the Muslims capable of using their immense oil wealth for anything except trying to destroy Israel and impose their various interpretations of their religion on each other?

Many on the political Left had become anti-Israeli but then many on the Left had supported Stalin, the Chinese Cultural Revolution, Castro's dictatorship, and other horrors, and I increasingly wondered whether I wanted to be part of a political movement that routinely confused every form of anti-Western authoritarianism with liberation. The later attempts by many on the Left to delegitimize Israel while remaining silent at the horrors of Muslim extremism, even silence at Putin's attempts to resurrect Stalinism, was to be expected. I joined the Ben Gurion students who watched a group of Communist Party Arab students protesting over some policy. Although Arabs and Muslims mostly had separate lives in Israel they sometimes met whether in the open air food markets or at an expensive Jerusalem hotel restaurant. And there were Arabs who were not orthodox Muslims and had their own customs, areas, and occupations, such as the Bedouins. Although there was no legal apartheid the communities were distinct. While Jews would go to Arab restaurants for more spicy foods, most Jews did not trust living alongside Arabs. A foolish American I knew had an affair with an Arab man and was told by his family to stop or die.

I stayed at the university dormitory and soon, although older, became part of a group of research students from Ireland, Germany, Portugal, Spain, and the United States with whom I would sometimes eat, attend concerts (mostly of the 19$^{th}$ century classical tradition—Bach, Mozart, Beethoven—that Israel had inherited from Germany), watch films (subtitled in too many languages), go to a discotheque, talk. They were mostly non-Jews who had come to Israel because of the strength of its science and computer faculties. They would eventually return to their own countries to teach at universities. From them I learned such basics of the modern world as how to use email to stay in touch with Adele, and watched doctoral candidates spend hours on-line with their dissertation supervisors in other countries. In contrast to the dirt and filth of much of the region, Israel seemed obsessed, even more than Americans, with hygiene. When I mentioned seeing a cockroach one night in the dorms I might have said an armed terrorist.

I was lonely and took up with an artist whose husband had died young and whose son was in the army. She was very skilled in bed, and I vaguely thought of her as a girlfriend, but we soon ran out of topics to talk about except her past, her former lovers, and the frustrations of her career. She bragged of bedding one of the younger British royals and I thought she was probably conned. She had exhibited in Paris and some of her work was in Israel's national art collection, but like many talented people the country was too small and too filled with other talented people for her to have the kind of success possible in New York or London. I told her about Adele and the relationship eventually became a friendship. When I returned to Israel a year later without Adele my artist friend suggested I come up to her studio to see her latest work, but she seemed less attractive to me than previously. A one night fling with an American poet, whom I had met in a Tel Aviv art gallery, the night I had asked Allen Ginsberg for Calcutta contacts, concluded with my packing my small bag and leaving her apartment in search of a taxi to take me back to the Ben Gurion University dormitory in the middle of night when after rapid unexciting sex she got out of bed to show me her poems and wanted to cook a frozen chicken with cauliflower. Persuading the night security guards at the University dormitory that I resided there was a more memorable experience.

Perhaps it had to do with the biblical concubines, perhaps it was the result of female high school graduates having to do military service where they were instructed about their bodies, sexual hygiene and birth control, perhaps it was the unsettled and continually changing immigrant population, perhaps as in much else the freedoms of the 1960s were rapidly assimilated and became the norm, but Israelis seemed to be non-stop doing it like rabbits are supposed to but seldom actually do. Friends simply accepted that their teenagers were having sex. Women would pick up men without hesitation. When I tried to teach Shakespeare's *Measure for Measure* I needed to explain the importance of virginity in Christianity and Western society in the past. My students saw no issue and claimed that the Torah taught human life comes before anything else. In a society with too many highly qualified people seeking the few good jobs (even a taxi driver might have five doctorates in five different languages), sexual attraction, advancement, and power were related. As family is central to Jewish culture (there are even biblical laws telling men how often they must copulate with their wives) the lack of fidelity among many Israeli couples was a problem and the government decided that paternity tests (which had showed that about one third of children had fathers not married to their mothers) could no longer be used in court as evidence for divorce. The Bible says increase and multiply.

Security was central to life in Israel. Before going into a supermarket there was an inspection. Even before going to Israel the Israeli Airline counter would be at a distance from other airlines and there would be heavy security. The airport outside Tel Aviv was carefully guarded and it was necessary not only to offer evidence of why you were entering Israel and what you were doing but on leaving the security was even stronger as they feared explosives being carried by naïve foreigners. The usual process is to have several guards watch from a distance while you are being questioned on your friends and contacts by a customs inspector. The first time this happened I broke out laughing at the absurdity of being a suspected pro-Arab terrorist. Why was I laughing? Another time when my taxi did not arrive at the dormitory and in the middle of the night I, knowing no Hebrew, had to telephone for a replacement. I arrived at the airport angry. Why was I angry? No wonder a Jew had invented psychiatry. Even Israeli high school students before they graduate are assessed psychologically as well as intellectually and physically as potential officer material by the army.

I regarded myself as a bird of passage at the university and never felt at home in a department heavily stocked with Linguists and their

American orientation. No one appeared to read the *London Review of Books* or the *Times Literary Supplement*, no one had a background of teaching in England and the Commonwealth. The one Oxford University graduate among the departmental staff seemed to be an outsider. I was then writing my book about V. S. Naipaul which included information about the uneasy relationship between Trinidadian blacks and Indians and how Naipaul had carried over this background into his reportage about decolonization elsewhere and liberal illusions. At Ben Gurion University Naipaul was someone who wrote for the *New York Review of Books*, not one of the major writers who had emerged from the last days of the British Empire and who had a conflicted relationship to England, English culture and India. I tried teaching Naipaul's amusing early short stories to my students who were as puzzled by him as the Alabama teachers in a night course I had offered.

Although Israeli universities, like much else in the country, originally followed a German model, they were now Americanized with classes ranging from the highest to surprisingly low standards. My seminar in modern poetry attracted brilliant students whose term papers could have been published. Russian immigrants who hardly could read English took my course in Jacobean drama. I might as well have been at some American campus giving a Shakespeare course as the main response was that I was marking too low and expecting too much. Perhaps I was. I had finished a small book about *Coriolanus* and its critics and was excited by such matters as Shakespeare's stagecraft and the psychological implications of the images that his characters used in their speeches. I expected students to share my interests and was disappointed that they only wanted to pass the examination. If some classes were applauded and others, I imagine, led to complaints, one brought me a close friend.

A student who was in the Shakespeare course came to discuss her low mark and mentioned that she wanted to go on to do a doctoral dissertation on a little known French Rumanian philosopher who was an early existentialist. She was amused that she has stumbled across the works of Benjamin Fondane while looking in a library for Jean de la Fontaine. I told her that it was not necessary to have an advanced degree to write literary criticism or scholarly articles and suggested that she start writing essays for publication. I met Dominique some years later in Paris while we were both waiting in line for a film and she told me that she had followed my advice and was indeed now a published scholar and part of a Fondane circle editing his works.

## III India 2 1988

We visited India three more times; as Fulbright Research Professor for six months, February to August 1988, when I started *Three Indian Poets* (1991), and for shorter stays the winters of 2005 and 2006. The six months in 1988 began as a totally different book project than it became. I continued to be fascinated by the relations between the Indian writers and the decades of Beats, hippies, and flower power drop outs who in discovering a relatively inexpensive place to have their scene also provided ways that Indians could publish and become known abroad. What was the history of this recent past, and how might it be made into a story? I asked for a Fulbright Research Fellowship for a year to research such a book and was disappointed only to be given six months.

Adele could not immediately come to India as she had to finish the semester at Ball State and was to give some classes and be interviewed in Israel. I flew into Delhi and began talking with writers about their time in the USA. For women writers that turned out to be the happiest part of their life when they could explore a new world before returning to marriage and the conventions of Indian society, a situation perhaps best symbolized by a female poet who said she had enjoyed reminiscing so much that she wanted to invite me over for a meal, but she had to find out first what night her husband would be away as she was expected to cook for him.

In Delhi I stayed with David Davidar in the extra room of his new apartment. David had moved on from editing a journal in Bombay to be in charge of Penguin India, a new locally funded company, established on Bengali money, associated with international Penguin. David with the advice of the famous writer Kushwant Singh was supposed to publish at the rate of a book a day, a seemingly impossible rhythm which he and Singh were somehow achieving. There was at the time no comparable international Indian trade publisher and there was a great reservoir of manuscripts in English and in other languages hoping to find print. The problem was to read and judge them, and David, besides handling the business side, was reading manuscripts every night and sending promising ones to Singh for his confirmation. That David had been part of the Bombay literary world and was friends with Moraes and others helped as he knew famous cartoonists who were ready to publish or write books, successful writers in such languages as Malayalam who could translate their own work, and journalists who felt they had a novel waiting to be written. David had himself been raised in Tamil Nadu, Kerala, and various schools in India, so he had an unusual familiarity with Indian literatures besides the writers he met through working with *Keynote*, *Gentleman*, and other Bombay publications. He also had formal training in the publishing business making him a clear choice for executive management as well as editor and production. Penguin had sent him to Harvard to study managing a publishing business.

David was such a success with Indian Penguin that Penguin International had its eyes on him. David told me about lunch with and the marriage of Pete Mayer who had headed Penguin in England (when he had published *Satanic Verses* and stood by Rushdie), and was now head of Penguin International from New York. I was in the same fraternity with Peter at Columbia College and we had travelled to England together on the boat when I met Adele; I had written an article of jazz and literature for *Lady Clare* that Peter had edited while I was at Leeds. David would be hired to head Penguin Canada with the idea that he would be moved on to CEO of Penguin International, an umbrella for Penguin in many former parts of the British Empire.

After some years in Canada a charge of sexual harassment brought David's association with Penguin to an end and he returned to India where he and Ravi Singh who had headed Penguin India and Pearson India formed Aleph, a new major publisher for up market books. Rupa Books, which invested in Aleph, would continue to publish mass market and Aleph would publish a small line devoted to literary excellence. Singh would resign when Aleph, supposedly under pressure from its investors, held off publishing books by the highly regarded American scholar Wendy Doniger that some Hindu nationalists tried to block, and announced publication of the poems of Narendra Modi, leader of the BJP and future Prime Minister, translated into English by a leading investor in the company. David was also a novelist, best known for the widely translated *The House of Blue Mangoes* based on the history of his family.

One night I was explaining to David my current project. The book would begin with the Bengali intellectuals who were studying at Harvard visiting Henry Miller in Big Sur California and listening respectfully to Miller's fantasies about India although he had never visited India (his "What India Means to Me" was published in a Bombay magazine January 1949). I would then move on to the Beats, hippies, and various American writers in India during the 1950s and 60s, their influence on Indian writers, and how the American magazines provided international publication for many Indian writers neglected at home. I wanted to bring in Ginsberg and Bengali writing, even the Hungryalists, and conclude with the more recent migration of Indian writers to the USA such as A K Ramanujan, Meena Alexander, and Agha Shahid Ali. David said it would be a hit, a new *Karma Cola*.

The trouble was that Gita Meta's *Karma Kola* (1979) was an amusing portrait of the mutual influence of America and India during the 1960s, whereas I had in mind a serious, straight faced, academic study. David was right, parts of my imagined book were comedy, other parts were dry academia, and I did not know, once it was pointed out to me, how I could reconcile the two tones in one book. I would keep working on the American influence on Indian writing, even going back to such nineteenth century authors as Mark Twain (who visited India), David Thoreau and the transcendentalists, to show continuing connections, but once the pieces were shown to fit badly I did not know how to put them together again. After a decade of further research, thought, and carrying photocopies from place to place I started discarding boxes that would no doubt be useful to others in their research but I now accepted I would never use.

I continued my research and interviewing and made some small changes in the 1989 reprint of *Modern Indian Poetry in English* including bringing the chronology up to date. The book would keep being reprinted every year or two. I added five new chapters—themselves a small book—for a "revised" second edition (2001) that continues to be republished. When I was first doing research on Indian poetry in English I was often told I was being absurd, I could not be serious, Indian poetry had to be written in a non-European language.

Oxford University Press unexpectedly rescued me from the immediate problem of how to write a book about the mutual relationship of Indian and American culture during the 1950s and 60s. Would I like to write a book on the Indian Novel in English to complement my book on the poets? When I replied no I was asked what I might write around 40,000 words as part of a new series of paperbacks. That was how *Three Indian Poets: Ezekiel, Moraes, and Ramanujan* was born. It too would go into a second, enlarged, edition when I added three chapters to bring it up to date.

Such a book would clear my conscience of sins I had committed in my first book on Indian poetry. I had left out Dom Moraes as he had not published a book of poetry for years and I did not know how I could fit him into the book I was writing. He was clearly one of the founders, along with Ezekiel, of modern Indian poetry in English, and I often referred to him while discussing the history but he seemed to have little to do with tendencies I was tracing. Now I would have a second chance, especially as Penguin had published a Dom Moraes *Collected Poems 1957–87* (1987). I could begin with his personal history that was so important as the subject matter of his verse, the colonial English speaking elite (including nationalist leaders) among whom he was raised, his years in England and other countries before returning to India, and treat him in his complexity as a founder who was also transitional and who went his own way and had his own problems. This was important to me as Dom had seen my book when it was published, noted the lack of any chapter about himself, even counted the references to himself in the index in comparison to those to Ezekiel, and was puzzled, even hurt. In justice I had to write about him at length, reading his poetry with more care and thought than I had in the past. Even now, many decades later, I think of Dom without having a clear vision as most of his best poetry was written in volumes after the publication of his *Collected Poems*, and as I am aware that his personal life could have with a slight different throw of life's dice led to his returning to England, America, even some other country, where he felt he would be better appreciated as a writer and a personality. But by the time I first met him he was already drinking too much and had been away from London, New York and Hong Kong too long for his renewal to have taken any place than in India, which it did.

I mostly neglected A K Ramanujan's poetry in my first book. His poetry appeared easy but was dense, difficult, often indirectly referring to Indian mythology, classical writings in Sanskrit, Tamil Naidu, and other ancient texts, while exemplifying Ezra Pound's advice to make it new. Pound, T S Eliot, Yeats, Frost, all the classics of 20th century English and French language modernism, had cross-bred with a rare knowledge of ancient Indian literature and culture. Ramanujan read everything, and yet the poems appeared to be about himself and his life, although as I got to know them better I could see that they were carrying on an argument with or within Hinduism. Luckily I had come to know Ramanujan, our lives kept crossing on flights from India, in Chicago, in Paris, in Jerusalem, and I started sending him drafts of what I was writing about his poetry. They would be returned with cryptic scribbles alluding to classical Indian texts, or polite corrections. We also often talked over the telephone. Ramanujan was lonely and a mixture of someone who needed to keep his distance and someone who needed company and attention. He once told me that he had succeeded at almost everything he tried (by which he meant his university career, his many prizes as a scholar, poet, translator, folklorist, theorist) but had failed at the most important, his marriage. He had been married to and divorced from the same woman twice. He had been offered professorships and visiting professorships at Oxford, Jerusalem, Michigan but kept returning to Chicago where his long time colleagues taught, and where his wife then lived, a source of comfort and pain.

I could not have written those chapters about Ramanujan without his aid. I did not have his immense knowledge, but who else had? No wonder so many Indian literary critics made fools of themselves claiming that he had stopped being an Indian poet when in actuality he was alluding to the classics of ancient Indian literature, using images and symbols from Indian spiritual texts, and addressing topics that had concerned Indian thought over the millennia, although the poems might be set in Chicago or appear to mock unthinking Hindu observances. Early in his life Ramanujan had studied Buddhism and that along with his reading in modern European thought colored his response to the themes of Indian philosophy. The seeming flatness and coldness of his poems was because, despite their personal subject matter, they were reflections about the nature of life; the poems and volumes are related to each other through titles, motifs, and recurring images. This would only become clear to me later; when I wrote *Modern Indian Poetry in English* I knew there was something that need study and explaining but I did not know what.

Although I had written a chapter on Nissim Ezekiel in *MIPE* I knew that there was more to be said. He seemed to me part of a Jewish intellectual tradition that values success and accomplishment in this world while being constantly aware of the moral and ethical dimensions. Nissim knew that fulfilling desire could not take place with clean hands, but he also desired clean hands. Much of his life was conflicted, and not having any belief he turned to a variety of methods of meditation and techniques he hoped would calm his anxieties. These ranged from experimenting with LSD to the various secular prayers he wrote, almost as parodies, in which he sought peace, but like Augustine of Hippo not quite yet. His verse technique also changed over the years from the early tight formal verse with which he is associated to a much looser, at times seeming off hand manner. Poetry came to mean less to him as an art or accomplishment than as a form of spiritual easing in which he gathered his thoughts and reflected upon them. Although he had become a leading figure on the Indian cultural scene he felt that he, unlike Ramanujan, was a minor voice internationally. When not at the centre of followers and admirers he seemed lonely, almost irritable by life being filled with incompetence, irrationality, lack of consciousness and conscience. Such feelings animated his poetry which I have come to regard more highly than in the past when he seemed someone who had shown aspiring poets how to write modern verse about India.

Before leaving New Delhi Adele and myself went to the American embassy as I wished to read files of one of their publications. I thought I should inform the cultural attaché of my whereabouts in case of requests for lectures, as two years previously just before I left some Indian universities had asked for me and I had to reply that it was much too late as the embassy needed months to arrange what they considered a properly sponsored tour with booked hotels and accompanied by an official. I should not have considered the prospect this time except I was envious of an American literary theorist who had been given luxurious support in Hyderabad when I was there and I thought here he is retelling a simplified version of "all culture is political" before an admiring Indian academic audience who seemed oblivious to the major cultural shift in our time—from the centrality of European and American literature to the new national literatures in European and other languages. It would be nice if Adele and myself had a few weeks of living well in India instead of in huts and student housing. This cultural attaché was not interested in my research or what the Indians wanted me to lecture about. He angrily told me that as I was being paid by the American government I should spend my time in India publicizing American culture. Increasingly American embassies seemed staffed by those more suitable for selling used cars in Prairie Oyster Creek Montana.

Whereas to enter the embassy I needed an appointment, had to sign in for a name tag, and was accompanied to my destination, when I left no one seemed to care as Adele and myself wandered about searching for an exit and found a back way that avoided the guards. Could we sell terrorists the back route?

Although Nicole was now in Paris I was aware of her commenting that we had not gone to Kashmir when she was in India. I knew Devindra Kohli from Leeds University and had met him one time when we were staying with Derry Jeffares. Devindra was then, I think, involved with some Australian student who suffered from poor eye sight and Jeanne had a foolish notion he might be her rescue. He was also writing a dissertation on Robert Graves under the supervision of Geoffrey Hill. Afterwards he had returned to India, married Vijaya, an Indian woman who taught English for the British Council, was teaching in some college in New Delhi where he had a home, and he had, foolishly I thought, turned down a Readership in the English Department of Pune University, and accepted a full professorship as Head of the English Department at the University of Kashmir, Srinagar. Although it offered little for my research I felt that Adele and myself needed a month seeing Devindra and environments.

At first we were put up in a post-graduate student apartment, a large room in a row of such rooms with a thin wall separating neighbors. Ours were noisy, argumentative, partying and we had no idea where to find food and other supplies, so we moved into the staff club intended for short term visitors where we had a room and would be fed several times a day and learned how to venture forth to see Srinigar and Kashmir, usually by employing a three wheeler to get us around the immense campus (part of which I understand has now been made into a "reserve"). Again it was like being back at the University of Ibadan again with British influenced suburban modern houses very unlike the native Kashmiri housing which was often wooden and irregularly piled up like some medieval street front, which seemed as archaic as the long haired smelly goats of their owners. There were also Muslim dwellings which could have been rooms in the Middle East, attractively decorated with metal engraved tea servers, beautiful flower vases impractically made of paper, and large photos of Anti-Western heroes that made you feel subversive even looking at them.

This was probably the last year to visit Kashmir and especially Srinigar as a tourist without fear. There were already police and army checkpoints, incidents involving gunfire, and tension building between the government and the Muslims, but Lake Dal was indeed one of the most beautiful places in the world, and travel on the many highly decorated boats that ferried passengers back and forth a delight. As the University was on the West side of the Dal we had to travel by the boats both ways when we left campus. We never stayed over night on one of the house boats, and we never flew by airplane to Ladakh nor even completed our taxi tour of the beautiful, dangerous, icy valleys in the mountains north of Srinigar as I started having headaches because of the high altitude and was given some pills by a doctor at a local hospital who told me to be careful. I was aware that I was often bad tempered and could suddenly feel aggressive. This, in spite of its beauty, was not place for me. The altitude was too high, the food uninteresting, mostly "Kashmiri" meant with some dried fruits and no spices. The Tibetan restaurants were even more boring.

It was not a good place for Devindra and Vijaya either. He was not local and not being a native meant that the university administration often treated him as an alien hired hand Whether it was a request for transportation or something needed by his department he was ignored, had to apply again, and generally was at the back of the line with any Kashmiri placed in front of him. I was both annoyed and saddened by his lack of power. I was not surprised when after a full insurgency developed and he and others fled, the university refused to recognize his contract, pay his salary, or later, his pension. It was claimed that he should have stayed on or returned from Germany where he and his wife struggled on yearly appointments until they realized that there would be no return to India.

I had seen versions of such violence and lawlessness while staying at the staff club. The university security guards were at times drunk and could be arbitrary with non-Kashmiris or concerning University regulations. When one of the Sanskrit epics was televised the guards and their friends decided that they wanted to see it on the staff club television set. Told they could not enter they brutally did, beating up anyone who opposed them, and they continued to watch episodes each week. The University administration refused to intervene—whether from fear or support I do not know. I just know that Devindra and Vijaya were out of their element and he had made a bad choice in not moving to Pune where Adele and I were to spend a pleasant month.

I gave a few classes to Devindra's graduate students of English and found them cheerful and welcoming unlike the usual aggressiveness of Srinigar. One female student, learning of my interest in music, invited me to a concert followed by a meal with her husband, a brilliant, internationally known, player of the santoor, which until recently had been considered a Kashmiri folk instrument used to accompany singers, and which had now been accepted into classical northern Indian music. The santoor, probably derived from the 72 strong Persian Santur, belongs to the musical family that includes the hammered dulcimer: in other words it is played with special mallets on its many strings which cover three octaves or more; some instruments have as many as a hundred strings.

The concert consisted of santoor accompanied by tabla. The musicians were seated on a white cloth, and Adele and myself were the only audience. It lasted about an hour and afterwards when the wife appeared she said that her husband did not like her to be present when he played; he claimed to be so inspired that he could not remember her name at such times. I talked to the drummer about a technique of sliding a note that I also knew from jazz and African drumming and that interested me in south Indian tension drums. He told me that it was thought vulgar by tabla players but audiences loved it. So do I.

At lunch the drummer departed. Brahmins do not eat with those whose hands touch animal skins. I thought by now I was used to the various, to me, oddities of Indian culture concerning eating and company. I knew that Muslim wives would hide in the kitchen and send out dishes, many Hindu wives would similarly disappear after serving the husband and his guests, and I had even been offered and had lunch in a traditional manner in which each person eats silently on his own without facing or talking to others. There seemed various social and spiritual rituals and hierarchies concerning food, but it never occurred to me that someone who had studied in St Louis and performed internationally with his accompanist would not eat with him. I was shocked. Years later, however, I realized that in jazz circles the leader is the leader and often does not socialize with the sidemen. Some big band orchestra leaders were thought tyrants. While there is a difference between establishing authority by distance and by caste hierarchy they are both ways of saying I am the one who pays you to accompany me.

I was supposed to be based at the University of Pune where we stayed for a month in a damp dingy ground floor room originally intended for I do not know what. There were visiting American Fulbright scholars who took rooms at an international hotel, but that was beyond our means. Our basement room would need do, although as it was often raining there were more uplifting enjoyable places.

We, however, often walked out, rain or not, to get a three wheeler to take us to the extremely enjoyable restaurants around Pune. The city was a gourmet's delight, especially for vegetarians. Long gourds, aubergines in many shapes, vegetables I had never before seen, were hollowed and stuffed with other vegetables and nuts in exotic sauces. Indian cooking is often regional and at its best unlike the repetitive tandoori chickens and curries, and rich northern butter and cream cuisine inherited from the Muslims. Why had no one told me about vegetarian food in Pune? And the charming restaurants with music?

And there were the ice-cream parlors with unheard of flavors, mixing spices, creams, nuts, saffron, cinnamon, into combinations I cannot remember but still recall with greed, an excuse for an afternoon in what was a beautiful clean city with large uncluttered streets and modern buildings, so unlike most of India. Even more than in Bombay, Calcutta and New Delhi I admired the way the British had left a tradition of good architecture fused with Indian styles. Many writers and artists from Bombay came here to live, it could be quicker to reach from the train station in central Bombay than many of that city's suburbs. It was also a city for music. In Calcutta there were expensive all night musical festivals with famous performers that lasted a week and through which most of the audience would sleep. Bombay had concerts in famous halls difficult to reach without an hour or more in heavy traffic. In Pune we went to an early morning concert of Bhajan, devotional music, and I felt that we had joined the select, those who discriminated.

We were especially friends with Dilip Chitre, poet, film-maker, painter, a friend of Arun Kolatkar, and one of the more significant modern Indian poets in both English and Marathi. Dilip had travelled around the world as a teacher, living many years in Ethiopia, had spent a year as a fellow at the Iowa International Writers Workshop, and was often central to what was happening in Indian arts. He had been in charge of the poetry archive at Bharat Bhavan, Bhopal, and arranged for a national poetry festival there; he co-edited with Kolatkar one of the first "little magazines" in Marathi; he edited *New Quest*; he made films; he had edited and translated parts of his *An Anthology of Marathi Poetry* (from which I had learned of the modernists who preceded Kolatkar in Marathi); and he was author of a great, semi-surreal, autobiographical sequences of poems, *An Ambulance Ride*, and of *Travelling in a Cage*, published by Clearing House which I had discussed in my book on Indian poetry.

Whereas Kolatkar participated in yet could hold himself aloof from Marathi culture with its local literary history and songs to the black god Tukaram, Dilip threw himself in head and heart together. He published books of translation of the seventeenth century Tukaram and of another, much earlier, bhakti poet. He was filled with energy, rage, could be obsessive, and I enjoyed his company. We adlibbed through four educational TV programs about "How to Read a Modern Indian Poem" instinctively understanding each other and what was of interest (ambiguities, syntax, near rhymes, assonances, syllabic count, sound and grammatical patterns) about the poems chosen at the last moment while a hapless young member of the University English Department protested in bewilderment that this was not how poetry was taught in India.

The English Department at the University was good and its students much better than in Kashmir. Its standards were high and the best graduates went abroad for further study. A problem was that some of the better members of the department appeared always on leave teaching elsewhere. The campus was attractive and clean, I liked the buildings, and I thought it an example of how the British had left a good educational system in their former empire that was rapidly being lost in switching to an American model of mass education.

Manohar Shetty had married and moved from Bombay to Goa where his wife was a journalist and he now edited the monthly *Goa Today* that lamented the hippies and other alien trash destroying local culture and traditions. Manohar was, Goan style, eating his curries with white bread instead of rice and through his wife friendly with those who governed. He was not, however, one of those nostalgic for a Portuguese Goa. He recommended to me novels lambasting Portuguese colonialism. His family was at first unwilling to accept his marriage to a Roman Catholic. It would take the birth of children before the estrangement ended.

We returned to the same house as two years previously, but there was a small white hotel not far from us with a miscellany of foreigners and westernized Indians. The attractive red headed woman was said to be a Scottish heiress recovering from drug addiction, but her Indian friend seemed unlikely to help. Perhaps I was jealous? As I was of the man, probably German, who claimed never to see a doctor and who cured himself by removing the offending area with a vicious looking knife he took from its leather holster. I lost my months of trying to give up cigarettes when at a party someone said "try this with liquid hash" and I thought how can it hurt, but a few days later I was once more chain smoking. On the road between Baga and Calengute there was now a fancy hotel with a swimming pool around which one could eat and drink and watch the manager trying to tandoori a large fish, his specialty that he would not entrust to the chef. We often ate fresh Goan shrimps and fish at a small house near the Calengute market. Was it only two years ago we had first come here, I drew water with a bucket from a well, and I worried about being bitten by dogs at night? The postman somehow found me and delivered a large registered package consisting of a contract from Ben Gurion University. It must have cost a hundred dollars to send and as it was totally in Hebrew I could not understand a word and ignored.

In Bombay we went with Nissim to a PEN meeting at which an Arab feminist lesbian writer was lambasting the West for racism, colonialism, patriarchy, and almost every thinkable evil. Then at question period after being asked by some Indians why she lived in London we learned that in her Muslim country she not only could not publish but she would be put to death as a lesbian and blasphemer. This did not modify her rant against the West. She seemed a perfect example of how reason and logic cannot change belief.

Menka Shivdasani was a founder of the weekly Poetry Circle. She was young, small, almost girlish, although physically confident. As we walked she told me that she had travelled by herself in Europe, worked as a journalist in Hong Kong. She seemed a new generation of Indian women. Then she led me to a dimly lit coffee bar where once we were seated she handed me some typed poems, the first began (I am quoting from memory) "We are seated in a dimly lit coffee bar/ and he is looking at my large breasts." I looked up; how had I not noticed? Afterwards I could not keep my eyes off them. Returning to the hotel I told Adele who thought it amusing.

Adele had visited Israel, was interviewed, taught some classes, been feted and decided that she would like to live there even though her visa would be stamped non-Jew. She, however, had to teach at Ball State for another semester as she had come to India on an unpaid leave. I said I would prepare the way, but returning to Ben Gurion after six months I immediately felt a chill. Although Adele had published three books and had another one ready for publication she was unlikely to be given a tenured Associate Professorship as an examiner found her books too short. I was not being courted as I was previously, indeed I was being broken from a leader into a follower. I would be expected to participate in a weekly departmental study group in theory and read the books of Jacques Derrida (which I had tried before and thought unreadable nonsense inflating perceptions any child knew), and so on. The department no longer needed a high profiled professor to offer advanced degrees. I felt demoted. I am also resistant to group pressure. I asked for my Professorship to be made into a Visiting Professorship and told Adele to take back her resignation from Ball State. I would return and work on my writing and editing next semester.

We spent summer 1989 in a one room studio owned by a friend in Paris. The same night that I was mugged, knifed and taken to a hospital emergency room, Adele was robbed of her purse and keys to our apartment. Locked out she spent the night listening to the telephone ringing as I attempted to tell her where I was. Late morning Nicole came to the hospital to bring me home.

I was depressed about how the prospect of our both teaching in Israel had ended and I almost did not go to the August 1989 Commonwealth Literature conference at the University of Canterbury, Kent, England. That would have been a mistake as for over a decade my life would be shaped and I would be working on books that had their origins there.

# Chapter 15: An influential Conference

*Paris, France 1990–91*
*Walcott 1987–91*

At a conference I first became aware that some younger scholars and critics were regarding me as, I overheard, "famous", which was ironic as I faced another academic year without a position. The conference was a celebration of the twenty-fifth anniversary of the Association of Commonwealth Literature and Language Studies existence and there was a book of essays and memoirs dedicated to Derry Jeffares who had founded the Association with a conference at Leeds University when I was teaching in Ibadan. I contributed the autobiographical "How With the Help of Derry Jeffares, I (An American) Became a Commonwealth Literature Specialist" or some such absurd longwinded title—a fashion at that time. My piece was shortened by one of the editors and now read in places like a jerky silent film played at the wrong speed.

I had become one of those who had founded what became a major new area of literary studies and someone insisted on taking my photograph with the West Indian poet John Figuerola and saying it would appear in a book for the next celebration. Very ironic as I had not attended the first, founding, conference, preferring to spend that summer in Deia, and while I had come to know many at Kent from Africa or India—the English poet John Haynes, Jayanta Mahapatra—I was indeed an American interested in what to me were new national literatures and tangential to those, often from England and the white Dominions, promoting the notion of Commonwealth Literature. I had no institutional support, no money from various Commonwealth funds, and my own connection to all that was happening in the decolonization of literary studies began in seeking the sources of jazz in Africa, interests in African music and art, and having studied for my doctorate at Leeds University when Derry Jeffares was Head of Department and placing academic job seekers in the new universities of Africa and the Commonwealth.

I had, however, for a long time been interested in Africa and this led to the Caribbean and then to India, and I had taught at universities in Nigeria, England, Scotland, Canada, and New Zealand, and increasingly regarded the Commonwealth along with France and Mallorca as my home, but I still had American citizenship, an American passport, filed American tax reports, and did not have the same relationship to the Commonwealth as someone from, say Canada or Australia, who could expect to go to conferences on Commonwealth money while being a cultural nationalist. To me such academics and some writers they favored seemed like the former clerks of the empire, the educated native elite who helped the British rule their colonies while claiming their own right to inherit power through representing the local culture. Although I disagreed with his solutions I agreed with Franz Fanon's insights into the psychology of colonialism and its effects on those who proclaimed independence. I remained skeptical of politics and those proclaiming that everything essentially was political. Literature, music, art, and dance had brought me to the new nations and they remained my interests. Despite my involvement and increasing presence I remained an outsider, neither one of the few Americans who were teaching African literature at an American university, nor someone from England or the Commonwealth likely to benefit from the network that had formed through ACLALS. I was an expatriate American who avoided Americans.

That I was questioning my place in Commonwealth studies was partly the result of my being jobless, but also because I was ready to move on to another stage in my life. This was the time of the *Satanic Verses* affair and it seemed natural to Adele and myself that the Conference should pass a resolution supporting Rushdie and right of artists to represent what they wanted. We were discussing this over drinks when I think it was David Dabydeen who said that such a resolution might further enflame anti-Muslim immigrant sentiments in

England. After a discussion that included Fred D'Aguiar, Caryl Phillips, Ben Okri, and others, David agreed to introduce a resolution supporting Rushdie and I became aware that here was a new generation of writers who were basically British and not the immigrants from the colonies and expatriates from the new nations that had written or been the subject of fiction by Samuel Selvon or V S Naipaul. I had so far written about those of my generation, Achebe, Soyinka, Walcott. Could Dabydeen and D'Aguiar still be thought West Indian rather than British? West Indians in England already had their own long history. My revision of *West Indian Literature* (1979, enlarged 1995), a book of essays I had thought up originally while at Ahmadu Bello University, and was now being asked to bring up to date, would need accept that it could not accommodate this development.

More important I would start reading, reviewing, and writing at length about these British authors and their social background. Where did they come from, what were they saying, how long has this been going on? I had not known at the time that Alastair Niven, in whatever official position he was then filling, had decided to support the inclusion to these writers in the Commonwealth Literature Conference to see what happened and I had risen to the bait.

## II    Walcott and Paris 1990–91

Derek Walcott had arrived at ACLALS conference in what I was later to regard as characteristically Derek: he flew into Heathrow and took a taxi directly to Canterbury expecting the conference organizers to pay the fare. Unlike the other writers he kept his distance from the conference participants and did not attend the conference dinner possibly because he was expected to pay. There were the usual claims later by women that Derek had propositioned them, which may or may not have been true as this might be a way to share in his limelight, or a woman not approached by Derek might feel diminished. I was more interested in an exchange between him and a young Marxist critic I knew who foolishly asked Derek why he had not protested against American involvement in Nicaragua and received what sounded like a distinctly religious reply about a poet's ultimate responsibility being to God. Although I had written essays on Walcott and edited a book on West Indian literature I suddenly felt I knew little about him and the West Indies. As long ago as 1974 Michel Fabre had published one of the best critical introductions to Walcott's poetry and his ability to make himself representative of the Caribbean. I was influenced by Michel as I was dependent on Robert Hamner's 1981 book. Suddenly, however, Walcott seemed more complicated and driven by different obsessions than racial politics and decolonization. Maybe I should immerse myself into the West Indies as I had earlier in Nigeria and Africa? Reinhard Sander had told me when I was editing *West Indian Literature* that my "Introduction" appeared, unlike the *Introduction to Nigerian Literature*, that of an outsider. He was right.

The next day at the University bookshop Walcott was looking at copies of the Modern Dramatists that Adele and I edited and asked me who I had in mind when I wrote that the series besides influential Nineteenth and Twentieth Century European and American dramatists would include Nigeria, obviously Soyinka, and Trinidad. I told him that I had without success sought someone who could write about his plays in performance. Walcott looked into his calendar and told me to see him in Boston during, I think it was, October when he would be free from travelling to readings and conferences. I did not know that Walcott had for years been wanting someone to write a book about his Trinidad Theatre Workshop. Although I had read his plays I had not seen any in performance, and knew little about the TTW and his long commitment to creating a significant theatre company in the West Indies.

Back in Muncie, where Adele was teaching French, I began looking into the Trinidad Theatre Workshop and Walcott's earlier involvement in the St Lucia Arts Guild. Nicole and I had seen his *Pantomime* in New York. I arranged to give a lecture at Amherst College to pay my way to Boston. Derek was teaching Poetry and Theatre in the graduate creative writing program at Boston University and had a large apartment within walking distance of the University. I later learned that his third wife, Norlene, legally was co-owner of the apartment and Derek would not divorce her until awarded the Nobel Prize in (1992) as he could not afford to buy her share. Although Derek seemed to earn more than poets do he also had many recurring expenses—his previous wives, his children, travel, and even his theatre productions sometimes cost him more than he earned. He also had an unusual life style, insisting on flying first class, taking taxis to wherever he could not walk (he never learned to drive an automobile), staying in luxury hotels. He was not someone who enjoyed luxury but it was part of his sense of his own worth. Much of Derek's life had been driven by his sense of uniqueness and talent, a highly gifted modern writer in English, from a small British colony, who was neither a "black" nor a "white", and who felt he should be paid what he was worth.

Our first meeting did not go well. I did not know Derek had a history of seeing what American universities would pay him and what he could get away with. I later heard stories of his being flown to readings, keeping his distance, and staying in his hotel before and after. I knew little about him beyond his writings and he knew nothing of me except that I had reviewed some of his books. He assumed I was an American university professor and he wanted to see what I wanted and could offer. I was puzzled when he spoke little about himself and I felt I was wasting my time. He would ask me how much Amherst was paying, was my fare being paid, and what Ball State might pay for a reading. I did not know that Derek would often size up someone this way. When I said I had to go to Amherst he asked when I was returning; I said I had not been invited and it seemed pointless. Suddenly there was a different Walcott. I must come back and stay with him and we would talk more.

And it was different. There was a meal of rewarmed Chinese food. He mentioned his girlfriend, Sigrid, who lived in Pittsburg. She was soon to become his constant companion. I lay on a couch in the front room reading my friend Michael Dash's recent translation of Edouard

Gissant's *Le Discourse Antilles* while Derek was entertaining his two daughters, both of whom were attending Boston University. I slept in the guest bedroom. Derek made some starchy breakfast and painted for a few hours, then tried to tell me about the Trinidad Theatre Workshop. He seemed surprised that I did not have a tape recorder, and perhaps that might have helped as he was lost in a flow of disconnected memories, apologetics, fantasies, names, plays, events, productions, as if he had been transported to another world, closer to the ungoverned associations of creating poetry than lineal, factual, historical prose narrative. At the end I felt I knew little but I had a string of names and plays to research. Derek obviously wanted to tell the story, and I would later find in the Trinidad Theatre Workshop programs he had tried to tell it and it was the backdrop for his great essay "What the Thunder Says" prefacing *Dream on Monkey Mountain*, but he was a great poet, whose mind worked by metaphors and associations, not a literal minded scholar.

 I suddenly had something I wanted to do but how could I afford it? The National Endowment for the Humanities had research grants that covered more than one year but they required several scholars co-operating and when I asked for the forms I was told it was too late as it was always necessary to have the staff examine and suggest revisions to the application. What else could I do but try? Not knowing what I was doing helped, I could be vague as hell. For the other scholars I needed to work with I contacted friends in West Indian and Postcolonial literatures and told them that I was using their names and if I was awarded a fellowship they would get a thousand dollars each for doing little. Meantime I carried on editing, writing, and trying to learn what I could about Derek, productions of his plays, and West Indian theatre. No wonder I could not find anyone to write about Derek's plays in performance, I would need start from scratch. Of course, one never starts from scratch, there were newspaper reviews, memories by actors, Derek's sketches and water colors for productions, a few photographs, articles by Roderick Walcott in St Lucian newspapers, Judy Stone—an actor, producer, and theatre critic, in Trinidad, had a scrap book of clippings she would lend me to photocopy and she would eventually write her own book about West Indian drama, but I would not know any of this until I went to Trinidad, St Lucia, Barbados, and Jamaica, seeking material.

During 1977–78 I had suddenly changed from unemployed professor to being offered both a Rockefeller Humanities Fellowship for the year to write my *New English Literatures* and being invited to teach at the University of Paris 3 for a year. Something similar happened again. Besides being awarded a two year National Endowment Fellowship to research and write a book about Derek Walcott's plays and their West Indian productions, I was unexpectedly appointed a visiting professor of English at the English Institute ("Charles Cinq") of the University of Paris 7 for 1990–91. I did not know how this had happened and it took me several years before I learned. I had written to Michel Fabre that I would like to return to teach in Paris and sent him my CV that he had passed on to his wife who led the American Literature section at another university. She silently inserted it among applications for a year's visiting appointments in British Literature and that committee without having any idea of how my papers were in the pile, or who I was, selected me. No I could not put it off for a year. I was told that there was no guaranty that I would be appointed the year after. I explained the situation to the National Endowment, told them that I wanted to take up my Fellowship in June when I would go to the Caribbean for four months (teaching did not start in Paris until October) and asked that the budget I submitted be spaced out over an additional six months, to finance my work after I returned from France.

Adele and I went directly to Port of Spain in June and rented an apartment owned by Trinidadians in Canada that Margaret Walcott (Derek's second wife) had found for us and I began interviewing (now with a cassette recorder) members of the Trinidad Theatre Workshop, searching libraries for materials such as announcements, interviews, reviews, working through the files of the *Trinidad Guardian*, seeing what was in the University of the West Indies, St Augustine, library. As with *Modern Indian Poetry in English*, first I would need to create a chronology, but here, besides productions, I would need establish cast lists, changes in casts, tours, places of performance, and such matters as lighting, acting styles, scenery, other and earlier theatre companies, and soon I understood that I had to trace the history of West Indian theatre and gain some idea of the cultural and social background. It was clear that Derek had attempted to create a world class theatre company in Trinidad that he had hoped to take to New York, London and other cultural centers. The ambition was enormous, there was even a dance ensemble within the company. It all seemed crazily improbable, how had this come about and how had it been financed? This was not a one year research project followed by a year of writing what I had found. I had stumbled into something which would take me five years and lead to spending another five years trying to understand how the Trinidad Theatre Workshop and Derek's ambitions fit into his life and the world in which he was raised. Rather than the broad untestable generalizations which were then becoming mandatory to English studies I wanted to know details and how they might fit together. Who were these actors and dancers, where did they come from, what motivated them, what were St Lucia, Jamaica and Trinidad like when Derek started writing plays for performance, who were his patrons? I had to research, learn about and understand the major cultural centers of the West Indies along with the Trinidad Theatre Workshop and Derek's plays.

Many people were helpful. Carol La Chappelle explained that she had studied classical ballet in London; Noble Douglas had a more African style centered on the hips and pelvis rather then Carol's European verticality that depended more on the vertebrae. As usual any information seemed to lead me back further. The Trinidad Theatre Workshop began as part of Beryl McBurnie 's Little Carib dance company. McBurnie was one of those who had sought the African roots of Caribbean dance and studied with Martha Graham, worked with Katherine Dunham and taught Pearl Primus. She was one of founders of modern dance, especially of the "primitive" modern dance that developed during the 1940s and 50s. So from the first in Trinidad Derek had dancing in his plays, and he inhabited the cultural and social world of the founders of modern West Indian high culture. The leading Trinidad Theatre Workshop actors, such as Errol Jones and Albert Laveau, were in acting companies before Derek invited them to attend weekly workshops. As was obvious by Carol and Noble, Derek's world was not only rediscovering its African roots—indeed that had already become the obsession of an earlier generation—it was trying to make use of its British and American heritage and trying to find its place in the modern world. I had seen something similar among Nigerians and Indians, but the specifics were different in each country as would be what was produced. Derek and Wole would see themselves as leaders of a new generation opposing the rapidly developing tyrannies of the previous decades of nationalists, but the societies they wrote about would obviously be different, especially in relationship to Africa.

I would feel part of Port of Spain that summer and when we returned again next summer after Paris. We attended evenings when the theatre community presents awards for the best plays and performances of the years, we went to 14th of July parties given by the French communities, I spent many evenings going through Judy Stone's note books, the great composer Andre Tanker, who wrote music for many of Derek's plays, took us to clubs to hear local bands, listen to the quattro (an older four stringed instrument favored on the islands), and explained the difference between Soca (more modern, mainly for dancing) and Calypso (older, includes verbal satire). I remember being driven the wrong way around a large city circle while the owner of the automobile drank rum and coca-cola. We danced at Calypso clubs where the husband of one of the actresses offered me at his expense my pick of the prostitutes (that Adele was with me did not seem significant to him), another actress stroked my thigh while confiding that since she and her husband separated she only masturbated, I bought a necktie directly off the neck of a bartender who wanted to know whether he should merely loosen it or could I tie a knot?

## 356   An influential Conference

July 27, 1990, disaster. I was interviewing for the second time Norlene Metivier, who was still legally Derek's third wife although they had separated and she had a child by a different man. Norlene mentioned that her boyfriend was jealous of me and so Adele came along. As Norlene drove us to her apartment on the outskirts of Port of Spain we were told at a stop that there had been a major explosion at the police station as part of a local black Muslim attempted coup d'état. At her apartment it soon was clear that there was a state of emergency, parts of Port of Spain were in flames including the lower floors of the television station, and part of the army had disappeared, while the police were fighting the rebels and looters. At first it seemed that looting and arson would be the main problems, but there was another development, the rebels had taken over the Red House trussed up the Prime Minister and his Cabinet with explosives and on a 6 PM television broadcast declared themselves the government. No one knew who was in charge and how to resolve the situation. Wole Soyinka arrived to give a lecture celebrating the end of slavery and had to whisked into hiding until he could be flown out.

We watched from the terrace of Norlene's apartment as smoke rose from Port of Spain, watched those in revolutionary army clothing declare on TV the need to burn down everything, watched TV stations presumably forced to play video's of Ali Mazrui's pro-Islamic documentary about Africa, and a calypso singer known as Lenin who sung of revolution and fire, and listened to someone telephone a radio station that he was tired of all the slow martial music and wanted something lively to which he and his mates could "jump". At an apartment block near to us the Indian families had come together, circled the house with their automobiles, and set up a system of armed guards strung out along the roads that were likely to be used by looters or enemies. Our apartment block had no central command but we filled bottles and other containers with whatever might be used against a raging crowd. There were the usual absurdities. Norlene would swim in the pool on the roof of the apartment house, wander around in her bathing suit while saying that she really should wear something less revealing around me. One day we risked going out to find food and we brought back the best, sweetest, most flavorful melon I have ever tasted. Norlene decided to puree half of it for her son, who of course, rejected it while I felt like stealing from a baby.

Then one day, with nothing yet resolved, it was announced that the curfew would be lifted for a couple of hours. We telephoned a taxi driver we trusted and asked him to take us to our apartment and wait while we would rapidly pack, then take us to the airport as we had heard that some flights were erratically departing. At our house there was fighting going on between police and rebels. The police were shooting into the trees behind the house and we hoped that we would not be struck by a stray bullet as we shoved everything into suitcases. Margaret Walcott had amazingly turned up to make certain we paid our rent. At the airport we searched for a fight and were told we might get on one to someplace the next day, but we would be limited to one small bag each. We found a room for the night and I trimmed all my photocopies to make them as small as possible to fit into a box which would need be sent separately from us as the American Embassy had declined to help; we were individuals not "official". The next day we managed to get a flight to Barbados and entrusted our bags and my research to someone who said he would try to send them on to St Lucia where we should contact the airlines to see who might have them. Shortly afterwards the attempted revolution collapsed when the Prime Minister agreed to amnesty the militants if they released their hostages. Had it only been six days?

The normality of Barbados seemed abnormal. We swam, ate good food, and I rapidly found what I could about Derek at the University library. I had earlier met Michael Gilkes who directed Derek's plays in Barbados, and who helped find materials. We would return to Barbados several times and I can no longer distinguish between the times. It was more likely next summer when I sought Walcott and Trinidad Theatre Workshop materials in Barbados, or on a return trip late August. More likely we booked the first connection we could to

Castries. We had lived through the Western Region Crisis and the attempt by Biafra to succeed in Nigeria, the assassination of Indra Gandhi in India, I had been knifed and mugged in Paris, but the arson and looting in Port of Spain was what I still remember.

Nicole and her boyfriend had been waiting for us in St Lucia for several days and she asked (confusing French for English) why we were "retarded". She had not followed international news. Several days later the bags with our clothing and a box with my photocopies arrived at the offices of a small West Indian airline in Castries.

We were staying at Villa Beach Cottages with the understanding that we would need move from the one air-conditioned cottage when Derek and Sigrid arrived. Derek always stayed in the same cottage and stored his easel, brushes, paints and paintings in a utility room. Nicole and her boyfriend occupied another of the handful of hut-like buildings. It was an ideal location, a small beach for swimming, a few coconut palm trees, a wild area next door, a Chinese restaurant within walking distance, and on the main road to Castries and before that some shops, a supermarket, and a place where fishing boats docked and it was possible to buy fresh fish before they were frozen and sent to Japan. The St Lucian national archives that I would use were also on an off-road.

Nicole was probably the only recent Bryn Mawr graduate who had a professional truck driver as a boyfriend. He was one of a group she was part of when she first moved to Paris and after she left the first lover he became the second. He read little but would wander off to beach hotels to hang-glide and other to us exotic sports. He was also an excellent driver for whom we rented a car in which he took us south to see Soufriere and other parts of the island that still had reminders that well into the 19th Century St Lucia was a French colony. We went dancing to local bands at Gros Islet and smaller places and ate in a restaurant whose owner illegally took his small boat to Martinique and back every day for supplies. Although you could clearly see Martinique from St Lucia the two islands were incongruously separated by politics and official languages; they shared a similar patois and listened to each other's music on the radio, but the island of St Lucia was an independent nation, officially spoke English, and followed British law, whereas Martinique was a part of France and required a French visa to enter. Castries seemed disorderly, lively, and suffered from the lack of planning I was familiar with from Nigeria, Ghana and much of India; Martinique appeared interchangeable with the Ivory Coast, the Riviera, and probably every other French and former French possession. A good place to live if you liked the provinces and required the same goods, books, and nightclub acts you could find in Paris or Nice. Did the mosquitoes also speak excellent French?

We left Nicole and her friend in St Lucia a few days before Derek and Sigrid arrived. We went to Jamaica where Adele and I stayed with Michael and Cheryl Dash while I researched tours by the Trinidad Theatre Workshop in Jamaica. Michael drove us to swim at Ocho Rios and Hellshire beach (in St Catherine Kingston), and I discovered how beautiful the island was. One night we went to a theatre to see a Jamaican play by Trevore Rhone, whom I interviewed, that seemed more professionally performed than the shows I had witnessed in Port of Spain but also more British in acting styles and body language. Derek was on to something Caribbean although American critics who had seen TTW called it a "folk" theatre.

The Mona campus of the University of the West Indies was like being back in Ibadan with a staff club, pool, tennis courts, and a multiracial staff. We were invited to dinner by Ralph Thompson, a businessman friend of Derek's, a poet and painter, who had been principle sponsor of two TTW tours to Jamaica in 1973 and 1975. We met the famous novelist John Hearne and saw the poets Eddie Baugh and Mervyn Morris who taught at UWI-Mona. Eddie and Mervyn had contributed essays to my West Indian Literature. As we were in Jamaica several times the details fuse in my mind.

## III  Paris 1990–91

Nicole had found me an apartment near the Place de Republique in Paris within walking distance to the English Institute at Charles V (named after its address), but no one in the Department seemed to know why I was there or what I would be teaching. Each time I knocked at a door I was met by puzzlement and told that maybe someone else knew. I think Geneviève Fabre, who was a leader of the American literature and culture section, was away at the time, but in any case she was long reluctant to admit that she had slipped my CV into the files of those being considered for a visiting professorship in the British literature and culture section. Anne Theieulle who had recognized my name among those appointed was away recovering from a broken leg. Except for my letter of appointment I really was on my own.

Eventually I found that I belonged to English not American literature (which at first it was assumed was where I belonged) and arranged to teach an undergraduate course in Commonwealth Literature twice a week, using an anthology edited by Anna Rutherford, and offer a weekly seminar in African and West Indian literature at the Masters level. Second semester I taught Caribbean literature to the undergraduates. Except for Anne when she returned I was left on my own by the Department and could have been on Mars. Everyone in the know arranged for their classes early in the week (there was a departmental meeting on Mondays); as Nicole was free from work on Mondays I requested my courses be scheduled Weds-Friday, unknowingly losing any chance of getting to meet colleagues.

I surprised everyone including myself by being a popular lecturer, students who had problems following lectures in English rapidly changed to other courses, those who wanted to learn about something different took mine. I not only attracted the brightest, those who could follow lectures in English, but also those from non-bourgeois backgrounds who normally felt out of place when hearing about the glories of English gardens, castles, tea parties. I talked about Nigeria,

Ghana, India, Trinidad, racial conflict, cultural conflict, decolonization, the disillusionments that followed national independence, and I expected my students to read the texts, be interested in them, and willing to offer their views as part of a class. Normally I would lecture for an hour, perhaps ninety minutes, and then ask students who were warned weeks in advance to take over for the next hour. Some were brilliant, some awful, but we got to know each other and often talked before and after classes. Some confessed that they were only enrolled at university for the various government paid privileges, such as cheaper movie tickets, student hostels and meals, inexpensive travel on trains. They were the ones that all the other lecturers thought unteachable, but they came to my classes and listened about the world outside of France. Most were from former French colonies.

The post-graduate seminar on African and Caribbean literature was less satisfactory. Few signed up and I never understood what they were doing in my seminar as they kept coming and going. Someone preparing to take up a government appointment in Africa attended for a semester. Michael was offering a doctoral seminar on V S Naipaul's *A House for Mr Biswas* at Paris 3, French style, explicating line by line. One of his students followed my seminar for a while. A few stayed the course (not intended as a pun) and wrote long papers at the end, but no one was outstanding.

Nicole had lived in Paris now for about eighteen months since her graduation. She had worked at a few art galleries where she was poorly treated and had to take one to court to be paid. She was now an Assistant at an important gallery owned by the Nahons, dealers in contemporary art, where she mostly welcomed people and introduced them to others who handled sales. She also helped hang and arrange painting for exhibitions and other tasks. It was permanent employment and she had many friends both from the gallery but also from a period she was a second language teacher. I became dependent on her, her French was native, she knew her way around French bureaucracies. She had French skepticism, could be rude and sarcastic, had learned to persist. When my appointment seemed to be endlessly stuck on some desk at some ministry she found out where and harangued the woman who kept claiming that I never sent in a form, that it had been lost, that I should send it again, until it was found sitting there on the desk among a pile of other forms that would never have been stamped and sent on to the next desk without Nicole's energy. As it was I did not get paid until January and continued to be paid for the next twelve months even receiving an increase in salary after the next summer so that according to French pension records, as I was to discover decades later, I had been employed for two consecutive academic years.

I had gone with Nicole to dinner where she told me that she had been asked to research some painters for articles and was thinking of breaking up with her boyfriend as he lacked ambition. On the way home we agreed to meet the next evening after her work. The next morning, November 19th. I received a telephone call urgently telling me not to go to the university but come to the gallery. On television that morning there had been news of a fire at the building where Nicole lived. I and one of her colleagues would need search the hospitals for her, if she was still alive. After a few futile days it was obvious that she and her boyfriend were not among the survivors. I telephoned Adele, made certain she had a friend she could stay with for a few days, and told her that Nicole had died in a fire. I would remain in Paris and she should stay in Muncie until Nicole's body had been found, there was an autopsy, and plans for burial. I also telephoned my mother in New Jersey who lamented that she was always "unlucky". I replied that no it was Nicole who was "unlucky" and I afterwards hardly ever spoke to my mother.

I made up my mind to keep giving my classes, although I almost blubbered my way, shaking and tearing, through the first lecture, and wandered aimlessly around what seemed an endlessly gray rainy Paris. I had several weeks of modern dance tickets for Nicole and myself. I never seemed to have the clarity to give them away to her friends, they went to waste. Adele came over for the funeral attended by hundreds of people I did not know and afterwards some of Nicole's friends arranged a party where I met others. In coming months I would be dependent on them. They would invite me to dinners, parties, shows. I was especially helped by the company of a young couple, she was American, he French, who often had me over for a night with their friends. At some point I realized that she also was lonely as her job, unlike his, did not take most of her time, and she wanted someone to talk to while shopping, waiting for her husband at a restaurant, or discussing what films and plays to see. Another of Nicole's friends, from Bryn Mawr, had moved to Paris and wanted someone with whom to have lunch, or tell about her adventures.

Tragedy is also a test. Nicole's death brought Adele and myself closer together.

I had decided to finish teaching the academic year but felt that I no longer wanted to be a wandering scholar-husband. Our marriage was more important than my career. I would in future stay in Muncie with Adele and find a way to support myself while writing academic books. I was supposed to return to USA at Christmas as I had arranged a Forum at the Modern Language Association in Chicago. I asked a panelist to take over and Adele would come to Paris for the Christmas holidays. We would over the next months criss-cross the Atlantic.

How had Nicole died? Adele thought the topic too painful and wanted to drop it. I was angry and wanted to know. This was more difficult than I imagined. The police would not investigate, it was necessary to have a lawyer ask a judge who had discretion about whether or not to take matters further. We could not address the judge, only a lawyer could and in the form of questions. I was impatient and kept changing lawyers. The third one managed to get a judge's attention asking the police to investigate. This turned out to be still more frustrating. Six people had died in the fire, but others had managed to climb the staircase and escape from the top floor. The front door to the building had been unlocked for several days and someone had started a fire at the bottom of the hallway stairs during night and the wind had carried the flames to the butane used on the (French) first floor by an Algerian and his girlfriend. This caused an explosion and the flame travelled upwards so those on the upper floors managed to escape but Nicole and her friend had died from the smoke before being charred.

The neighborhood where Nicole lived was rapidly going up market, and she and the other students and immigrants paid rents far under market value. The obvious explanation is that the landlord hired someone to set the fire planning to get rid of his tenants who under French law could not be moved unless there was a major renovation, but the fire went out of control because of the butane. In America or England the police would probably begin by seeking clues about the arsonist and then who hired him. The French police wanted to know whether it could be a crime of passion, might the landlord have enemies, his bank records showed no debts and so uselessly on. The fire might have been started by a drunk or drugged vagabond. I later learned that I would have done better to have hired a criminal lawyer and sued the landlord in court for damages. I thought it gross to seek to profit from Nicole's death but the French would have understood that as serious.

I continued to find Charles V cold and unsupportive. At Nicole's funeral I was told that I could use the Institute's Fax machine without paying to contact Adele; a lecturer at first would sometimes turn up as I tried to negotiate the labyrinth bureaucracy necessitated by a death, but Nicole's art gallery had assigned someone as a support who met me every day, ordered taxis to take us where we had to go, counseled me against committing suicide, and arranged for Nicole to be buried at Pere la Chaise. In France there is even a social hierarchy of cemeteries and Pere la Chaise is thought a place for famous artists, writers, and rock stars, as well as the wealthy.

Soon after my arrival I began planning a conference in the New Literatures in English. It is difficult now to explain my obsession but that is what I then did. I had regularly chaired or spoke on panels at Modern Language Association and ACLALS conferences. That was part of my profile, my career. It also represented my response to what was or should be happening in English studies and the world. The Forum I had petitioned and had approved for this year's MLA conference, that I decided not to attend after Nicole's death, concerned "Decolonizing Theory". The previous year I had chaired a panel about Nadine Gordimer (the start of the book I edited *The Later Fiction of Nadine Gordimer* (1993) and a special session on Salman's Rushdie's *The Satanic Verses* at which to my surprise only one panelist defended the author's rights, whereas the other three attempted to defend those who claimed he was offensive to Muslims. The era of liberalism and tolerance was already being replaced by political correctness and those finding excuses to appease new tyrannies. During the special session I had watched several heavy olive skinned suited men eyeing each other as well as the audience and was preparing to tell my speakers to duck under the table before I understood that these were security guards not terrorists.

The logical time for the Paris conference was Easter holidays when foreign academics would be free to attend. The head of department, however, planned to be away and told me to choose another weekend. When I replied that nothing else would work she said she was unhappy with my decision which meant I would be swimming upstream on my own as far as the Institute was concerned. I was eventually also told by others that conferences took years to organize as they knew from their own (usually limited) experience, and that in any case the tradition of the Institute was to invite someone from England for a morning talk followed by tea and discussion. I only heard such objections later; at first no one at Charles V paid attention to what I planned. The French woman then in charge of readings at the British Council offered financial support and decided to invite some British writers from outside the white mainstream. We agreed on Wilson Harris, famous in Commonwealth literature for his strange magic realist novels about Guyana, and Ben Okri, for the concluding evening of the conference which would be held at the British Institute. Okri had recently been awarded the Booker Prize for his magnificent novel *The Famished Road*. He was interesting as he brought together the various literary modes associated with different Nigerian ethnic groups, especially the Yoruba, although he was not Yoruba. I invited Olive Senior, a Jamaican poet, fiction writer, and sociologist, who lived for periods in England and France house sitting, and Frank Moorhouse, a major Australian novelist who was in Paris for the year while researching what became *The Grand Design*. He required Swiss libraries. I met him through a French couple who had worked for a long time in Australia and held regular parties for Australian cultural figures and those with Ozzie interests. I think Frank had rented the studio behind their house. The London office of the British Council, I think at Alastair's suggestion, sent to the conference E. A. Markham. A West Indian writer who lived in England, but had spent time in France and who would eventually buy an apartment in Paris. Markham came from Montserrat where there was no African heritage and the colonial influence was Irish rather than British. Denis Hirson, the South African poet and translator who lives in Paris, added to what had become a distinguished and international group of writers reading at the conference.

I now had my writers; academics from many countries were submitting titles for papers or asking to chair panels, there was even a New Zealander. There would have been more except that a visiting colleague who I assumed would help had seen her future at Charles V and was telling foreign academics that the conference would not take place. She was almost right. Unexpected problems started to occur, but Michel Fabre decided that his research centre at Paris 3 would come to the rescue. When I was told that Charles V would be closed on Saturday he found a place for the Saturday panels. I thought as the schedule was so tight to arrange a lunch on Friday for writers and panelists and there was an excellent restaurant close to the campus that would be perfect and was willing but they would not to wait for the University to pay, which might take a year. Michel said that if the funding was paid to his research centre he would see that I was reimbursed. When it became clear the conference was going ahead the women who said that it must be changed to another date ordered that the Charles V imprimatur be added to that of Paris 3, making the sponsorship dual. I gather that Michel and Geneviève (although I never discussed the situation with her) were used to such creation of hurdles by colleagues and felt that I was naïve to assume that there would not be resentment and difficulties.

Thursday evening Olive Senior, Archie Markam and Frank Moorhouse read before my combined classes. Afterwards we went for a meal together as I had not known them socially. Friday was intense with multiple panels of the kind familiar to those who attend MLA or other large conferences. On Friday evening there were the British Council's readings by Ben Orki and Wilson Harris, which brought in many outside the university. The conference continued less formally on Saturday with discussions of exile and a reading by Okri.

I was not amazed, nor amused, when my fellow visiting American who had told others that the conference was cancelled started asking panel chairs if she could replace them—she clearly wanted to insert herself and gain credit from a conference she had tried to destroy—but I was stunned when another colleague complained that I had not invited her to read a paper or chair a panel as this was the first time she had ever spoken to me. I had, however, made several younger friends among the junior staff and assistants. Charlotte Sturgess, British, had at one time been part of a dance company in the Netherlands, was editing a collection of essays on Female writers, and we would often lunch together and discuss what was needed as part of the conference. She would drift off to Vancouver, return to France as a provincial high school teacher, recommend me for a visiting half year professorship at the University of Angers, and herself become a full professor at another French university. Vera Dickman, a South African, examined possible hotels to recommend and put me in contact with others in Paris from South Africa. She edited the conference papers and I will need to talk about her more later; she and her husband Mike became long term friends.

Anne Thieuille, the member of the English Institute who recognized my name among those appointed for a visiting professorship, had returned to work and I found that we had Africa in common. She had taught in Sierra Leone for a year, and while there had met her boyfriend (now her husband) a small lean Englishman who worked in oil fields. He was an odd choice for a Parisian. He hated Paris, wanted to live in the country, hated French food, and seemed to live on a diet of roast chicken and potatoes. She was tall, not attractive, and would claim that while female academics were studying, other women learned how to dress, use make-up, and choose the right lipstick.

We often dined together, exchanged stories, and she would tell me what the department was thinking and how I was stepping on toes. After my conference it was felt that they should appoint someone to teach post-colonial literatures but they had in mind a translator of South African literature that they knew who would be a team player and not disturb feelings. The Institute was created by those who in 1968 rejected the importance the French gave to the State Doctorate and publishing. Although they were obviously the new establishment, they still thought of themselves as rebels and kept fighting for the right to appoint their own staff, promote as they desired, and rule their own roost. Anne was amused that this meant hiring friends and former students and I could have pointed out that having promoted one of themselves to a full professorship on the basis of six articles published in their university magazine was reason enough to require external assessment, but these were her friends, she was one of them, and Anne was the only company I had among my colleagues. I wish they really had been a team, but all that I saw and heard was vain, unhappy, self-seeking, individuals who were proud to have come far by flying low. The one male member (again no pun) of the British section had published some books locally and stayed in the shadows except when ruthlessly chairing committees. When he retired he made certain that his highly qualified, well published, mistress did not replace him, but that his position was given to someone without any known history in the area beyond regularly taking tea with him. Many of the French are raised to be silent, watchful, worldly. Anne was different. She had a strong sexual drive, had chosen her boyfriend mostly for his sexual appeal, and one evening when dining with an African writer I knew she drank too much and made clear her availability which the next day she laughed off as haven't we all slept with the black, yellow, purple, green. Adele and I several times rented her apartment in Belleville while she was away. It was during those times that I learned about oriental restaurants in Paris.

I suppose I have always felt myself an individual and leader rather than part of a group, but when appointed as a visiting professor I tried not to have opinions about others as I would not be there long. I had, however, been welcomed when I taught at Paris 3 and still had friends there. I also respected most of their professors as established literary scholars. Charles V was different, there was no one in the British section with a reputation, they were lucky to have turned the rebellion of 1968 in permanent positions in Paris, and I would have felt like an excluded member of a high school sorority except that I was emotionally enclosed by Nicole's death and I had a lot of editing work that kept me busy.

Besides the book about the Trinidad Theatre Workshop I had taken on several books to edit and they often presented problems. The proofs of one book, I think it was *Postcolonial English Drama* (1991), were sent by Macmillan, my usual publisher, by courier, but when the company could not locate the concierge they failed to leave a message or any sign until, a week later, I telephoned Macmillan to learn who had the proofs. The company said come to the airport to get them. I refused and it took another week for delivery. I now warn publishers that proofs couriered to Paris might be better sent by slow mail.

These proofs had no punctuation, no sign where quotation marks began or ended. This was the copy editor's resignation. And I did not have the manuscript of the book in Paris.

I was used to problems editing books. Adele and I had to guess what a contributor intended to say in an essay in the book on Nigerian literature and we had rewritten the entire chapter in our words. For *Literatures of the World in English* I had to find another essayist after the first felt insulted when I objected to the his chapter beginning

"Australia started as a prison colony and is now a cultural desert." His replacement was a well-known nationalist critic who at first refused to mention such modernists as Patrick White and A. D. Hope and accused me of belonging to the Sydney English Department school (whatever that was) with which I gather he had been carrying on a literary argument. Some twenty thousand words, and several apologies for his drunkenness, later I managed to piece together a seven thousand word chapter and declined co-authorship, but I had not previously had to guess where to place full stops, commas and quotation marks. It was a new experience that I would not recommend.

The year at Charles V was my last visiting professorship except for the semester at Angers five years later.

After my year teaching in Paris we returned again to the Caribbean for the summer, staying in Trinidad, St Lucia (where Derek gave us one of his water colors), Barbados and visited very beautiful Grenada where I found some details of a TTW tour, enjoyed the sandy beach, heard The Mighty Sparrow at a Calypso tent, and met the brother of the Guyanese novelist and literary scholar Ron Dathorne (who I knew in Ibadan and who now was a professor in Kentucky). The brother was relaxed, amusing, and we discussed the 1983 American invasion that put an end to the chaos that resulted from the takeover of the small island by Cuba backed Marxists. He felt it was unfortunately necessary. Shortly after Granada became independent the New Jewell Movement seized power and had ruled in an arbitrary but tolerable manner under Maurice Bishop until an extreme faction ruthlessly eliminated Bishop and his followers, imprisoned others, and invited Cuba support. That was enough for the Americans who used the excuse of Americans students at a medical school on the island to invade. This seemed to fit in with what I knew about the Caribbean where Washington appeared ready to accept authoritarian governments of any political pretensions provided that they did not become a base for Cuba or the Soviets during the Cold War.

We stayed at the same hotel in Barbados as we had in 1990, but this time Adele left for a conference in Japan and I went on to Jamaica by myself, once more staying with Cheryl and Michael.

# Chapter 16: My Derek Walcott Decade

*Boston, Trinidad, St. Lucia 1992–2003*

For a decade Derek Walcott was central to my writing and sometimes my life. I wrote two major books about him and his career as poet, painter, dramatist, and I supplemented Adele's underpaid position teaching French with National Endowment for the Humanities two year fellowships. Somehow during those years my life changed from being an odd ball unemployed university professor to, in some academic circles, a famous literary critic and editor who had help start new areas of literary study, bringing to attention the writings of Africa, India, and the West Indies, knew major writers from around the world, even, I sometimes, heard "a legend".

I suspect this later fame would not have occurred, although it was partly based on my Indian books, if I had not written the two Walcott books. During a period when it was still common to write introductory books or edit a jumble of conference papers as a themed volume I had followed up my detailed study of *Modern Indian Poetry* with two scholarly books about Walcott. Whether you liked my books or not, and many reviewers did, I was treating writing from former colonies with the same scholarship and critical seriousness previously reserved for British and American literature. My books were sometimes reviewed in the *Times Literary Supplement* and for a few years I was included in the international edition of *Who's Who*.

I find it difficult to write about this period of my life. Although it was highly productive, and my publisher now was Oxford University Press, I was often troubled by my relationship to Derek who sadistically could change from encouraging to discouraging according to his own perceptions of his world and the accompanying moods. When I began the book on the Trinidad Theatre Workshop I was taking on a history that Walcott had long wanted and my main problems were that Walcott at times felt he should have been writing it instead of me, that he was annoyed not to have somehow earned money from it, and that his second wife Margaret distrusted me as she felt I was seeking to uncover a personal secret (which I soon discovered everyone knew) and, I was later told, she distrusted whites. Derek could be charming when he wanted but he was used to power over those he knew and befriended and I kept meeting those who had fallen out of favor and were severely hurt in the process. Derek's word was meant to be law and while working on the two books I saw how someone who Derek relied upon and I thought as part of the inner circle would conflict with his plans and suddenly become an untouchable. Untouchable seems appropriate; if you were rejected by him you would be rejected by others around him until he changed his mind, if he did.

I felt the force of being an outcaste while working on the second book, the biography. Supposedly the reason was that I had not obtained permission and paid to use a photograph in the book on the Trinidad Theatre Workshop and I suddenly learned that Derek had hopes of a pictorial biography from his own collection. I had also been naïve in not knowing all the ins and outs of permissions about photography and art works, and while I was learning rapidly the legal side of using illustrations in books I did indeed foul up in one case, reproducing a cover of a Trinidad Theatre Workshop production without getting permission and paying a fee. The real problem I think had little to do with paying fees but concerned Derek's legal situation about sexual harassment, which will need come later in this chapter.

Paying fees, however, became a theme as I studied Walcott. Derek had struggled for years to earn his keep as a poet, indeed supposedly decided when young to become a poet rather than a painter as it would be comparatively easier to support himself as a writer. Much of his life can be seen as trying to earn a comfortable living in the arts, as a poet, essayist, reviewer, theatre director, trainer of actors, stage designer, lecturer, university professor, winner of awards, and painter. From early on he quibbled over fees, over amounts and late payments, over contracts. The arts in the colonies were mostly thought of as hobbies, ways of improving social status, and while a few people earned money abroad as artists it was uncommon in the Caribbean. Walcott had to change the perception of a poet or dramatist from someone self-published or anthologized without payment to a professional who earned a living, a decent middle class living, like any professional. He had known from an early age that he was unusually talented, and had the potential for greatness, and he wanted the money that would enable such a life as well as recognition of his ability. Only with financial independence could he be artistically independent and himself become a sponsor of other artists rather than himself dependent on patronage. The letters, I found as I was working through the papers, letters, and diaries in the Walcott archives at the University of the West Indies, St Augustine, Trinidad, often referred to disagreements over finances, whether reminders of small unpaid amounts for contributions to anthologies or the broadcast of his poems, or, later, claims to ownership of theatre scripts and projects that others had had a hand in at the inception.

Walcott's motives seemed to me more complicated than he recognized. He would, sometimes enthusiastically, sometimes not, begin a project, a film or theatre project and edge out the other person, having gained financially from the relationship. Then the project might disappear. Derek's career in the West Indian theatre was something he wanted written, yet he no doubt felt he could do it better, or should benefit financially, that the facts were his, although he could not put them together in a scholarly fashion. I became aware that I could easily become one of those he had initially welcomed, helped make contacts, and then dropped, and the project would wither. I heard stories about Derek being generously paid to give campus readings and then staying all the time in his hotel room working on manuscripts instead of meeting those who had made his visit possible. I knew academics who had made appointments to meet Derek in St Lucia and found he was elsewhere at the agreed upon dates. In several of his unpublished works Derek's characters remark that in the alphabet "I comes before U". He believed in the importance of survival, the life force, and mocked the weak, the inhibited, the gentlemanly. When my daughter died he was impressed that I returned to St Lucia the next summer to continue research; that to him was the West Indian, the New World, spirit unlike, he claimed, the British who would retreat into self-pity and inactivity.

As I researched Walcott's life I learned that many of his traits resulted from the conditions in which he was raised and were especially influenced by his mother, "teacher Alix". He was, as he often wrote, descended from white slave owners and from slaves. His maternal grandmother, Caroline Maarlin, had been passed around by white Methodist estate owners on St Marteen and eventually moved to St Lucia. Her daughter, Derek's mother, was brought separately from St Marteen to St Lucia at an early age, educated by the Methodists and eventually became a primary school teacher, later head of the Methodist primary school and prominent among the English-speaking community on what was essentially a French and Franco-Creole speaking island. In a society in which "shade" mattered she was conscious that she was partly white and bragged of the features and color of her children and of her family connections on St Marteen which she sometimes visited. She mostly taught English to those raised as French or Creole. She hoped to marry white but settled for Warwick Walcott, born of a relationship between a St Lucian plantation owner, a bastard part of the famous Walcott family of Barbados, and a local brown woman, descendent of a slave.

Warwick was raised to be a gentleman; he painted, read, listened to classical music, belonged to a literary society, and was thought overly polite and unassertive. Derek would at times mock him as impotent. He died, supposedly because he would not jump the queue for an operation he needed in Barbados, a year after Derek and his twin brother Roderick were born and was often held up to them as a model cultural attainments. His death, however, left Alix and her three children (a daughter had been born two years before the twins) without the financial support she needed to become part of the mulatto upper-middle class. Over the years she had to struggle for security while asserting her superiority. She would be part of the small circle invited to celebrations by the British governor of St Lucia for whom she would take charge of sewing of clothes to supplement her earnings as a teacher, but the sewing would be done by others. Some details of her survival are less amusing and probably influenced Derek's callowness towards others, especially women, and his feeling for the relationship between social connections, money and dignity. He was a mixture of charm, generosity, spontaneity, calculation, egotism, selfishness, and guilt. He was proud of his ability to judge, lead, and use others, and he could deceive himself as to his motives. There was a basic insecurity that drove him and undermined his feelings of success; he was always climbing while pretending that he was above such ambitions, and he felt that others were advancing themselves by association with him. Distance was not only for self-protection, it was also a weapon.

He had been a child prodigy—especially so as thought of as black rather than white—helping teaching his mother's students, earning a scholarship to the one preparatory college on St Lucia. St Mary's College was run by Irish clergy and intended for the elite. It was extremely good and among its graduates were the island's future professionals—who would go abroad for a university degree and then return to a comfortable life—many of whom would rule the island after it became independent, and a Nobel Prize laureate in economics, Arthur Lewis. Those, such as Walcott or V. S. Naipaul in Trinidad, who won island scholarships to schools modeled on British "public" schools were brilliant and likely to be in competition for the even more difficult island scholarship to study at a British university, usually Oxford or Cambridge.

At St Mary's Walcott met Dunstan St Omer, a black Creole painter who became a best friend and part of a small group that tried to create a new culture on the island by fusing the modernist revolution in the arts with local black Creole subject matter. Cultural and racial politics were part of an aesthetic, social, and moral challenge to the dominant values of St Lucia. Besides Derek, his brother Roderick (who also attended St Mary's College), and Dunstan St Omer, there was the older Harold Simmons, himself a Methodist from an important family, for a time a civil servant, who devoted himself to the arts and died a drunkard, an illustration to Derek of what was likely to happen to an artist isolated on a small island. Harold Simmons had loaned Walcott the latest poetry books from England, given art lessons to Derek and Dunstan, played them classical music on his gramophone, taught the necessity of creating a high culture that celebrated the local while following the life of a bohemian intellectual and artist. Walcott would later immortalize him in his long autobiographical poem, *Another Life* (1973). I had assumed that Simmons was white, but later while doing research on St Lucia realized that he and his family were light skinned mulattos, part of the same English speaking Methodist mulatto elite that was a notch below the white British on St Lucia and socially above those, usually black, who spoke Creole. Their rebellion was against the puritanical French Catholic clergy that dominated the official culture of the island. Paradoxically Derek, Roderick, Dunstan, and Simmons celebrated in English local black Creole culture in contrast to the imported European values of the French priests, although Creole was a French patois. But they themselves were giving modernist "form" to local content in contrast to the polite provincial accomplishments that then passed for the arts throughout the Caribbean.

Derek become famous at an early age for his poetry, clearly influenced by awareness of James Joyce, T S Eliot, Dylan Thomas, Auden and the so-called Faber poets, and his poetry was published in the new literary journals, such as *Bim* in Barbados, that began appearing in the West Indies, or broadcast on Caribbean Voices, one of the BBC Overseas Services radio programs for the colonies, or when he could afford it, self-published in pamphlets. The new university colleges, the overseas broadcasts, were among England's ways to turn its former colonies and dominions into the British Commonwealth by the creation of local elites and institutions for education, culture, and self-government.

Although praised as a poet by those who read English literature, Walcott's rebellion against the local priests and colonial bourgeois culture was confrontational in the theatre since others including Roman Catholics were involved as actors and audience and the plays were reviewed in the local press where there were often prominent controversies with the two Walcott brother's justifying themselves and offering a view of art as speaking for another truth than religion. Derek and Roderick became leaders of the St Lucia Arts Guild, basically a theatre group that performed their plays in which fisherman cursed god for the dangerous life of their work and in which sex outside marriage and with more than one woman was celebrated as part of local traditions. The cultural assertion that accompanied the demand for political independence in many of the former colonies took shape in St Lucia as a brown English speaking urban elite rediscovering and praising rural black folk life and customs in opposition to the priestly values and French language that dominated the island. Even teacher Alix helped make costumes for the celebration of folk traditions. During the colonial period St Lucia had passed back and forth between the British and French but was allowed to be French although governed by the British. To celebrate the folk was to empower black Creole culture through brown English-speaking leadership. Much of Walcott's tense relationship to the West Indies can be understood by similar color and cultural contrasts. He would want to be the spokesperson for a community of which by color, education, command of language and being a poet he was different. When the brown professionals inherited government with independence he could speak for the intended inter-racial community of white-brown-black but when Black Power asserted itself he became a green eyed "red" man who left for the United States.

Roderick and Dunstan would remain on St Lucia (although Roderick would eventually move to Toronto) while Derek went to the new University College (London) of the West Indies in Jamaica, where he would join the first group of students in the Arts and be treated as more an honored guest than a scholarship boy. He felt he had already learned through Simmons and St Mary's more than his university courses taught so he concentrated on languages and writing his poetry and plays, many of which were published in a series of theatre texts by the Extra-Mural Department of the University. He soon married the secretary of the Extra-Mural Department who had typed his manuscripts and he later became a father. Faye was a middle class brown Jamaican easily impressed by the young man who brought her his scripts to type; married they would quarrel as she felt he was unfaithfully chasing other women. (Many decades later even over the telephone I was aware that she remains very feminine, seductive, and without a clear vision of her future).

Theatre was thought a way to bring together the different peoples and cultures of the region as it moved towards independence. When it was decided to stage an immense pageant play (with Rockefeller Foundation support) to celebrate the West Indian Federation (that only lasted a few years before the various islands turned into separate nations) he was the choice as writer. Besides becoming famous at an early age he was meeting many of the actors and other artists who were becoming well-known in the region and with whom he would remain in contact as theatre became, along with poetry, central to his life.

When the capital of the Federation became Port of Spain Trinidad he and many others moved there. He joined a lively cultural scene that included dancers, painters, actors, and others, black, brown and white, some of whom were already internationally known, such as Douglas Archibald, a dramatist who had plays on Broadway, and the dancer Beryl McBurnie, who had performed and taught "primitive" dance in New York. Before Derek went to Trinidad the Rockefeller Foundation sponsored a period during 1958–59 in Greenwich Village New York City where he was supposed to study theatre. This was a lively and influential time in the arts and he read and saw what was happening in the late 1950s. It influenced his view of modern culture and that it was now centered in New York and especially the Village rather than in London towards which West Indians usually gravitated.

Faye saw her future in New York and they went different ways. Their son was left with his mother Alix in St Lucia and eventually Faye was edged out. In Trinidad Derek married a second time, and Margaret Walcott would remain central to his life although there was eventually a third wife, Norline, and other companions. Margaret and Derek were at first the fairy princess and prince of Port of Spain's cultural life, invited to all the prominent events, often mentioned and photographed in the local newspapers.

He wrote regularly on the arts for *Trinidad Guardian* and soon using those from other theatre groups and McBurnie's Little Carib, formed his Trinidad Theatre Workshop which he hoped to train as a world class company of which he would be director, producer, main writer, scenic designer, and everything else including trainer of the actors. The Trinidad Theatre Workshop even had a dance company that combined European ballet with the more pelvic movements of black and Caribbean dance. The Trinidad Theatre Workshop would be a great love of his life, an illustration of how local artists and subject matter could be brought to the level of high European culture. Derek was driven by the need to prove himself and the West Indies by creating an equivalent to the European national theatre companies which then often were led by brilliant avant-garde directors. He also hoped to earn a living by it through such fees and royalties as training actors and performances of his plays. His ambitions for the Trinidad Theatre Workshop resulted in him investing earnings in its productions and tours.

This was the group whose history, participants, productions, acting, music, stages, tours, scenery, finances and reputation I researched and wrote about in *Derek Walcott and West Indian Drama/ Not Only* a *Playwright but a Company: The Trinidad Theatre Workshop 1959–1993* (1995). Although it had struggled to survive in a small basement theatre in Port of Spain it had toured the West Indies, performing in Jamaica, St Lucia, Guyana, Barbados and even, with the backing of Joseph Papp, New York's Central Park (1972 ). Walcott had illusions of it regularly being part of the international theatre festivals that were becoming common in the summers. He was also becoming famous as a dramatist, the Negro Ensemble Company's production of his *Dream on Monkey Mountain* was awarded an Obie in New York (1971)—it was taken by them to the Olympics in Berlin; he was commissioned by the Royal Shakespeare Company to write a modern version of Tirso de Molina's *El Burlador de Sevilla*, which eventually became Walcott's great musical comedy *The Joker of Seville* (1974).

Such success however was a sign that his dream of a great West Indian theatre company would shatter as it came into conflict with his own increasing fame. There was no place for such a company speaking West Indian English and performing plays with Caribbean subject matter and concerns in New York or London. He wanted his actors to follow him abroad but that was a dead end. One of the best, Albert Laveau, tried, had a moderate success, and returned to Trinidad knowing that there were many talented and more motivated black actors in the United States. The break with the Trinidad Theatre Workshop was marked by Walcott's taking up with Norline Metivier, one of the younger dancers who replaced Margaret as his wife. Margaret had invested much of her own life in the Workshop and Walcott's principle actors were on her side. Moreover this was a period of Black Power assertion in the Caribbean, especially in Trinidad, and Walcott was continually criticized for writing within the language and conventions of white rather than the often illiterate black ghetto culture then in fashion. The political situation in Trinidad became dangerous as the government, after putting down a Black Power revolt, passed laws against public expressions of dissent and one of Walcott's acquaintances who criticized the government died in mysterious circumstances. Faced by a Black Power movement and sympathy for the Cuban revolution, the government in Trinidad asserted its own support of black culture and enforced it. It was time for Walcott to go.

His leaving for the United States with Norline was the theme of *The Star Apple Kingdom* (1979) volume of poetry. He would teach poetry part time at Columbia University, be invited to Los Angeles as a playwright, become friends with the major writers then living in

Greenwich Village, and become part of the American literary scene, then a visiting and later a tenured professor at Boston University where he was appointed to what had been Robert Lowell's position. Besides teaching poetry he started a Playwrights Theatre that eventually led to his reconciliation with those who had continued to produce plays as the Trinidad Theatre Workshop. With Boston University's sponsorship the TTW would reassemble as a company, perform in Boston and once more begin touring the Caribbean with Walcott's plays.

I had been lucky in the past teaching in Nigeria when Soyinka, Achebe, Okigbo and others were bringing into existence a modern Nigerian literature; I was again lucky to have decided to write my first book about recent Indian poetry in English at a time when the major writers who created it were still flourishing and willing to tell me about their struggles as no one had given them such attention previously; I was a third time lucky to have been asked by Derek to write a book about the Trinidad Theatre Workshop just as it was being resurrected.

Derek and those involved with the Workshop had quarreled, broke up in 1976, went their separate ways, but increasingly courted each other like separated lovers and were finally reconciled. At first, after leaving, Walcott would not permit the Workshop to perform his plays and it was no more than a production company, a rubric allowing the principle actors to produce plays by others. Then they found they needed each other. A 1986 celebration of the 25$^{th}$ anniversary of the Trinidad branch of the University of the West Indies when Walcott was awarded an Honorary degree meant that the TTW actors, dancers, and musicians were brought together to perform extracts from his works. Several people associated with the Workshop kept in contact with Derek and produced his plays with his permission with their own companies but using Workshop regulars and Albert Laveau as Director. Walcott even formed his own company to produce a few of his plays in Trinidad.

The award of the 1992 Nobel Prize in Literature soon brought Derek and Workshop together again. Besides the performances of extracts from his plays at *Celebration*s in St Lucia and Trinidad there was one during 1993 at a major theatre in Boston and Walcott's new Playwright's Theatre opened with a production of his *Pantomime* with two of his Workshop actors. Andre Tanker who had become Derek's Trinidad Theatre Workshop musical director arrived from New York where he was involved with another West Indian play. There was a party after *Celebration* at which Galt MacDermot played piano, Paul Simon performed with a South African singer and an immense electronic musical box; Henry Louis Gates Jr from Harvard was also there. I could understand Derek's reluctance to have his acquaintances mix with his famous friends as I was embarrassed to hear someone call attention to Simon's small size. I mentioned to Derek that I thought the party surprisingly successful and he praised Sigrid's choice of the restaurant and the various small plates. For all his fame and experience of the world he still felt an outsider to the sophisticated.

Boston University, which tried to compete with Harvard and MIT, had several famous professors, but realized that Walcott's many talents made him an exceptional prize and invested in bringing the TTW to Boston in 1994 for a short season of plays at the Playwright's Theatre and at a larger theatre for a production of *The Joker of* Seville. This was the Trinidad Theatre Workshops' longest tour in North America, its thirty-fifth anniversary, and it was video-taped by Boston University. There was a cast of 30, and the director for the performances of *Joker of Seville* was Greg Doran of the Royal Shakespeare Company. There were also performances of *Dream on Monkey Mountain* at the Huntington Theatre using the marvelous lighting scheme originally designed by John Andrew when the company had to improvise with cellophane and oils and which now seemed almost impossible in its complexity. *Pantomime* was shown privately at Playwright's Theatre along with four days of workshopping *Steel*, a play Walcott was hoping would soon be produced.

During the two trips to Boston there were the usual dances, parties, meals together and other events that Walcott used to keep his company together. The rule had always been that when on tour there should be no private dinners, nothing to which the company was not invited. I could see the point of this when an actor arrived late after a long night with a local woman he knew. It all seemed very Trinidadian when a boat ride around Boston Harbor kept erupting into a dance, or when a meal at a restaurant ended with a "lime", where everyone sings together and the person mentioned at the end of the refrain is expected to improvise something personal and topical. Everyone knew the routine and had their party pieces prepared except me. One of the female dancers had a routine about wanting to "Princess Di" me which meant that I was next up. Stunned I inaudibly mumbled something about surviving the recent attempted coup d'état in Port of Spain and wanted to hide under the long table. Both tours had music by Andre Tanker's band for dancing.

I never lost my interest in music, especially in drumming. I was pleased when staying at a small hotel near Derek's apartment the manager told me that he had formerly been a musician and produced a small snake skin headed hand drum on which using his fingers he played some complicated rhythm that he claimed was basic to Greek music. It was the kind of imposition of more than one rhythm that Tanker's rhythm section appeared to find natural although the sources might simultaneously be Indian, Brazilian, Soca, Venezuelan and jazz.

I thought Andre, who had on my visits to Port of Spain shown me around the local music scene, a natural for discovery by a record company, and told Bob Koestler about the group's visit to Boston, but Bob was too busy at the time to fly from Chicago and he passed on the information to Rounder Records. They contacted Andre but there was some disagreement—I seem to remember that he wanted to record his latest suite and they wanted something more rootsy or folksy such as calypso or Soca—and nothing came of it. This seemed to be typically Trinidadian in its self-centered stubbornness and feeling that local fame was enough. Tanker was a great musician and had a great band but there no paying record industry in Port of Spain as recordings were always illegally copied and the only real market was to West Indians who lived abroad. But then Andre had studied music in the UK and been brought to the USA to provide music for some West Indian plays so perhaps he thought there would be other opportunities. That was not the feeling of some of the younger cast members who disappeared and became illegal immigrants before the tours returned to Trinidad.

My story of the history of the Trinidad Theatre Workshop possibly contributed towards its new international fame and to recognition in Trinidad that it could no longer be ignored as an eccentricity. The government accepted that it needed a regular home and acknowledgement of its role in contemporary Trinidad culture. It was to be advertised to tourists whereas in the past it was shunned by black nationalists as not part of the folk art they promoted. Walcott had always claimed it was necessary to earn one's place in Caribbean society, rather than assume that race determined authenticity; almost everyone in the West Indies was an immigrant or descended from immigrants, whether slave, slave owner, refugee, or castaway. His theatre company had survived many changes in government, cultural fashion, attempted revolutions, and earned its place by producing plays, training actors and dancers, by fighting for and eventually becoming accepted as necessary to the arts of the island and to Trinidadians who hoped to work in the theatre abroad.

II

It was primarily as a poet that Walcott had become famous and it was his poetry rather than theatre that caused me to be interested in him. His earliest verse had appeared in small Caribbean magazines and self-published pamphlets and when he first appeared in hard cover books in England and later in New York, he already was an experienced poet, who had put behind him his experiments and had polished his rough edges. Each volume had a theme which was picked up by many of the poems, was autobiographical or alluded to his life and situation, was located in the West Indies and showed his familiarity with the traditions of English language poetry while increasingly making use of West Indian speech or culture. It was not only excellent poetry, and part of a new world of poetry from the formerly colonized, it also expressed social, cultural, and linguistics subtleties and complications in a manner found in major writers. That Walcott's poetry had energy was expected from someone from a former colony, but the range of attitudes, the voices, the varied tones, the seeming ease in form and cadence was a sign that Walcott was as good as any poet then writing in England or the USA.

Port of Spain was in the 1950s and 60s a place to go to avoid the austerities of post-War Britain. It had a lively arts scene stiffened by those who had come for the sun and the cultural excitement that was part of the last stages of colonialism before European empires withdrew. Indian born Alan Ross, known as a poet, owner editor of *London Magazine*, and a writer of books about cricket, came to cover the England-West Indies matches and returned to England with Derek contributing to his then famous magazine. He also arranged for Derek's first books of poetry to be published in England with Jonathan Cape. There were favorable reviews and Walcott was soon recognized as a major contemporary poet. But the real breakthrough resulted from a 1962 visit of Robert Lowell and Elizabeth Hardwick to Trinidad on their way to Brazil on a trip paid by the Congress for Cultural Freedom. Robert Giroux had heard about Walcott from Tom Maschler, Walcott's British publisher and asked Lowell to see him. Lowell returned to the USA singing Walcott's praises as the only major "Negro" poet and brought him to the attention of Robert Strauss of Farrar, Strauss and Giroux. He also introduced Walcott first major reading at the Guggenheim in New York. With Lowell's backing and publication by FS&G Walcott was becoming part of a cultural establishment that was happy to have a well-read black exotic writer of great talent. At first there was some skepticism, the *New Yorker*, for example rejected poems that Farrar, Strauss and Giroux, acting as Walcott's agents, had sent, but after *Dream on Monkey Mountain* had won an Obie, and the poetry editor, Howard Moss, read a volume of Walcott's verse, it encouraged him to submit and over the decades published many of his poems before they appeared in hard cover. Soon it replaced *London Magazine* as his main poetry market. Although denying any such conscious intention, Walcott often made friends with those in helpful places, and *The New Yorker* besides paying him well for his poetry, often reviewed and mentioned him. The New York literary scene can involve a small number of people who move between publishing houses, editing magazines, arts sections of newspapers, foundations and other cultural institutions; Walcott was becoming one of their favorites.

There were other aspects of Lowell's friendship. Lowell was important to the Congress for Cultural Freedom, a grouping of usually leftish anti-Communist intellectuals that the CIA indirectly funded during the Cold War, sometimes through the Fairfield Foundation which helped Walcott attend the International Congress for Poets in Berlin, September 1964, at which he met Wole Soyinka, J. P. Clark, Chinua Achebe, Leopold Sedar Senghor, Amie Cesaire, and others who were creating what we now think of as postcolonial literatures. Walcott and Soyinka were to remain good friends, often citing each other about the importance of freedom for the arts in contrast to the authoritarianism that was becoming common in the decolonized world.

It is easy but wrong to assume that Walcott, Soyinka and others were part of an American Cold War plot against Communism, but the period after the Second World War was artistically and culturally exciting and both the West and the Soviets wanted to influence it. I had African friends with offers of scholarships from the Russians, French, Germans and others. The CIA seemed to have unlimited funds to back anything provided that it was likely to be for freedom rather than Communism, and there were great publications such as *Encounter* that I wished still existed. Writers from what was then termed the third world were especially encouraged and supported through scholarships, visits, and publication.

The indirect support of the CIA for the new post-colonial literatures in their early stages is well-known and I have often been asked and refused to write about it. I am asked because I was one of the group of foreigners directly hired by Nigerian universities who received no outside funds and we looked on with envy as American professors arrived with large salaries who were provided with excellent housing and medical care for themselves and their families. I met, although did not known closely, those, such as Robert July, who were influential in deciding which writers should be given grants, fellowships, and travel expenses. Later I heard rumors about Walcott. Walcott and his Trinidad Theatre Workshop had received much support from the Rockefeller Foundation and he later moved to USA, thus...

I had been interviewing those involved, such as Gerald Freund; I applied for and was given a small grant to consult the Rockefeller Foundation Archives in Tarrytown, New York. The Foundation had traditionally supported the little theatre movement in communities in the USA along with research in the sciences, especially medicine. After the Second World War it began cautiously to provide aid to universities abroad including the early development of a few university drama departments such as at the University of Ibadan and it thought the University of the West Indies. Wole Soyinka had already established himself as a playwright before returning from England to Nigeria and wrote a play celebrating the independence of the African nation. He was not only a dramatist but it was assumed that he would join the Ibadan University Theatre Department once it was established. He was offered a Rockefeller Foundation Fellowship to research drama in Yoruba ritual. This followed on from the then commonly held view that the origins of tragedy were in religious ritual and his research became the basis of his own Yoruba-centered world view that is central to his early plays and many poems. Yes Soyinka being anti-authoritarian was anti-Communist in terms of the Cold War, but in Nigeria he was a prominent supporter of the Action Group, which claimed to be

Socialist. Wole was more influenced by Nigerian politics and by G. Wilson Knight's interpretation of William Blake, Lord Byron and Nietzsche than American global politics.

The Rockefeller Foundation reluctantly became involved with the University of the West Indies because of the pageant being prepared for the start of the West Indian Federation, which was also a way to support local theatre groups that would be participating in the event. Walcott, who was already known as a dramatist, had been asked to write *Drums and Colours* (1958), and became part of the small circle of people who it was hoped would start a Theatre Department at the University of the West Indies linked to a local Little Theatre or performing arts center in Trinidad. It was soon obvious that Walcott and the dramatist Errol Hill were the only ones with a clear idea of what was needed and as Hill left to pursue advanced degrees at Yale and took a professorship in Nigeria, Walcott and his new Trinidad Theatre Workshop would become the focus of what little support the Foundation provided to a Little Theatre movement in the region.

The person in charge of the West Indies desk was J. P. Harrison, who arranged for Walcott's fellowship to study in New York. His successor was Gerald Freund, who became important in the Foundation world during future decades and had a significant role in Walcott's future. Freund had to persuade, and sometimes failed to convince, others that fellowships to train Trinidad Theatre Workshop actors and grants to the Workshop were not an expansion of Rockefeller policies beyond what it had traditionally done in the past. He felt that foundations should discover and sustain original talents and he thought that there was insufficient awareness of how difficult it was for artists to survive. He managed to persuade his colleagues to award Walcott a three year grant 1967–70 so that he could quit journalism and devote himself to writing and theatre. When Freund moved to the John D and Catherine T MacArthur Foundation he fought those who initially wanted to limit Fellowships to those in the Sciences. Walcott, Henry Gates Jr, Robert Penn Warren and Joseph Brodsky were among the first given five year Fellowships (1980). The $250,000 changed Walcott's life, allowing him security he did not previously have; he purchased a large apartment in Boston where he was moving to replace Robert Lowell as a Professor at Boston University.

I several times talked at length with Freund over the telephone shortly before his death in 1997. He was still a foundation administrator—he had at times moved in and out of university positions, now in charge of Whiting awards to writers and an advisor to other granting bodies. I found him charming, helpful, friendly and was surprised that he was interested in my research and asked whether I needed financial help. I did not and I decided not to arrange to see him in New York, nor did I telephone the person he told me had run the Fairfield Foundation. Walcott probably was innocent of the ultimate source of two tiny grants he had received, but Freund was more in the know.

As I worked on Walcott's biography I found little evidence of any CIA support after the Berlin meeting except for a latter Fairfield Foundation small grant when Walcott, often in debt because of the mortgage on his house, his investment in plays, and travel, needed money, but he did benefit in other ways after being introduced by Lowell to his friends, the anti-Communist left intellectuals who were becoming the New York cultural establishment. Many of them were Jewish and later termed the Family as they seemed inter-related through such publications as *Partisan Review*, *Dissent*, *Poetry*, the *New York Review of Books* and even *The New Yorker* when Howard Moss became poetry editor. It was the social-cultural-political scene to which Woody Allen's earlier films allude and often parody. (A Woody Allen joke: if you cross *Dissent* with *Commentary* you get dysentery.) Many of the Family taught at universities in New York or close enough to attend parties and cultural events, and Roger Strauss, who had inherited money, was central.

While Walcott was outgrowing the limits of being a West Indian writer he never left behind the dream of bringing his Trinidad Theatre Company to the world's stage, nor did he become a black American. He remained a West Indian whose perspective was based on St Lucia,

Jamaica, and Trinidad, but had enlarged to include career and publication possibilities offered by the United States. When I met him, although he was now a famous poet and a professor at Boston University, he still would speak of Americans as simpler, cruder, less complicated, less cultured and less sophisticated than Europeans and West Indians, a cliché he had inherited from the British. But then he was often ironic and cynical:—the blandness and supposed innocence of Boston made it the capital of Canada. The West Islands was where he was born, raised, educated, formed, and lived until he was in his mid-thirties and still thought of as home although he was aware with each return that he had become a wealthy tourist who was no longer part of the poverty and culture that had been his subject matter. His poetry spoke of this growing gulf between himself and his past; he later appeared more concerned with observing America, the friends he made there, his travels, until with *Omeros* (1990) and *Tiepolo's Hound* (2000) he was once more writing about the Caribbean.

There is a difference between the earlier poetry which seems engaged and written from within a society even when lamenting his estrangement, and the latter poetry which although much more technically proficient seems like that of an observer reflecting upon life and friendships. That an angry young man should become more settled and a greater craftsman seems normal, but there is less emotion, more sentimentality, more intellectual clichés in the latter poetry although the verse itself became almost too complex as metaphor builds on metaphor and statement dissolves into image, sound, comparison, analogies, music. It is perhaps only in *White Egrets* (2010) where he approaches once again the variety of registers and dimensions that his poetry lost after he moved from Trinidad to the United States, but even these poems for all their again found strengths seem pale in comparison to such earlier volumes as *The Gulf* and *The Star-Apple Kingdom*.

Even before I handed in my manuscript of the Walcott biography I knew that there would be trouble; it was much too long, too loaded with factual detail, and not enough story and characters. I had tried to cut the first half down to a manageable length, but that seemed to make it one dimensional, a narrative about Derek, and leave out so much of what I found of interest, such as the history of the Methodists in the West Indies, the role of Roderick Walcott in the St Lucia Arts Guild, the role of the Rockefeller Foundation in preparing the colonies for independence, the social and cultural differences between Jamaicans and those from the Eastern Caribbean at the new University of the West Indies. At least for the first half of the book I had a story-line, which took Derek from St. Lucia through Jamaica and Trinidad to the United States, but I was not clear what the story was after he moved to America except it culminated with the award of the Nobel Prize and the events that followed.

Now I can see that his seeking the security of a full time professorship as a creative writer, his continuing support by Gerald Freund and Robert Straus, his finding a place among the American writers and intellectuals, his close association with Seamus Heaney, Susan Sontag, Joseph Brodsky, his desire for the Nobel prize, was itself a story, and not simply a series of facts. Although Walcott had come to know and often work alongside of many of the American and British literary and theatre establishment his close friends remained a few poets, such as Brodsky and Heaney, West Indians from his past, and a very few others, such as Galt MacDermot and Greg Doran who had shared his projects. And they were also from the Commonwealth, MacDermot, being Canadian. Walcott's outlook was still that of a West Indian who had travelled north and east as part of his career, and while avoiding the sentimentalizing of "home" he always hopped to return, a very successful prodigal son.

I had asked my editor at Oxford University Press for another year in which to revise my manuscript or better to let me publish the West Indian part as a separate book as I was still finding and adding details about Walcott's American years and I needed to see the manuscript from a distance after a rest, but no she said it had been advertised and sold to Oxford companies overseas, so I handed the text in hoping it would sort itself out somehow.

I could sense dissatisfaction, no offer to drop by for a meal, no congratulations. This clearly was not the best seller for which she had hoped. Then it was sent to one of their best line editors who however was not the best for this book. Her sense of style was that of an old fashioned school teacher; she was usually correct, but her suggested changes would have slowed down even further a book that was already too slow and that I had tried to speed up by making the style concise. I felt embarrassed rejecting her work, but it had to be. Worse was to come. Alan Ross mumbling had handed me a photograph of Derek that I understood as having been taken by someone now forgotten. It was chosen for the cover of the book and Alan wrote asking why he had not received credit and payment. I had to write a calming explanation. Then after advance proof copies had been sent out for review to major publications and one review had already appeared someone discovered that the biography lacked a "legal reading" and the newspapers and journals were told that publication was halted for the present.

Oxford University Press had little history of publishing books about living persons and with British libel laws being so arbitrary (even the factual could be libelous if it damaged a reputation) and awards so high there were fears that Walcott might sue. It did not matter that I claimed to be totally factual and Walcott had in principle approved of it, it had to be read by a lawyer. I had mentioned in passing the widely reported claims of Walcott harassing students for sex. Had I checked with all the British newspapers and international weeklies that Walcott had never sued them for libel for reporting such claims, could I really assure Oxford University Press that Walcott would not feel that my book harmed his reputation, and, unbelievably, did I not understand that saying that Walcott and V. S. Naipaul promoted different "myths" of the West Indies could be used in court against my book? What was this word "myth"? Literary critics should not use it.

Somehow the manuscript survived and the book limped out many months after the announced publication date: months I would have liked to reduce the length and clarify the story, had instead gone into legal quibbles and consequent delays without any fundamental change in my text. It at least was reviewed in England; the *Times Literary Supplement* reviewer slaughtered it over several pages; but then he often reviews Derek's books and wanted a different story. The American edition appeared unnoticed sometime between Christmas and the New Year, and had no reviews of which I am aware. In all the biography was a disappointment. It was not advertised and sold only 3,000 copies of the expected 6,000 and never went into a paperback edition. For academics used to print runs of 500 to 1,000 copies 3,000 might seem satisfactory, but most of my books had sold more. My *History of Seventeenth-Century English Literature* had, for example, sold 8,000 copies before going out of print. I do not know how many copies of my *Modern Indian Poetry in English* sold but I imagine well over 10,000 copies. I once calculated that Adele's first book on Albert Camus probably sold 25,000 copies between the various British, American editions and translations. My agent wanted to take back the rights and sell an updated version to another publisher and I might like the opportunity to revise and shorten the book, but unless specifically mentioned in the contract the days of rights reverting to the author are over; publishers hold on to rights in the hope of orders for expensive print-on-demand copies.

I flew to New York to view the Paul Simon—Walcott *Capeman* three times during its long epic rewritings and restagings and at each performance I could only see a mighty flop. When I warned Derek that he had to give up some of his ideas for the play I was criticized by Sigrid who claimed I should be supportive. That was really the last time, over a meal with some of his West Indian acquaintances, that I felt any relationship to him. I paid for my visits by writing an article for an Indian newspaper that had an internationally aware Sunday cultural section and I was tired of finding ways to fund my Walcott research. I have only on a few occasions seen Derek since the publication of the biography and while he has sometimes been friendly and other times barely acknowledged my existence we both remain guarded. I feel the decade I spent working on him is an episode of my life that was interesting but I was glad when it ended and despite my agent's desire for an updated version of the book I do not wish to return to those days. A professional biographer once told me whether or not you have the subject's permission when you start your research you will end disillusioned and unhappy. I gave all of my Walcott files to another academic writing about him so that I would not be tempted to do anything more about him or West Indian literature.

    I briefly had thought of bringing together for students Walcott's reviews and articles on West Indian writing during the 1950s through the mid-a1970s from the Caribbean newspapers and literary journals in which he then wrote. I thought them an excellent introduction to the region's literature of the period as it was appearing and more written from within the society than Derek's later North American essays where, for example, he simplified V. S. Naipaul into a racist who disliked black people, whereas earlier he spoke of him as contributing a much needed ironic and satiric vision to local self-regard. Living in Trinidad he knew of the terrible violence that had developed between blacks and Indians during elections and even controversies about whether Trinidad should be independent as the Indians feared black majority rule. Naipaul mocked both sides and what self-rule brought. Seen from outside the Caribbean such skepticism appeared reactionary or racist, but there was a point to it. The views in newspapers and journals by Walcott and others during those decades should be brought together by academics trying to understand the realities of the post-colonial world. Some of the best critical comment on Walcott remains those of the writers and intellectuals who reviewed him as his work appeared. But someone else will need re-do the work. I was glad when Gordon Collier, an Australian teaching at a German university, objected to my bringing out a book of Walcott's West Indian criticism as he planned eventually for an edition of the complete journalism. I no longer had to face trying to get Derek's permission or that of the many publishers of the reviews and articles; I would not need to try to negotiate with Derek the royalties for what he, not I, had written. I was free.

A decade was more than enough. I wanted a new life.

Of course you are never free of the past. One year in France Derek's poems were set for higher level examinations and as usual there was a conference as a way of generating academic studies for a book of secondary readings and while I no longer had anything to do with the academic world I was asked to be involved and I reluctantly agreed to give the opening address. Derek was invited and the predictable happened. First he wanted for himself and Sigrid to fly first class. On arrival he complained about the three star hotel and the conference organizer had to find money for a four star hotel which was still not satisfactory. Then Sigrid's purse including their money and documents was swiped while shopping, etcetera, etcetera. I telephoned to ask if he needed help, but he was already being helped and I felt free once more.

With *Derek Walcott*: A *Caribbean Life* (2000) I had now published two major books on Walcott, an updated edition of my V S *Naipaul* (2003) and an enlarged version of *West Indian Literature* (1995 ) with five new chapters. I was thought an expert on the literature of the region. But I was a white American unable to get any job at an American university. I had written two books on Indian literature, both of which I had revised and enlarged to bring up to date, the first of which was the standard work on Indian poetry in English. I was a recognized scholar, perhaps in the view of many Indians then the only major scholar-critic, on the subject. (There later were others) My first book was recommended in Indian universities and schools. No American university wanted my expertise. I was still being offered visiting distinguished professorships abroad, but I would have been living on the streets in America if Adele had not supported me for twenty years and if I had not been awarded National Endowments for the Humanities fellowships. Adele was now seventy years old and tired of teaching among uninteresting colleagues where the only achievement was undergraduate student evaluations, tired of freezing during Indiana winters, tired of living in a small mid-western city with little to do, where we had few real friends with common interests. Our life had for two decades remained overseas. Even the drive returning from Indianapolis to Muncie seemed harder each year. It was time for her to retire and for us to move to Paris were she had always wanted to live.

# Chapter 17: India III
# Two More Visits 2005–2006

I would meet Menka again soon after returning to Bombay in 2005. We flew from Paris where we had lived since 2003 and once more stayed at Bentley's hotel which had increased its price and reputation in the past decade and a half but otherwise not changed. Menka sent her chauffer driven immense black shiny luxurious Scorpio SUV 4 x 4 to take us to a gathering of poets she had arranged in her house in a wealthy suburb known as the New Bombay. Christopher Levenson, a Canadian academic and critic of Indian poetry, also would also be there with his wife.

Menka had married Raju, a publisher and was co-responsible for the supplements they produced for *Times of India* and other journals. She explained that she could publish books of poetry but none of her distributors would handle them. My describing her *Nirvana at Ten Rupees* as one of the best volumes of Indian poetry published during the 1990s (in the revised *MIPE*), was, she claimed, the start of her present reputation.

Menka was instinctively a leader, someone who would translate, edit, anthologize, and organize. She was obviously wealthy, had a good looking, bright, energetic, loving husband, who treated her as an equal and was willing to help in the kitchen and look after their daughter although they had servants. The house was spacious, well equipped, and while it was at least an hour's drive from central Bombay, it was part of a community for those similarly young, attractive, well off, living modern lives. She had published a second book of poetry and would contribute to various publications, but she was now a leading participant of the new cultural establishment and a business woman.

My standard for poetry was Arun, Nissim, Dom, an era of giants who were dead or aging badly and whose work I had stumbled upon because I wanted a trip to India. *Modern Indian Poetry in English* and *Three Indian Poets* had made me known among those who read and wrote poetry in India—Adele noticed copies of *MIPE* in the bookshop of the Taj Hotel, but I was more interested in seeing old friends and enjoying ourselves than I was in discovering poets. On the return, the chauffer drove us to the Banganga music festival on Malabar Hill. It overlooked the harbor, had an active temple, and it took over two hours to get there from New Bombay. Throughout the music although unrelated to it, there were prayer bells from the temple, and while we enjoyed the evening getting transportation home was an impossible problem until we spotted a taxicab leaving people and their luggage at a hotel in the neighborhood and we persuaded the driver to take us instead of someone he had promised.

Menka she was busy with her job, with meeting women friends for cocktails, and cultural events. To catch a glimpse of her meant visiting the expensive Hotel Oberoi, in Nariman Point, South Bombay, and having something else to do in that neighborhood to justify the cost of the taxi. Luckily Arundathi Subramiaman had an office at the National Centre for Performing Arts across from the Oberoi. Arundathi was tall, powerful in appearance, and beautiful as well as one of the better new poets. It was difficult to know how good at first as her poems were written for performance and that can both seem impressive and suspicious. As I followed her writings from volume to volume I began to see she was indeed one of the very best poets in India, and that what I at first took for lyrical exclamations were connected, part of a vague spiritual autobiography in which the over-animation of Bombay was contrasted to the quiet and calmness she sought through meditation and yoga. I had a distrust of Indian mysticism and spiritualism but Nissim and now Arundathi showed that good poetry could be made from the search for peace.

Bombay was changing into a city of skyscrapers, international restaurants, cinema complexes, fashionable shops. It had touches of Singapore, Hong Kong, New York, Paris. We went to a play in the experimental theatre at the NCPA and in the garden of the David Sassoon Library saw a Hindi production Shakespeare's *Midsummer Night's Dream*. I was surprised at the high quality of acting but many of the actors were television performers. Adele had some bright colored pure cotton dresses made by a young designer with a boutique in a large modern building. I had two cotton suits and trousers made by a tailor on Cuffe Parade and shirts at the Taj. The suits and trousers lasted, the shirts from Maharaj's did not. The suits are so out of style that whenever I wear them someone thinks I am dressed in the latest and compliments me. I sought and bought and still own some "mules", shoes without backs, that I need because of a heel spur. Leather belts were much less expensive than in Paris. For the first time we were tourists in India. We went to Pathar Hall in Churchgate to hear a visiting Trinidadian "chutney" band and danced in the aisle and on stage. One of the female performers recognized that I knew Trinidadian dancing and kept encouraging me as if I were a star, which in a way I was. Indians were not familiar with this hybrid of Soca and Indian music and did not know how to dance Caribbean style. I could then get low or go belly to belly with the best.

We went to a reading at the Poetry Circle and afterwards to a party at the spacious apartment of Jane Bhandari, an English poet and interior designer whose Indian businessman husband died a few years before. The view from the apartment window was the best I had seen in Bombay and Jane had a computer in the middle of the room on which she searched for men. I remember a conversation to the effect that she could have an affair with men for their body or their intellect. She avoided older and married men. I knew it was now accepted for women to meet men through computer dating but Jane was the first among my acquaintances to speak of doing it. Some years later she remarried, lives in the same apartment and still writes poetry.

The car we were in had to stop because one of the passengers wished to pick up a picture she had framed. After a time waiting I got out to see what was happened and heard the woman hysterically screaming that she was always being overcharged because she spoke "shitty Hindi". That was how I met Janet Fine who for the year or two before her death was one of the more interesting of the international crowd in Bombay. She was the daughter of Benjamin Fine, who wrote op-eds for the *New York Times*, and her sister was a professor of French at NYU and her brother headed an important research center at MIT. She and, she claimed, her entire family had an interest in erotica and she assumed so did everyone else. When we went to lunch at her apartment every decoration was pornographic and her purple bedroom appeared ready for an orgy. She was on the editorial board of an Indian pornographic magazine which I looked at and thought tame; and she had her own publishing house devoted to scholarly editions of Indian pornography, badly translated, unarousing regional versions of the *Kama Sutra*. She took us to a Muslim market where a few stalls were devoted to crudely made pornography. We were clearly expected to purchase something and after looking through masses of junk we found two watercolors of Ganesh having sex with a thin young maiden, which we were told were painted in Ahmedabad during the 1970s. When we bargained on the price Janet seemed shocked. Had we committed an unlikely faux pas or was she hoping for a percentage? She told us to keep the paintings wrapped until we left India as they could start a riot if Hindus saw them.

Janet was a social and cultural columnist for a Bombay newspaper. She was invited to many openings of artistic events that she wrote up. Perhaps I was not aware of it twenty five years previously but there was a large group of foreigners in Bombay who were part of the cultural scene. Janet had an interest in films and I remember her leaving us at a restaurant so she could visit another section to talk with a Bollywood director. Someone told me that she had when she first came to India appeared in a few films and was supposedly the mistress of some director until he felt he needed a fresh foreign face. She was always on the move from Bombay to Egypt, Greece and the USA, often to film festivals. I was sad to learn a few years later of her death. She had a strange high pitched almost hysterical voice, but was friendly, welcoming, likeable, and seemed vulnerable despite her travels and involvement with erotica.

Anand Thakore had started Harbour Lines, a small press to publish poetry by himself and a few friends. When I favorably reviewed his first book I thought it very competent, a bit old fashioned in being song like but that was because he also is a classical Indian singer. When I met him at Jane's party he invited us to a poetry reading at his apartment, celebrating the return of Vivek Narayanan from the USA, and drew a map and wrote such elaborate instructions in literary Hindi that people thought it amusing. Chandri villa was in the Gowaliatank area of Bombay. As we neared the villa he unexpectedly pulled open the taxi door and claimed it was a robbery. I was to become used to, even appreciate, his sense of humor.

Chandri villa had a cultural history, and had been in the paternal side of the family for generations. The quit India campaign was started by Gandhi during 1942 in a field across from it and Anand's grandmother was among the protesters who were jailed. Several apartments were sold and Anand and his father kept the rest. The meal was Spartan vegetarian, chapattis and peppery lentils. This was far different than the Westernized India I usually visited. Anand had studied English and Sanskrit and told me that he used to resent Bruce King who's *Modern Indian Poetry in English* he had to read; who was this American to tell him about Indian poetry, especially Arun Kolatkar. He showed me an essay on Kolatkar he had written.

Bombay, now Mumbai, was for me a chance again to meet Adil Jusswalla and Arvind Krishna Mehrota who returned every winter to see friends and family. I noticed that at Anand's house once more I was being introduced to a new generation of writers. Then Gieve Patel arrived and Gieve was treated with the respect given the famous. Gieve invited Adele and me to a meal a few nights later and we went to a restaurant he liked that served meat and had many Arabs among its cliental. This may have been the first time I saw women in heavy burqua and I was shocked. Gieve was, like everyone we knew, translating bhakti poetry, but mainly concentrating on his painting as he had an exhibition in New York.

The Goa we knew had become much too popular and every space had filled in with hotels, guest houses, restaurants, bars, discos. The hippie, dropout, beaches of Baga, Calingute, Anjuna had become overcrowded slums for mass tourism. As well as the small stalls everywhere selling Indian clothing and souvenirs and Anjuna's Friday night market, there was now the immense well-lit Arpura Saturday open air market. The line of cars waiting to enter was long and police or guards directed them with flashlights. This was a movie India packed with food stands, musical instruments, and all the Indian bags, saris, tee shirts, and art that any tourist could ever want. This was not the young and bohemian crowd we knew from Goa in the past; these were mass tourists who had flown to Goa on direct flights from Europe. While waiting for our driver we talked to an Indian family from Birmingham who were stuffing their bags with Indian decorations probably made only for export. There were hand crafted pots instead of pot.

Manohar Shetty had moved to a lovely house in Dona Paula outside Panjim and was now, probably through Vickie, well connected locally. He arranged for us to stay south of Calangute at a small hotel beyond Candolin which had a swimming pool and was normally rented for the season, as was most of Candolin, by the British, especially retired Brits who found Goa less expensive and a better place to be during the winter than the English midlands Except for the shops and taxi drivers it was like being abandoned in a tasteless British suburbia. There was even a local open air club with a dance floor that featured out of tune, hopeless, supposedly professional visiting British "artistes" who badly copied hit recordings, entertainers of the kind I associate with working man's clubs north of London. Why do I find such accents and dress offensive? I really have become a British snob with the same ear for distinctions of class, education and culture. The best part of the club was the Goan drummer who was said to play with a northern Goan symphony orchestra and who did not need any of the other musicians; whether soloing or backing he was what we danced to. One night a young British woman impressed everyone by making small patches of her skin, an area here, an area there, shake and shimmy; she had studied some yoga technique for body control. I felt defeated. (Yes, a bad pun.)

From Candolin we went north to visit Nick Arnold at Arambol Beach where he had moved after Anjuna. It was crowded with every nationality, especially Israelis, and we met him at a café with chess boards and American sandwiches and cakes. He had a cell phone and regularly checked for email at the café so he was easy to contact. I thought the area hell especially as it was necessary to walk through a crowded long narrow lane to reach the packed beach, but for Nick who wanted to live like a European but had little money it was a colony of backpackers, druggies, drunks, bohemians, and a substitute for Deia. Nick was his usual self, seeming half hung over, skeptical, polite, and waived to those with whom he usually played chess. He was, suffering from cancer, broke, bought a bottle of local rum to drink himself to sleep each night, and would die a few years later having pushed heavy furniture against the door of his room so no one could break in and rescue him. A Harvard degree and inheriting a million dollars had led to this. He was co-author of one story in a magazine and may have published some untraceable pornography.

I had been warned that a sunken ship, visible from the beach, created dangerous currents just behind our hotel but paid no attention to it until one day I was caught in a powerful riptide and I had to struggle to keep standing or I would drown in the waves. I thought I heard a terrible crack and felt an awful pain in my left thigh but somehow struggled back to the beach and collapsed until three or four men saw me and carried me to a beach café where a visiting British nurse decided I had a fractured pelvis. With Adele's help I made it to our hotel where I was told that I had been warned about swimming and should stick to the pool; eventually a local doctor came and would have been more helpful if he had recited prayers. He bandaged my groin with a smelly, nasty, hot salve that made my testicles shrivel and looking guilty and nervous he injected me with a steroid that produced a blood clot.

Before the doctor had managed to kill me we flew to Bangalore where Jeet Thayil and his new wife Shatki Bhatt took me to a local hospital, found a well-trained doctor who prescribed medicine to remove the clot and told me that I had a badly damaged muscle that would take years to recover (it never did, but it was not a fractured pelvis). He wrote me a letter to show airlines that I needed special seating and gave me the name of a Harvard trained doctor in New Delhi where I was going next. In Bangalore we stayed at the house of Jeet's family, a family of writers. I could not climb stairs and Jeet's mother and father let us stay in their room on the ground floor. Jeet's father was a distinguished journalist who had been the first newspaper editor after national independence to be tried for sedition—for criticizing a government minister. He founded an edited *Asiaweek*, which is why some of Jeet's poems are set in Hong Kong where he lived when he was young.

The father was also a prize-winning historian who spent nights writing and listening to news in his study. He was an advisor to the *New Indian Express* and had a chauffeured limousine available. He had at one time been part of the Indian presence at the UN. His wife, who enjoyed cooking and entertaining, was still unhappy that he had given up his UN position and its high salary and pension to return to India, although by Indian standards their life was among the elite. Besides Jeet who had worked on *India Abroad* in New York, other journalists in family contributed to or edited such publications as *Time Out Bangalore*. I remembered that Jeet's uncle had translated Baudelaire into Malayalam.

Jeet had met Shakti Bhatt while both were working for *India Abroad* in New York. She replaced the beautiful but dumb druggie Jeet had brought from California to his Muncie reading who caused him and Shakti to change apartments as she continued to follow, hoping he would return to drugs and her although drugs had damaged his liver and would kill him if he had not stopped. The self-discipline needed to give up drugs included the rediscovery of religion, which is why many of his poems refer to Christianity. He had carelessly waited too long to renew his American visa which forced his return to India, a situation alluded to in some poems in *English*.

Shakti was pretty, thin, with what seemed a surprisingly large head for her small body, articulate, highly competent, and up to date in clothing and attitudes. She was usually smiling, which she did well. She had been sent to Florida for an MA and was knowledgeable about literature. Shakti was also from a journalistic family. Her father worked in Mumbai as a journalist, her mother Sheela was a well-known reporter, columnist and Editorial Director for rediff.news in New Delhi. Jeet was several decades older than Shakti and divorced. Shatki and Sheela were close and the mother had flown to the USA to discuss the marriage as she had wedded an older journalist, from whom she lived separately.

Bangalore had become a mixture of modernity and slums, often the two sitting next to each other on main roads. There were fancy restaurants, places to dance, shopping malls with the latest names and styles, alongside filthy shacks with tin roofs and matching customers. There was even an occasional cow blocking the bumper to bumper traffic slowly edging through the city towards urgent destinations. Contemporary blue jeans and tee shirts, designer clothing and drugs, were for the wealthy young trying to avoid men in polyester shirts and rags, smoking bidis and spitting red juice. Even a restaurant bar we went to meet Ramachandra Guha had two sections, one popular and the other more expensive although there seemed little difference between what was served. But then I had come across such distinctions in restaurants in Bombay with the downstairs less expensive and less decorated than the rooms above. Modern urban India is no longer solely about caste, money is class.

Jeet arranged for us to meet some of the local writers such as the poet, novelist, editor Anjum Hasan and Vijay Nambisan and his novelist wife and medical doctor Kavery Nambisan. Bangalore was a good place to live, with a temperate climate, a modern hospital, various universities and institutes with high educational standards, and especially attracted those who retired or wanted to rest between jobs. I missed the famous Kannada dramatist Girish Karnad who was leaving to direct a film. Although I economized my time by reading only English language authors it was impossible to ignore the translations of Girish Karnad's plays as he is the best known Indian dramatist. We talked over the telephone and I learned that he had known Ramanujan when they were young and had advised him to send his poetry to Oxford University Press in England which published his first volume. In a nation divided between Marxist, other radical and ultra conservative Hindu historians Ramachandra Guha had gone his own way as a skeptical liberal uncommitted to dogma and had been criticized for his attempted objectivity and celebration of what had succeeded in India. I was impressed that he had written scholarly books while not attached to universities and that he was often appointed abroad as a visiting professor. I had vague thoughts of writing a book or at least an essay on Ramachandra but I was not an historian and was overwhelmed by the extent of the material I would need master.

Shakti claimed that her mother would like visitors and we were met by Sheela's chauffer who somehow immediately recognized us after our airplane arrived in New Delhi. Sheela Bhatt was impressive. What was our program from day to day? She hired a driver to take us around the city to galleries and markets (where we bought a couple of paintings by an Orissa folk artist), and who would try to see us off for our return journey to Paris. Sheela was a Guajarati language journalist who was not educated in English and would file five or six articles at night which would be put into better English by her staff. She was a Jain, a strict vegetarian who insisted that food be freshly made and if it sat and was no longer warm it had to be given away to the poor and more made freshly. Oddly, one time she took us to a hotel restaurant and ordered kebabs, a treat she very occasionally gave herself. I was impressed by her energy, knowledge, abilities, and personality. She was used to being in charge, respected, known. Her field was national and international politics, and she was one of those Indians who often travelled abroad to Italy, France or England on a combination of business and pleasure. From her I learned that as a result of globalization people were having a modern life in regional languages and in the cities beyond Mumbai and New Delhi. You could watch television, go to a shopping mall, use internet, discuss world news, become wealthy, buy the latest fashions, without English.

We visited Susan Visvanathan at her staff apartment at Jawaharial Neru University. I had met her previously in Paris but it was only after by chance spotting her first book of short stories, *Something barely remembered*, that I started reading and took an interest in her work. Although I knew she wrote fiction I had thought of her as a feminist sociologist. Through Sheela I met Jeet's friend, the poet Vivek Narayanan, who had been raised in Zambia, I thought his poetry, which I read in manuscript and later reviewed favorably, interesting and mentioned him in a rather formless lecture I give to the graduate students of English at Jamia Millia Islama University in New Delhi. My argument, if it could be called that, was that research should be based on personal interests and chance discoveries rather follow well-trodden roads. One wrong road was a Thai restaurant I had heard about and insisted that Keki Daruwalla eat there with us. It seemed to take his driver hours to find it and as soon as we entered the basement of a smelly house it was obviously a mistake as was the food.

The drive to the airport was long, the wait for the airplane was longer, until Air France eventually confessed that because of a strike in Paris we would not be leaving that night, and we waited still more before we were told that we would be bussed to hotels in the city and a flight the next morning. By the time we reached our hotels it was almost time for breakfast and back we went again. Everyone has nightmarish stories about the difficulties of airline travel but mine are mostly about Nigerian Airlines and Air France.

## II  2006

We returned to India for a few months the next winter, 2006. An Indian friend in Paris was from Pondicherry and it sounded like something new to experience. The flight from Paris went through Bombay to Chennai and once we eventually reached our hotel we joined others in the lobby complaining that after an exhausting trip we expected to have the room we had reserved long in advance, but the desk clerk said we should have sent payment which none of us had been told. We would need wait for hotel guests to check out before we could be housed. The nearest hotel, up the street, was already filled and there was nothing to do but wait and hope not to die from thirst or mosquitoes. We were only in Madras for a few days and tried to make the best of it before going to Pondicherry. We wandered around a film studio, went to a concert of southern Indian classical music, and were invited to a meal with Vivek Narayanan's parents who turned out to be traditional Hindus who served without joining. They regarded tomatoes and many other fruit and vegetables as alien invaders unfit for pure tongues. How had they survived decades in Zambia?

On the way to Pondicherry we stopped at Mahabalipuram, where we had visited in 1987, to once again see its temple by the sea, the carvings in the rock, and to have a brief swim. The driver from Pondicherry was connected to the Aurobindo Ashram that hired him to tourists and to which he gave his earnings. Our Indian friend in Paris had family ties to the Aurobindo Ashram and "The Mother", a French woman of Jewish parents who was Aurobindo's disciple and was elevated by him to the incarnation of Female Divinity. Our friend's mother had moved to Pondicherry and remained part of the circle of followers, educating her children through the Ashram's schools. There was a rule that once you left the Ashram you could not return so our friend who went to New Delhi for graduate work and became a Marxist was now an outsider although most of his family remained behind and he visited them at least once a year. I listened to his nephew telling me about modern history as he had learned it at the school and it seemed to consist solely of events in the lives of Aurobindo and the Mother. He had no idea that Pondicherry had in the ancient past traded with Greece and Rome, or knew anything about The Mother before she came to India as part of her divinely inspired destiny. Late nineteenth- and early twentieth-century Europe was filled with many such self-proclaimed mystical leaders, but how had Indians fallen for one? They had enough of their own gurus.

Pondicherry was disappointing, a mixture of a small "white" orderly clean French town close to the beach with French restaurants, food, and cultural places left over from late colonialism and a dirty, chaotic, loud, energetic Indian city. Our hotel was on the edge of both and it was amazing the differences travelling a short distance could reveal. I could tell when our taxi or three-wheeler was leaving the French section by the loud blaring Indian music, the bright lights, the many people and cars.

Although we had several names to look up and were always active I found Pondicherry more a sad provincial cul de sac than a gem left by the French. The French language publishing house had been revived but there seemed little worth publishing. I had an introduction to the recently retired professor of English literature at Pondicherry University who took us to a restaurant for a meal but I found him uninteresting and dated like the food. David Dabydeen had told us to look up an Indian West Indian who was starting a museum for ancient artifacts when Pondicherry was a Roman trading post. The topic of ancient trade was exciting but the objects were minor, without interest, and I felt sorry for our friend hoping to rediscover an interesting past without the means for a major excavation. William Wilson, an excellent black painter we knew in Paris, created a book for children while a guest at an artist colony outside Pondicherry but when we visited it had not recovered from the 2004 tsunami.

Another friend from Paris told us to look up her friend who lived with an Indian artist. We were invited to a party that turned out to be a regular French language social and cultural event that excluded Indians not fluent in French. We were welcomed as fresh blood but I was happy to leave the pettiness of Pondicherry for New Delhi.

So much happened in the weeks we stayed at Sheela's that it seems like months. Jeet and Shakti had moved in with Sheela while seeking a house in Delhi. We had one room, Jeet and Shatki the other, and Sheela slept on a cot in the main room. Then Janet Wilson came to visit India and took over the cot while Sheela moved to a friend's apartment. I was embarrassed but Shakti and Jeet thought it amusing. Before Janet arrived I unexpectedly found myself writing two essays for a special Indian issue of Janet's *Journal of Postcolonial Writing* that

Jeet had agreed to guest edit (42.2, November, 2006). I had suggested the issue and Jeet to Janet so this was becoming inbreed.

The first, longer essay, "To be or not be diasporic", brought together ideas raised in my recent writings in which I wondered how in an era of globalization, international education, travel and jobs, the lowering of boundaries, and inexpensive air travel, it was still possible to speak of national literatures. Jeet and Vivek Narayanan seemed to me examples of authors whose sense of place was where they wrote or set their poems. Rana Dasgupta was an even more extreme example. Born and educated in England, having lived for periods in France and the USA, he settled in New Delhi because of an Indian girlfriend whom he later married. His first novel was set at a Tokyo airport, his second moved from eastern Europe to the USA. For them and many other writers "home" was a choice unlike "exile" and "return" in the past. The other essay was a review of two recent late novels by V S Naipaul. As contributors included such already well-known young Indian writers as Amit Chaudhuri, Tabish Khair, Ravi Shankar, Siddharta Deb and Rana Dasgupta I was in good company. The issue is still often in demand. The publisher claimed my "To be" essay was among their top ten "hits" for the journal.

Jeet was responsible for renewing my interest in Indian literature. After the rush to complete the second editions of *Modern Indian Poetry in English* and *Three Indian Poets* I felt I was unlikely to have anything further to say. The second editions of both books used up what I knew. Then in 2004 Jeet had asked me to contribute to an Indian section of *Fulcrum* 4, an American literary magazine. I used the recent deaths of three major Indian writers, and a survey of their last works, as the theme of "2004: Ezekiel, Moraes, Kolatkar", which Jeet republished in his Penguin and Bloodaxe anthologies of contemporary Indian poetry. Now, in Delhi, I was meeting through him young writers and hearing him, Shakti, and their friends discussing books, especially novels, that were new to me and that out of curiosity I began reading and reviewing. I was also learning I had become a famous literary critic in India. *Time Out* mentioned that I was visiting Bombay and

New Delhi, a friend of Shakti's mentioned that in school he had read *MIPE*, and a lecture I gave was attended by several writers including a surprising attractive Susan Visvanathan, surprising because when I saw her previously she appeared a harassed gray haired mother. Now her hair was curled, colored, she wore a short skirt and she had lipstick. How could this be the same woman? She said she was in remission from her illness and had been at her favorite hairdresser.

Also in the audience was Sarayu Srivatsa, Dom Moraes' final muse who had co-authored several travel books with him as well as being the main source of that immense burst of poetry after he left Leela. Jeet introduced me to her. She indeed had the large breasts that attracted Dom but explained that they had been already reduced after her son told her about an operation. We would see her later in Bombay, where one night we went dancing at a jazz club. She asked me to write Dom's biography, but as his life was international and would have required years of research, the co-operation of his other former wives, and much money, I had to decline. I recommended Jeet but thought no one would probably take such a difficult task.

I did not know it then, and it was another eight years before I published *ReWriting India* (2014) but I was at the start of what would become my third book on Indian literature. Keki Daruwalla had said wasn't it time I wrote at length about him and he was right. The publication of two major books by Arun Kolatkar before his death meant I wanted to see how they fit with *Jejuri*. I at first found Amit Chaudhuri's concise tight lipped fiction puzzling and began asking myself what he was doing. The American republication of Upamanyu Chatterjee's *English, August*, along with the controversy over his later novels, made him of interest. I had not read Pankraj Mishra, started to read his books and was hooked. Then there was Susan. I had previously written about Jeet and Tabish Khair in the revised *MIPE* and since then they had published much more. A group of authors for a book was starting to form but I did not know at first know what might hold it together until I came across remarks by Chaudhuri and Mishra that they had been influenced by such poets as Kolatkar and Mehrotra.

Suddenly I had a possible theme, what the poets had accomplished and why and how they had influenced later writers.

We purchased a large terra cotta of Ganesh drumming that Adele had to fix by gluing together the parts after we returned to Paris. At a market we bought two framed "folk art" pictures; folk art was clearly becoming less innocent. A trip to Dehradun to stay with Arvind had to be cancelled at the last minute as the night before Jeet and Shakti had taken us to an overly popular restaurant and I had an embarrassing case of Delhi Belly. We did go to see the erotic temples of Khajuraho with Janet, stayed at a tourist hotel with swimming pool, enjoyed what was clearly a prosperous town and its restaurants—one even had a tree house with table—and I discovered that I was aging, not from the sexual positions (although I do not imagine I would enjoy it with a horse) but because I was afraid to walk up the stairs to visit the insides of the temples. It is difficult to reconcile the eroticism of many Indian temples with the inhibitions of contemporary India where until recently a kiss in a film might cause a riot, but as so often the past was far different than the bleached version claimed by nationalists. Tantra was now claimed to be a gateway to the spiritual.

Shakti and Jeet were a loving couple; it was a pleasure be around them. They were discovering a world together whether it was late night kebabs at a crowded stand on a dark side street, or the latest imported fashions for the young and wealthy. They had friends like themselves about whom they gossiped and shared stories. Life was fun. We were invited to a dinner at a house owned by the parents of a publisher and Jeet and Shakti were the ones in jeans, many of the others wore what Shakti mocked as "designer khadi", the adaptation of the woven cloth that Gandhi recommended Indians to wear instead of Western clothing. Jeet and Shakti were proud of being Americanized unlike the Marxist cultural intellectuals who were the Indian establishment. But they, especially Jeet, knew and appreciated the Indian art in the house and remarked on one of the guests being among the first to write well on the subject.

Shakti was bright, clever, ambitious. She would no sooner be interviewed for and appointed to a top position at one of the new branches that international publishers were starting in India than she would start complaining how holdovers on her staff were cramping her decisions, and start eyeing other companies as further steps up. With her personality, decisiveness, and nose for trends and Jeet's knowledge of the Indian literary scene, they made a formidable couple. Shakti's unexpected death the next year was a shock to everyone. For a time Jeet seemed unable to recover, then published a volume of poems associated with her, and threw himself into his secondary career as a musician. The autobiography he had been writing somehow was transformed into the novel *Narcopolis*, a finalist for the Man Booker and winner of the DSC prize.

Manohar arranged for us to stay three weeks at the Nanu Beach Resort, Batalbatim. South Goa was less crowded with back packers and their culture than the North, although still dependent on foreign tourists. Some Indian families were staying at Nanu, but there seemed to a preponderance of Russians. Adele, who had been brought up swimming in lakes and who had never liked diving into waves was happy to have swimming pools instead of going to a beach, but the all you can eat breakfast and dinner buffets were spiceless and there was little to do in the evenings. One afternoon we walked south along the beach to Benaulim to visit Sudhir Kakar, a novelist who was also a well-known psychoanalyst He often held workshops at French universities where Business schools would pay him enough for a weekend to fly to Europe. He and his German wife lived in a charming Goan colonial bungalow with a garden, and we were given bags of black peppercorns from their vines. Although a small village with only a restaurant or two it appeared a paradise for work and we thought of renting an apartment for the next winter, but instead I was invited to the Rockefeller Bellagio Center and we never returned to India. This would be our last visit to Goa and Bombay.

The Bombay literary scene I had known was passing. Nissim's office at PEN was now occupied by the poet and art theorist Ranjit Hoskote. Lunch with Leila Moraes was worse than embarrassing as she insisted that she had given up drink and that the clear liquid she gulped from her glass was medicine and her cook tried to sell us his book of recipes. The guest room was now for paying guests and we being old friends should stay and board; they obviously needed money. She insisted that she was still Dom's wife and that he had never legally divorced her before moving in with Sarayu Ahuja, conveniently forgetting that Dom had never divorced any of his wives before marrying at a public ceremony the next one, including Leila. He was a major poet but a multiple bigamist. Sarayu was Dom's literary executor. She lived in Bandra which was a long drive from Colaba and we probably would not have met except through Jeet who had been close to Dom. She was intelligent, had been an architect and journalist, and Dom was still the major event in her life. I asked her about her husband and was told that he had been happy to support her and Dom so he could have his own life without hindrance.

Before flying from Bombay we stumbled upon an immense brightly lit Muslim wedding with perhaps a thousand guests, live music, and distinct areas of meat and vegetarian foods. It was the kind of multiculturalism that India was intended to have and which was formerly common to Bombay before Hindu and Muslim extremism made rare such enjoyable celebrations, and the city became Mumbai.

# Chapter 18: Home

*Paris, France 2001–15*

We bought an apartment in Paris two years before Adele retired; it is on the Canal St Martin, a lovely area that was early in its ascent to one of the more desirable and fashionable areas of Paris. Although tour boats regularly go up and down the Canal even some Parisians do not know of the Canal's existence although it flows from outside the city through Parc Villette down to the Seine. It goes underground near Place de Republique, flows under the boulevards Jules Ferry and Richard Lenoir, before reappearing at Porte de l'Arsenal and emptying into the basin at la Bastille. The distance from where we live to la Bastille is not far, good walkers tell me that they do it in an hour. As I usually roam around the apartment nude in the morning and the boats need to stop at the lock in front of our building while the water level is being changed I unintentionally might be a tourist attraction.

The Canal has an interesting history. Napoleon built it to help bring more water into a parched Paris. It was used throughout the nineteen century as a waterway with barges transporting industrial products from the small factories along the quays. There are still some attractive factories remaining, architecturally outstanding with their small, almost classical, proportions, colored bricks and climbing vines in contrast to later unadorned monstrosities. But then it is difficult to live around the Canal and not be aware of a few remaining near bye art nouveau buildings with their marvelous glass decorations, of some rare cubist apartment buildings, and of how since World War II the French governments seem unaffected by taste. Left to its own, without popular protest, French governments would have turned Paris into a high rise jungle of dormitory apartments for the workers who increasingly are forced by the rents out of the city into the suburbs as foreigners such as us keep Paris a cosmopolitan city.

By the early twentieth century the Canal was losing its function as a waterway and had become notorious for crime, prostitution and murder. Drawings of the period inevitably show it as barely lit, shadowy, dark, dangerous, although one area was becoming a symbol of futurism; the elevated iron subway above the automobiles (metro line 2) at the junction of Stalingrad and Jaures still appears a surprising feat, like the Eiffel Tower and I have seen animated films in which something like it with airplanes or spaceships is symbolic of modern life in a dystopia.

The Canal also was a place to get rid of victims. A decapitated body was found at the lock just outside our window, the inspiration for Georges Simenon's *Maigret et le corps sans tete* (1955), which was later made into at least four television films (English, Japanese, and twice in French). This period of the Canal for the down and out was sentimentalized in Eugene Dabit's *L'Hotel du Nord*, awarded the first *Prix du Roman Populist* (1931), and portrayed in the famous Marcel Carne *Hotel du Nord* (1938) starring Louis Jouvet, Arletty, and Annabella, although the film, accurate in its details and atmosphere, was shot in a studio where the scenes imitated the reality. The *Hotel du Nord* was allowed to deteriorate and would have been pulled down except for an Englishman in love with its myth who bought and restored it. It is now a good place for a tourist to eat and sleep. After World War II the Canal became an area for immigrants, and the working class that owned the buildings started to leave for the suburbs. Our building still has a contract saying that the apartments cannot be used for prostitution and other immoral purposes. Then the government decided the Canal should be covered and become a motorway.

This was not to be. Artists started buying the old apartments and reconverting them. Bobohemia was on its way as those with blue jeans and artistic tastes followed the painters. Soon the Canal was fashionable, increasing featured in magazines. The Greens became powerful and stopped the government from selling land with old buildings for new apartment blocks; instead, after often years of confrontation, there were parks. The Canal was next closed to cars on Sundays and holidays when it became a pedestrian zone. Cafes, restaurants and many shops decided to stay open on Sunday (they instead close on

Monday). Eventually the government realized that the Canal rather than being an eyesore had become a feature of Paris, its banks lined with new cafes, with young lovers and guitarists during the summers, with a financially respectable young middle class and some rich people who preferred it to the more stuffy expensive bourgeois districts of Paris. Recognition of this change was (as often in France) a master plan making the Canal an historical zone (we spent a fortune having our building washed and restored to its nineteenth century look), planting wild plants along the unpaved banks of the Canal to prevent people living in tents, and bringing order and enlightenment to the disorderly. By now most artists could no longer afford it and moved on, but we have more fashionable fusion restaurants than any other section of Paris, and even to live within walking distance of the Canal is thought to be mentionable and worth a higher rent.

We had come to the Canal by chance. Our daughter Nicole died while I was teaching for a year in Paris and Adele was teaching in America. Lonely, depressed I wandered seeking something to interest me, and was surprised to find the Canal and youngish obviously artistic people at a café. I sometimes returned, learned that artists were converting dilapidated housing into lofts and living space and I thought if we ever did move to Paris this is where we should live. More than a decade later Adele had a one seminar sabbatical which we decided to extend to a half pay year off and looking for a place to rent we discovered an apartment for sale that had been modernized. The agent said it would be a good place for a writer; a frenzied negotiation followed with the businessman owner who had modernized the apartment and wanted to sell it as rapidly as we wanted to buy it: he needed the money to move to the suburbs with his wife and daughter.

Although the bargaining was intense I liked him. He had an artistic interest in the reconversion which he planned (some of which such as the attempt to create a shortage space in the ceiling did not work), including a false ceiling made from a shiny white plastic material, removing the doors to the main room so that space flowed through arches, large double windows. He told me the happiest days of his life were before he married and was an accountant at the Folies Beregere. He was part of an older French working class that an American would consider middle or lower middle class. The family had owned the building and his brother had an apartment upstairs that he sold months after we arrived. They had had office jobs and now were leaving the building as they retired and the Canal became a place for immigrants and next was boboized by artists and their bohemian friends. A nun, part of the original family, still owns one of the apartments that she rents to a Portuguese couple who help keep the building clean. The apartment next to our own was the property of a woman born there where she had lived all her life, over ninety years, except for a period when she was posted to Vichy during the Second World War. Now foreigners, such as ourselves, were moving in.

French friends tell us how lucky we are to have a beautiful already modernized apartment in such a lovely fashionable area. I had no idea how lucky we were. Adele had lost a lung to cancer, I was grossly overweight, we were both in our seventies; I thought of the apartment as a place where we would die during the next decade. It is now more than a decade later and these have been the happiest years of our lives. We are active going to concerts, films, jazz, restaurants, operas, almost every night and we have many friends, almost all much younger than ourselves, and including some American jazz musicians in Paris. Occasionally we are invited to jam sessions, usually when someone is having a party, or when there is a celebration among the black American community in Paris. Very occasionally I will have the guts to sit in on drums, but I know that I no longer have the skills and muscular reactions to be a musician. We still have a reputation as good dancers, and people send us photographs they have taken of us dancing. Friends from England, Nigeria, India, the United States and now especially New Orleans, visit us when they are in Paris. I continue to write and edit books, Adele wrote a short biography of *Camus* (2010), reviews Francophone African and Caribbean novels, and contemporary French fiction. She remains beautiful, sociable and charming. In recent years we too often groan with arthritis and other pains, but life remains enjoyable, something to which we look forward. We go to New Orleans for a month of jazz, dancing, food, and friends during the spring; in summer we go to the island of Hvar, Croatia, for swimming, friends, fresh air, to lose weight, for very fresh food (including fresh fish every night), concerts, and some years dancing and films. That will need wait until next chapter.

After we settled in Paris we were often away for a month or more in winter, first for two visits to India, then a month in Italy, and there were several trips to London and Oxford. In Oxford Janet Wilson had back issues of *Gravesiana* and other journals focusing on Robert Graves that helped my research for the short biography I wrote about him (2008). With Jeet's backing, I applied for and was invited to a Rockefeller Fellowship at Bellagio, Italy (2008). This was my second, and by the rules, final fellowship there. I had first been a Fellow when returning from India December 1984 to work on my *Modern Indian Poetry in English*. I wrote a draft of my chapter on K N Daruwalla for what became *ReWriting India* (2014) and an article on the novels of Kamala Shamsie as the book was not clear in my head and I hoped to have space to discuss some recent Pakistani literature and why the writing in the two nations had differently evolved after Partition.

The Rockefeller Center at Bellagio had now become a place for politically correct minorities. New York wanted to create an international mix and had turned selection of Fellows over to another body but the result was a mostly all-American group of minorities except us, being Paris based, and two Poles and a Brit, the only aliens. The Poles and Brit escaped into the village every night and found things to do, such as a film club. Adele made a fourth at a Bridge table. The food had changed since my first Fellowship. Then it was Italian and delicious. Now it was supposedly healthily free of spices, sauces, fats, starches, and taste. It may have been the most boring food in Italy. In a shop I found a large bottle of Tabasco sauce that I used every meal. Most nights we were brought together for a quick drink and brief conversation and sent to bed, a second drink was frowned on, a third impossible. One night we were allowed to stay up and the cats could play a little. We surprised everyone by dancing to some classical music, but then we danced to an organ grinder in the streets of Como outside the Cathedral. On an excursion we met a mid-western American woman who invested internationally and now lived above Lake Como. She was glad to hear English, nostalgic, and wanted new friends, but one of the Fellows so strongly advised against becoming involved with what she claimed was an obvious fraudster that we let the opportunity pass, although I thought our advisor wrong about everything. I hear that the Rockefeller Foundation now mostly uses the Centre for expensive conferences on feeding the Third World.

When we returned to Paris we discovered that in response to hurricane Katrina the French invited several musicians from New Orleans to residencies at the former convent near us which was now used for visiting foreign artists. We soon met the clarinetist Evan Christopher and the pianist Tom McDermott. The pianist David Torkanowsy and the singer Leah Chase were also in residence, as was an American who claimed to be a film director and who kept having photographs taken of us dancing for the film she was making about Evan and Tom. The shooting continued in April when we were in New Orleans. After she ran out of funds she received further grants including from Sundance and PBS but the film was never finished. She left boxes of clothing, negatives, and I know not what in our cave for several years until I got rid of them.

In October we went to a London University conference celebrating the fiftieth anniversary of the publication of Chinua Achebe's *Things Fall Apart* as it was an opportunity to meet old friends from Africa; a young academic told me he was there as perhaps it was his last chance to see the "famous old farts". A friend asked me about a mutual acquaintance and I said I cared as much about her as a "bat's fart" and he said "That is very little indeed". I do no recall having heard the expression previously. Had I just made it up? Had the young man's words been the source? While it was good to see many "old African hands", it was clear that others, some with little connection to teaching in Africa, were now the establishment. The field had advanced beyond introductory clichés and while the political quibbles continued a few scholars had done impressive specialized research.

I thought London nasty after Paris, the hotels bad and expensive, the transportation was even worse than in the past, the city noisy, dirty, windy, chilly, damp, the distances far. It required energy. The theatre tickets were expensive. Had I really years ago wanted to live here, hoped to have a job in England and often visited? Had I actually not many years ago commuted to London to interview writers for *The Internationalization of English Literature* (2004)? Suddenly Paris seemed our welcoming retirement home; it was smaller, warmer—food and transportation were better, and it required less stamina.

In December, 2008, Adele was operated upon for ovarian cancer that was detected before being advanced, but in hospital she had a blood clot and began medicines to prevent clotting. A year later there was a second operation and six months of chemotherapy. We did not go to New Orleans but did return to Hvar. Later she suffered from fainting, ministrokes, developed a heart condition, had a pacemaker installed, and was told that with her one lung it would be dangerous to have further operations. We never returned to India and would seldom return to London although we continued to travel to New Orleans and Croatia.

At first our life in Paris revolved around old friends such as Michel and Geneviève Fabre, Francophone African writers Adele had reviewed, the black French artist William Wilson and a few jazz clubs where we danced. Adele had published a short story by Ben Sehene, translated from French, in *From Africa* (2004) and we often went to restaurants with him and his French girlfriend Liesel. Ben's family had earlier left for Uganda and then Canada, but he returned to Rwanda after the massacres and even followed the killers into the Congo where he was jailed with them. He had written a documentary novel about the period and his experiences, but he wanted to become a full time writer—V S Naipaul was his ideal—and never mastered another topic. Eventually he began drinking heavily and became so erratic that we stopped seeing him. Occasionally I still get a drunken telephone call late at night.

Liesel's parents were French fascists who met in prison after the Liberation. Her life was to be largely a reaction against them. Her mother remarried a man of letters who chaired a well-known literary prize but found little time to read the novels so it was Liesel who read and reported to him her opinions. Like Ben she wanted to be a novelist, and I imagine that at least for a time she helped him. Although she worked as an editor at publishing houses and wrote some booklets on social topics no one wanted her longer fiction that was stereotyped and wooden. Her relationship to Ben seemed very French to us as she was married and the husband would even invite Ben to eat with them during holidays and only insisted that she remain at home for family occasions. When younger she desired to be a ballerina, but after years of lessons she was told that her build, especially her bosom, was too large for a classical dancer. She still was a patron of the Opera's dance company.

After my March 1991 Conference on New National Literatures I continued to see Archie Markham when I visited London as I thought he was a good writer although he spoke too much about himself and his opinions. After he moved to Paris and purchased an apartment we often went to meals and concerts with him. I was surprised that Marilyn Hacker, the feminist poet and translator who had given me a number of useful leads when I was working on my Walcott biography, enjoyed Archie's company and regarded him as a friend. Archie did think of himself as a feminist and among his repeated topics of conversation was the mistreatment of women, especially his father leaving his mother. Then Marilyn remarked that she had once been married to a black man and had a child by him and Archie's role in her emotional economy made sense. Archie kept talking about eventually unpacking his books, having book shelves built, and he was becoming part of the Paris Anglophone literary scene often giving readings. I was surprised when he turned up with a former creative writing student from Sheffield and she was surprised that I gathered seventeen people who mostly did not know each other to celebrate Chinese New Year in a restaurant that offered a ten course banquet, dueling lions, and a dance band. (A friend claimed the women were dressed like aristocrats, matrons or concubines.) Archie died of a heart attack in his Paris apartment and would not be discovered for over a week, after which no one knew who to contact. Another friend of this period was the French poet and translator Claire Malroux, but after her longtime partner died she devoted much of her energy into putting his writings into order and she seemed afterwards to be reclusive except for her prize-winning volumes of translations of American and Canadian poets.

For several years Paris jazz clubs were central to our social life. Sylvia Howard, who we had met in Indianapolis, sung at Sunday brunches with a small group of musicians, including the bassist Peter Giron and an excellent Italian pianist. Sylvia was one of several singers who had come to Paris as part of a black American musical show and remained. Other musicians, such as the trumpet player Rasul Sidik and the guitarist Juju Child would sit in; we danced. Silvia, Peter, Rasul, John Betsch, Mra Omra, Katy Roberts were among the musicians at Sept Lezards, a restaurant with a jazz cave in the Marais. Again we danced although there was even less space. The musicians loved Sept Lezards and were angry when the government closed it supposedly to build an entrance, which was never built, to a park or monument behind the building. Probably some influential neighbors had complained about late night music. Mra Omra, a trumpeter from Philadelphia who married a French woman, managed a small club in the Gout D'Or, a then dangerously unlit immigrant area. Katy often played piano and Rasul would sit in on trumpet and we danced. It was like being back in Nigeria or Harlem, but the city closed the club as part of its redevelopment of the Gout d'Or. As jazz in Paris, especially as black American jazz musicians, moved from club to club we followed.

We tried to dance several times a week which included Brazilian and Cuban clubs where the music and temperature were often hot, and Adele learned by experience to guard her purse. I especially liked a club on rue Moret that had Cuban food along with excellent musicians on Sundays. After it closed another less exciting Cuban club opened which was soon replaced by a jazz club which was temporarily in fashion and I do not know what is there now. Mra Omra's excellent Brotherhood of Breath sometimes played at the old Duc des Lombards and had Bobby Few on piano until he was replaced by Katy Roberts. We used to dance at a blues club in St Quen, just outside of Paris, that often changed its name, and as the evening progressed became wild as one of the owners would climb on a table, and drunkenly offer bottles of brandy and scotch. The music was good, often a jam session led by the famous American trumpeter Boney Fields, but the woman co-owner bought out her male partner, everything sobered up, and she eventually sold the place that was no longer of interest.

We still dance, but now in a tiny area, at the modernized shiny Duc des Lombards, currently the best jazz club in Paris, indeed the most likely place to hear famous and new American musicians along with the Israeli musicians now prominent in New York. We are usually the only dancers and perhaps because of our age, perhaps because we continue a tradition of dancing in French jazz clubs, we only pay for the first concert and stay on for the second set. I am told that we are a "legend". Touring New Orleans musicians often perform there. We have danced to Evan Christopher's European group, Henry Butler with Steven Bernstein's Hot Nine and the Stanton Moore Trio. We invited James Singleton and David Torkanowky and David's girlfriend to lunch and I was surprised that David who seems distant and superior on stage was like a house trained puppy, taking dishes to the kitchen and washing them between courses. I still do not understand how they were going to fit in another meal to which they had been invited before the show that night. David arranged for our tickets for the first weekend of Jazz Fest and hospitality passes to the Executive Committee's dining room and (more important) toilets. When Adele said she had trouble climbing David told her that his mother had travelled around the world as a flamenco dancer and now was so arthritic that she could no longer walk across the kitchen. I looked her up and indeed she had been a famous dancer.

We used to go to the Jazz Club at the Hotel Meridien-Etoile at least once a week but the policy changed and the music is more often jazzy than jazz; there are seldom visiting American bands and the club is no longer called the Lionel Hampton Jazz Club. It is only now open four nights a week, instead of six, and promotes food and cocktails. The Caveau da la Hachette has good dancers and traditional and swing period jazz, but there are long lines to get in, the tiny floor becomes too crowded and the stairs to the cave dangerous. Mra Omra bought and reconverted a building in the suburbs for a jazz club but whenever he tries to open "Espace Lady Day" the police claim he lacks a further permit or authorization and I imagine he has neighbors who will see that it remains closed.

Rasul is from a black family that for several generations has married white and become lighter. He complains that he is too dark to be white and too white to be black in films. His family has usually been involved in New York black politics, but he left for St Louis and Chicago to be part of the Free Jazz scene and feels close to the Art Ensemble of Chicago. He is a well-known musician who has been a sideman on hundreds of recordings but lacks the discipline to be a leader. His revolution may never come. When he eventually paid to record a CD under his own name it was a throwback to a late 1950s Free Jazz ensemble with too many musicians playing at once. He is highly respected, musicians claim that his sound is instantly recognizable, and I have been at large concerts where although he was a sideman the crowd chanted his name whenever he soloed.

Katy is white, an excellent pianist. And wild. There are stories about her studying in Ghana, her automobile having a flat tire at 2AM in Harlem, her falling off a snow covered roof during a party and breaking her leg and hip (why she was there is another story), being jailed after a concert for driving in the wrong direction,. She is however more "together" than Rasul. She, unlike him, has her French papers and teaches at a music conservatory. She rents an apartment; he is homeless and hopes others will lend him an apartment or bed. He cannot live with Katy as she says he will suddenly start practicing trumpet long after everyone else has gone to sleep. Indeed whenever invited to anything he turns up very late, sometimes even after it is over.

I can chart our aging by how New Year's Eve has changed from our dancing through the night to Latin bands at La Java and New Morning (which now has electronic music for the occasion), dancing and eating at African clubs closer to our apartment, dancing at an Algerian restaurant even closer to our apartment, and in recent years a non-dancing multi-course banquet at a Thai restaurant even nearer at which we are the usually the only non-Thais invited except for a few French husbands. I know we can stagger home without aid, or if needed it will be offered. We used to dance (and once danced New Year's Eve) at Opus, a former British officer's club, not far, further up the Canal. Besides live Latin music it then had Bal Musette once a month on Sundays. It has a new name, the hole in the dance floor is repaired, and the music is often soul and funk played by DJs, but it still has the same bouncer who recently asked where we have been since he last saw us.

That was on a Friday when we were returning from the opening of La Philharmonie concert hall at Parc de la Villette where we listened to William Christie and the Arts Florissant, an ensemble specializing in Baroque music that we have tried to hear whenever possible since several decades ago when I thought their production of Mac-Antoine Charpentier's *Medee* at the Opera Comique the most profound dramatic experience I had ever experienced and became a fan of Baroque opera. The black masque ballet in Act Four was memorable. When I was a student in New York I discovered and enjoyed chamber music of the Baroque and Classical periods but opera and the trained human voice seemed to me caterwauling unlike the blues and jazz singing I enjoyed. Adele likes the singing voice and eventually I became converted. I now attend Baroque operas whenever I can, provided I can afford them although I still much prefer to see actual productions. I seldom listen to Adele's many CDs of operas and singers. A week later we heard Agostino Steffani's *Niobe Regina di Tebe* at the Theatre des Champs Elysees, played by the Boston Early Music Festival Orchestra and sung by Karina Gauvin aided by the great countertenor Philippe Jaroussky whom Adele needs to hear several times a year. I understand that this was only the second production of Steffani's 1688 opera since it was rediscovered a decade ago. We probably go to forty concerts a year and I am surprised when friends do not. I like the old Cité de la Musique concert hall now called Philharmonie 2; the new grand Philharmonie 1, built to compete with the large concert halls of Berlin and Vienna, I find cold and avoid. I assume it is a bad joke when officials claim that Philharmonie 1 was built to attract the poor immigrants living outside the city boundaries.

We do not and never have had a television set and go to several films each week, especially at the Forum des Images where we often see two films a night. Membership for seniors is less than an hundred euros a year and allows for unlimited films and also reduced prices at the UGC in Les Halles. There are intense periods of film going early September for L'Etrange Festival (a private artistic enterprise now sponsored and better supported by the government) and in late May, early June, by films from the Cannes Festival. The Forum presents a retrospective of Quinzaine des Realisatuers; Un Certain Regard films are shown at the Reflet Medici. After seeing perhaps 10–12 new films within a week I really do not care that other cinemas have the Semaine de Critique and for a few days it is possible to see films from the main competition which will later be shown commercially. It was at an L'Etrange Festival we were being drawn by Paul Poutre, who interviewed us, thought it amusing when I said the films had become more mainstream since the government took over, and who became a friend. He seems to have continual girlfriend troubles and at one time was limiting himself to lesbian parties to avoid involvement.

Because of increasing problems being on our feet for long periods we seldom go to large art exhibitions. The small Museé Luxembourg is about my limit. We went to a small display of Romare

Bearden's Matisse influenced Odyssey art works at the new Global Gallery that Columbia University opened. I first came across them by way of Derek Walcott who had used Bearden's 'The Sea Nymph' for the jacket of *Star-Apple Kingdom*. At the Romare Bearden exhibition we met our friend John Betsch, a drummer who was part of the New York East Village loft scene before moving to Paris. Besides often working with Steve Potts, John is also a painter. We also rediscovered Claire Caillard and Berry Hayward, Early Music musicians whom we often used to see at jazz concerts before their son, the saxophonist Timothy Hayward moved to the USA. I was surprised Claire and Berry have choral workshops at Ateliers du Chaudron where they sometimes have concerts with Steve Potts, Peter Giron, and John Betsch.

Although I graduated from Columbia College I usually avoid Columbia events. I remember going to hear Harry Connick Jr at New Morning when he was supposedly on leave from being a student at the College; but he was part of the New Orleans jazz scene and I might have gone anyway. Another time we went to a drinks party given by the Dean of Engineering at some fashionable bar, and I became aware that Columbia was now indeed a multiracial, multicultural, multinational university unlike when I was a student.

My third and most recent contact with Columbia until the Bearden exhibition was attending a festival of Indian literature to which several writers I know were invited and some of my friends with interests in India were in the audience. Since the publication of the second editions of *Modern Indian Poetry in English* and *Three Indian Poets*, Indian writers, their translators and the few scholars of contemporary Indian literature are among our friends in Paris or when they visit.

At Columbia's Reid Hall Jeet Thayil and Karthika Nair were reading. I had known Jeet since my first trips to India, Adele invited him to read his poetry at Ball State, I had reviewed several of his books, stayed with him and his wife in Bangalore and Delhi, he had republished one of my essays in anthologies of Indian poetry, and I had written a chapter about him in my recent *ReWriting India*. By now we were old friends and he told the audience that I was one of the few serious critics of Indian literature. I had met Karthika Nair through him when he asked in an email message whether I knew she was living near us in Paris.

Karthika, also originally from Kerala like Jeet, is remarkable as she is afflicted with E B, epidermolysis bulosa, a terrible condition from which people die young and which means that she needs be extraordinary careful that she eat nothing solid or rough and which requires her constantly to go to hospitals for blood transfusions, operations, and treatments. In spite of her condition she learned French well enough in India to be sent to Paris to study Arts Management, be hired as an administrator at the Cité de la Musique, be made second in charge at the then new Centre National de la Danse (the person who was appointed Director supposedly wanted an assistant who knew less about the subject than himself), and having learned about contemporary dance and its contractual and financial sides she quit to work with two major dancers and their companies, Akram Khan and Sidi Larbi Cherkaoui. She has written texts for the former and managed the latter. She also published a book of poetry (*bearings*), a prize winning Indian tale for children in both French and English (*The Honey Hunter*) and her second book of poetry, *Until the Lions*, based on the *Mahabharata*, even before being finished was predicted to be one of the best books of 2015.

We see her regularly and I like the challenge of preparing different soft foods as I think she normally eats poorly. Despite her condition she looks smart, well dressed, fashionable. Karthika knows that she is likely to die while young and wants to accomplish as much as possible. I and others are amazed at her continual trips to India, China, Korea, England, Belgium, Italy, to see her family, on writers' retreats, to discuss the travel and accommodation of dance companies, to consult on tours, to read at literary festivals. If it were not for her I probably would have lost my interest in contemporary dance, which I once again often see.

We also see Laetitia Zecchini, although she is often so busy that she will disappear for weeks without any sign or word. Laetitia's father was a foreign reporter for *Le Monde* and she was raised in New Delhi and Washington before being sent to Paris to prepare for university entrance. After writing a doctoral dissertation on two English and two Hindi writers she was appointed to the Centre National de la Recherché Scientifique and has become France's leading expert on Indian literature and Postcolonial Theory. I met her by way of Janet Wilson and later a Parisian Indian literary journalist who has dropped from our sight. As she was working on Arun Kolatkar I put her in touch with Arvind Krishna Mehrotra who was literary executor of Arun's English language poems and translations from Marathi. Arvind was impressed with the essays she sent to him and she wrote the notes for the *Jejuri*, *Kala Ghoda Poems* and *Sarpa Satra* volumes in the Bloodaxe *Arun Kolatkar: Collected Poems in English* (2010). Then she published an impressive French translation of Kolatkar's *Kala Ghoda Poems* (2013). We had visited Oxford when Arvind had been invited to give lectures and at a dinner, given by Janet and Kevin, Laetitia met Matthew Feldman co-editor of a series of books, "Historicizing Modernism", that Bloomsbury publishers would soon take under its imprint. This resulted in Laetitia's *Arun Kolatkar and Literary Modernism in India* (2014), one of the best studies of Indian literature. Among the "Acknowledgements" Bruce and Adele King are "the best dancers, jazz aficionados and gourmets in Paris ... I cannot begin to enumerate the ways in which I have benefited from their intimacy with Indian literature and Indian writers". Laetitia was part of the Reid Hall program interviewing Amit Chaudhuri, who I had met decades previously at a Cambridge University conference and who was one of the eight authors given extended analysis in my new book.

Laetitia is tall, beautiful, has dark eyes, long dark hair, loves clothes and shopping, good restaurants, dancing, meeting people, and travel as well as her work and took the photograph of me eating with chopsticks at a Chinese New Year celebration that is on the inside back cover of my *ReWriting India*: Oxford airbrushed out the dancing lions that hovered behind me. She married late August, 2015, at a large wedding outside Vannes for which we returned from Hvar.

In the audience was Simone Manceau, who had translated several of Amit's novels into French, and who we had originally met in New Delhi at a party given by her friend Ritu Menon, publisher of the feminist Women Unlimited, whose books about Pakistani literature and Urdu Indian writers I sometimes reviewed. Simone was originally from North Africa—her first language was Arabic—and is a former university professor of French in the USA She has an immense apartment in Paris, a house in the south of France, and often is travelling with her husband Eric, a semi-retired economist. I never feel I know her well but she keeps surprising me, once bringing me a free-lance TV writer who was her "orchid doctor", who made my orchids blossom, and another time inviting us to a meal with a famous tap dancer.

Also in the audience was Vera Dickman whom I met two decades previously while I was teaching at Charles V and she was trying to find where she might fit within the French academic world. An extremely attractive actress in South Africa she had come to France to sit at the feet of the famous writer and painter Henri Michaux about whom she had written a French dissertation and then discovered that she was unlikely to be hired to teach French at a French university. She was bright, super-energetic, interested in everything, and without the abstract categorizing mind that French education forms and requires. Eventually she found her niche, as a teacher of translation, foreign languages and cultures, at which she proved excellent and soon became head of a large department at Telecom, an organizer of conferences, and went to meetings around the European Union to explain practical ways of learning about foreign languages and cultures.

She edited a book publishing papers from my "Return in the New English Literatures" Conference, curated exhibitions of Henri Michaux's paintings and wrote articles about him. That the three volume Pleiade *Henri Michaux Oeuvres Completes* includes many notes by her means that she has gained a scholarly prominence superior to the French academics who dismissed her as hopelessly untrained. When she ate with Karthika, and us one night of the conference, Jeet was impressed by her having known and written about Michaux whom he admired.

Her husband, Mike Dickman, also achieved notice by an unusual route. A well-known rock guitarist in South Africa he followed Vera to France, dropped out from performing and spent his time meditating, studying Buddhism, and taught himself ancient Tibetan of which he has become a respected translator. He teaches Tai-chi, sits in with improvising rock groups at local bars, and, as I learned at his immense sixtieth birthday party, has become an international guru. He sometimes performs at readings by Denis Hirson, an excellent poet, editor and translator, and also a South African, who is their friend. Denis is also president of Ateliers du Chaudron in Menilmontant where we go to occasional jazz concerts led by soprano saxophonist Steve Potts; we dance when we find space.

The opening of the Bearden exhibit was on a Monday. Tuesday, we saw a new production of Samuel Beckett's *Fragments* directed by Peter Brook at the Bouffles du Nord. This was the second or third time we saw *Fragments* directed by Brook and the best. The actors seem to improve each year, become more and more living, amusing eccentric, characters rather than performers of a text. Yet they embody the pointlessness and pains of life while the ways they respond are comic, engaging, and part of the desires, hopes and pleasures of continuing. Where most writers are loquacious in trying to give significance to a godless, purposeless world, Beckett understood that reduction to the comedy of survival said it all. The combination of the minimalisms of Beckett, Brook, and the Bouffles du Nord is in itself perfect, and the theatre is within walking distance of our apartment so we go there often. It helps that the theatre is in a neighborhood of inexpensive Indian and Sri Lankan Tamil restaurants, some of which are pure south Indian vegetarian. Although Bouffles is no longer Brook's theatre it carries on in the 1960s spirit of minimally staged plays and music, stripped down embellishments—there are even pillows for seating on the performance space.

A few nights later we had dinner at a small restaurant, "Comme sur une ile", with our friends Dominique Guedj and Marika Rosen. It was the first time Adele and I had eaten food from the Isle de Maurice, and we enjoyed it. The first course included the comedy of the others claiming that my smoked Marlin must be Cod (Merlu), the second involved an extremely hot pepper sauce that without tasting I added to my already spicy curried shrimps and rice.

Dominique is my former Israeli student who had become a published scholar on the works of the Romanian French philosopher Benjamin Fondane. For a few years she seemed directionless and talked of becoming a Spanish language actress, of opening a studio for learning the tango (she even went to Argentina for a month), then quit her job as a hospital receptionist so she could study Chinese massage and exercises. Her affair with a Chinese immigrant and her trip to China is long over but she now rents a place to give massages and travels around Paris teaching qigong. Dominique looked happier, better dressed and more attractive than ever. She had always been flirtatious and amusing and although she must be at least twenty five years younger I think we both have fantasized a fling; she told humorous stories about her son who was finishing art school wanting to be a publisher of erotic cartoons and planning to go to Lyons with his girlfriend. Dominique who used to laugh at herself for acting like the stereotypical Jewish mother protecting her son and his dependency on her from unsuitable women, now seemed free of such worries. "Let him starve."

Marika was in Paris for two nights for a medical checkup before returning to Nice where she moved last year. She was from a small Sweden town, a tall Sweden beauty (Laetitia often termed her "flamboyant"), and was studying chemical engineering when swept off her feet and impregnated by a married man who divorced his wife to marry her. She raised his two sons by the previous wife along with their son as they moved from Sweden to Norway to Italy and Paris as he became a European Community advisor on nuclear projects. She had unused energy and travelled twice to Ghana to study drumming, became expert in Japanese flower arrangements, a serious cook, a respectable amateur painter who won prizes and fluently spoke several languages. She loved to tell stories about Swedish foibles such as everyone being so enamored of Donald Duck on television shows for children that they had to change church times on Christmas Day.

She is one of the people we met through music and dancing. We regularly went to a blues club behind an Algerian restaurant where on Sundays there was a crowd of all ages, often in what the fashionable would consider vintage clothing except these were well worn like their owners, dancing to a great world music band called Paname Tropical. The musicians had in the past lived in French colonies or former colonies and the star was one of those drummers who could get everyone wildly dancing without needing anyone playing melody or harmony. It was comparable to the groups I had danced to in Africa or the Caribbean. They had a job at the Swedish church in Paris but we needed an invitation and that is how we met Marika who had persuaded the blond female pastor to hire them.

Her husband had died a few years previously and a gay Caribbean former tenant was escorting her around parts of Paris she was unlikely to go on her own when she heard Paname Tropical and also became a follower. After the band had a dispute over pay with the blues club we went with her to several restaurants featuring bal musette outside Paris; they eventually broke up. The leader now teaches saxophone at a music school and the band has become a legend. By then we had become good friends with Marika, went to each other's parties, and when she developed serious hip and leg problems it was Dominique, with her hospital connections, who found Marika the right specialist for the infection. Marika who remained restless sold her summerhouse to buy an apartment in Paris (her husband had invested in the summer home but rented in Paris), then sold the Paris apartment to move to Nice where she finds it easier to walk for exercise as well as it being warmer. Her cat also likes having a large garden and other cats with whom he can fight.

Gabriella and Nancy are among our closest friends. Gabriella, a Southern Italian, is very feminine, clothes conscious, always looks great, is an excellent cook, taught French in the USA and worked as a senior administrator for Skidmore and NYU programs in Paris. Nancy is tall, thin, almost American Gothic, raised in Kentucky and Indiana, and is or was a militant feminist who lived in communes, shared apartments with prostitutes, worked at bars, and was for years employed by the American government as a folk culture administrator. Their marriage a few years ago was the only lesbian wedding I have attended. The actual roles between the two women are the reverse of what I would have expected. The femme Italian is strong, aggressive, a fighter. She speaks casually of the various mafia and worse organizations that continue to operate where she was raised, and where she often returns to see family. Although she seldom cooks I learn from her: mix raw garlic with warm green beans, cook vegetables with pasta (—a Sicilian film jokes about cooking potatoes with spaghetti but yes it is done.) When I first met her I was puzzled by the lack of vibrations between us, the odd attraction that men and women have for each other regardless of age or position; then after seeing her in the company of some masculine women I understood. We became just friends.

Nancy is often ill, does most of the cooking, needs company, talks incessantly, reads novels, follows the news, and avoids confrontations. She is an excellent dancer, and is always trying to get Gabriella to follow her lead. Nancy likes large gatherings for meals; a birthday or anniversary is reason to invite others. Her annual Thanksgiving dinner has about twenty people, mostly French who somehow know and remark that Americans would not begin with champagne or have a cheese course, although everything else is based on what Nancy's mother would serve. As, however, she is a strict vegetarian the turkey must be cooked in a neighbor's oven. I only once had a real Thanksgiving dinner before I came to Paris—in Muncie a friend invited us—so this is as exotic to me as to the French. I notice that a few streets from our apartment an uninviting French bistro each year for three nights offers a Thanksgiving turkey with stuffing and trimmings. Maybe like Halloween this will become a French custom, an excuse for still another holiday off work.

I like to be around attractive, young, bright, well dressed, amusing women (Laetitia has joked about my harem), and the men we know are usually musicians. I am a good cook and serious when shopping for food. French women have several times remarked that they cannot cook so well and would be frightened to invite us for a meal. I was sorry to see Marika move to Nice and Dominique become so busy with her new career as they are both excellent cooks without inhibitions about inviting us.

Like many who moved to Paris food is important to our lives; it is not only something to eat. At first we belonged to a Parisian branch of Slow Food, but it was soon obvious that the French members knew less about food and were more easily satisfied than us. I did meet an English woman married to a Frenchman who had paid trips around the world researching such basics as potatoes and tomatoes for beautifully photographed books of international recipes sponsored by French produce associations. He was in advertising and would not let her prepare anything except leftovers since the results were not gorgeous. We decided not to renew our membership after an especially dismal meal and evening in Montmartre, that I thought no self-respecting naïve tourist would recommend but which was packed with celebrating groups of Parisians. As the majority of Parisian restaurants serve already commercially prepared food I say they are good enough for the French and leave it go at that. The better restaurants are discovered and patronized by foreigners like us. Our favorite restaurants in Paris are "Astrance" and "Abri", where we started going soon after they first opened, and where we are welcomed. To those outside Paris restaurant going the names may mean little; those who have telephoned month after month seeking reservations will be envious. Adele translated Philippe Delacourcelle's first cook book into English (*Spiced: Recipes from Pre Verre*, 2007), but we have not yet gone to the new restaurant at the other end of Paris which the brothers opened after selling Pre Verre. Our life changes each year. January, 2015, I decided not to have a birthday party; instead Vera invited us, Karthika, Laetitia and Philippe to a "cheap and nasty" Indian restaurant of my choice.

There are many others who we see, should see and some whom we promise to telephone to meet, but between writing, reviewing, appointments with doctors, seeing those I mentioned, evenings dancing, and many jazz and classical concerts, and films, there really is no time. Janet's *Journal of Postcolonial Writing* will suddenly need a review article for a special New Zealand or perhaps a general issue and could I help? Janet will be asked to start a new series of books for a German publisher and I will think, this is my chance to republish some of my essays and reviews, and a new, unexpected 600 page book will emerge and urgently shove aside the autobiography I was writing. Retirement to Paris is not restful, for that we go to Hvar.

# Chapter 19: Second Home

*New Orleans 1988–2015*

If we did not live in Paris I would live in New Orleans. London and New York are attractive, and there were times in the past when I lived in them and desired to return, but they are expensive, large, exhausting, speeded up cities, no place to begin again. You need the energy, blind ambition of youth, the comfort and ballast of a continuous past, or be moneyed, for such cities. New Orleans had and has jazz, especially New Orleans traditional jazz, New Orleans rhythm and blues, New Orleans brass bands, New Orleans klezmer, New Orleans Canjun, every kind of music. It has excellent food, a restaurant scene, movies, gossip, larger than life characters, history, places to dance and great dancers. It is a beautiful city, especially the French Quarter, Faubourg Marigny, Bayou St John, with a multinational culture. It is a place of tolerance, easy living and opportunities. New Yorkers like to brag if you can make it here you can make it anywhere; if you cannot make it in New Orleans you cannot make it anywhere. It is filled with unusual people, musicians and non-musicians, who have drifted in, stayed, and flourished. Depending on entertaining to survive, it attracts entertaining people as well as entertainers. New Orleans can be violent and risky, but it is also a large open stage in which those who might have dreary bit parts inflate and gain attention.

The people I know beyond New Orleans are writers, artists, musicians, journalists, university professors, arts administrators, parts of the cultural world. Except for musicians I know few such in New Orleans. When we were walking one night Sammye happily shouted that she was "drunk as a skunk". She turned out to be a compendium of lively southern expressions, such as "[ra]coon assed", that were new to me. She was originally from a small town in Mississippi and abandoned by her first husband who left her with a gasoline station. The New Orleans lawyer who became her second husband was married but after having satisfied his need for gasoline, oil, and other matters, divorced his wife, and brought Sammye home, made her his secretary and married her. He was successful and owned houses and oil wells. When I met her she was waiting for renters for one of her properties across from her house. The next afternoon we met her again, with her husband and we discussed renting one of her houses the next year. We never rented from her, but once exchanged our Paris apartment for her house for a month. As Sammye regularly keeps open house in her carnivalesquely decorated garden for anyone with a sense of humor who enjoys drink and talk, our having the gates locked as we went out every night was a disappointment to her neighbors and friends.

Instead of renting from Sammye we rented a house and the next year exchanged our apartment with Marge Lea, raised in the small Mississippi town, but who left her husband in California for New Orleans. Marge obviously enjoyed men, that was one of the reasons for leaving California, seemed to make money, she claimed to earn more than the husband she left, drank too much and eventually gambled away her money and house, failed to declare and pay tax on all her income and the IRS had a lean on anything she earned. As soon she earned money she spent it wildly before the government found it. She had a younger, attractive, boyfriend, who supposedly took her car when he left, although by then whatever she said was mixed with fantasy. She died of brain cancer. She would talk of her sex life, how long the average man lasted before ejaculating, and remarked that Adele and myself danced and still regularly enjoyed sex because unlike her and her boyfriend we were not heavy drinkers.

Johnnie, the third of the friends, remained in Mississippi with the chicken farm that her run-away husband left, but she and Sammye often visited each other, and while she was less amusing she shared a taste for sexual jokes. Drinking a coca-cola she offered me the cherry -"it has probably been a while since you had one". She was the serious one of the three. One April we rented a small house from another of Sammye's acquaintances. When we opened our door there often would be crowds of strangers with bottles and food on the steps, our introduction to stoop parties, supposedly a New Orleans custom.

Through Sammye we also met Rae with whom we stayed four years during jazz festival month. Sammye and Rae both had lawyer husbands who invested in property and oil wells. The lawyers were their second husbands although Rae once confessed to having had a third husband whom she refused to remember, count, or talk about. Her husband's death left her with a beautiful house in the heart of the French Quarter and five smaller houses behind in the "plantation". Rae and her husband had lived well, travelled to London and Paris, and were part of the social scene in New Orleans that seems to have been the only world she knew and when she was no longer part of it she had no idea what to do. She was from a wealthy California family that sent her to finishing school in New Orleans, which was the extent of her education before she married, divorced, married, became a widow. Now she remained house bound, crippled, and increasingly weaker because of her refusal to exercise, walk, leave the house for anything except to see doctors that along with endlessly going through bills and old documents was her life. She bragged never to have worked, never to have shopped for food, never to have done housework, never to have made a meal or coffee or tea, never to have washed dishes; her servant could do it for her. She had one servant, Alfred, a black man who followed an unusual version of Christianity, and upon whom Rae depended. She was perpetually concerned about money, hated to pay taxis to take her to doctors, and required Alfred to drive her car although he had no driver's license and was uninsured. He shopped, made her meals, prepared the sequence of medicines she was supposed to take, and was expected to be always available. If she fell in the middle of the night going to the toilet she would telephone Alfred to rescue her or if she was unable to reach her telephone she would lay there all night until he arrived in the morning. I once found her on the ground and tried to lift her on to her feet, but she was so hysterical, screaming all the time to telephone Alfred, that she pulled me down although I must have been three times her weight.

Somehow Sammye had persuaded Rae to rent us a room in her house. Rae hated the idea, had never let anyone rent her guest room, and began by making so many conditions that I told Sammye it was a mistake, but Sammye made Rae back down and to everyone's amazement Rae decided that she loved us, that we were like family, that we must keep coming to stay with her. She spoke of Adele's beauty and claimed I married Adele for her legs. Rae had been an attractive woman when younger, and would praise my shirts or jackets, but she had no idea that couples shared cultural interests, wrote books, had intimate as well as social lives. She used to laugh at the tough times she had given her husbands, refusing to live with them for years out of sheer contrariness. We won her affection because we dressed and danced well and we would sometimes take her to monuments of her past. She especially enjoyed going to a jazz club where she could watch us dance, although she would before long need to return home. One of her husbands had been a professional Arthur Murray instructor who on their wedding night had danced with someone else more his level. She kept saying that we danced like professionals and we would send her copies of photographs we were sent of our dancing. Although she appeared to live on grilled cheese and ice cream she liked to be taken to the latest restaurant where she would pronounce on the food she would not eat. We hardly had anything in common but when she died in hospital I felt sad and missed her.

 Marisol was the restaurant we were going to when we met Sammye and her husband. Opened in 1999 it was within a few years recognized as one of the best restaurants in America. Marisol was where the hip went to when visiting NOLA. It was located at the bottom of Frenchmen Street on the Esplanade in Faubourg Marigny, a spot since occupied by Mexican or South American restaurants one after another. Peter Vazquez who had come to New Orleans from Pennsylvania by way of cooking in Virginia, married Janis, a local artist. They were not ideal to run the most far out fusion restaurant in New Orleans at that time. Peter made aggressive comments in interviews about not serving anything that resembled traditional Creole and Cajun food, while Janis kept her distance from customers who were not her friends. Peter created unfamiliar dishes, while Janis' manner offended those locals who decided to give the place a try. On-line there were complaints about waiting over an hour for water or bread and eventually leaving because no one seemed to be interested. Janis felt taking orders from customers was the job of the server who thought it her job. As business fell there were less expensive dishes and even all-you-can-eat luncheons where Peter obsessively cooked great fresh food non-stop for a few customers. It is spectacular to have over 40 cheeses, each with its own house made accompaniment, but Marisol should have been in New York, San Francisco, London or Paris, not New Orleans at that time. Hurricane Katrina in 2005 was the excuse to close.

It also allowed Peter to become even more maladjusted. He appeared unwilling to work again full time as a chef. He provided homemade food one night a week at a wine bar. There were on-line notices by him that he would like a job cooking in Thailand (just what Thailand needs, an American learning to cook Thai food); he would like to be a security guard in London and indeed he had a job at a protection agency in New Orleans. He has an obsession with weaponry and is large, but he also has a bad leg and limps. He and Janis separated and they argued about his jobs, as she felt the way he could support her was to return to cooking which he refused. He unexpectedly became co-chef on Sundays at an obscure Italian restaurant on a strip mall far outside of New Orleans that everyone wanted but never found the time to try. The meals sounded great but I do not know anyone who actually drove there. Then there was a strange pop up restaurant one night a week at a place that opened for breakfast and lunch, followed by an always different Sunday menu at a Deli. Then for three years a hush-hush weekend on-line restaurant where you could order what he felt like cooking provided you came to get it, followed by an appeal for contributions to fund a tiny take away counter in Algiers Point near his house. I know cooks can be eccentric and will suddenly close a successful restaurant to travel around the world, become full time painters, or decide to cook in exotic foreign lands to learn the cuisine, but Peter is eccentric even in comparison to other eccentric and temperamental chefs. He wants to cook what he feels like cooking when he feels like it, and mostly what he wants to cook requires a personal following of those with a taste for the exotic. New Orleans may be one of the few places he could survive.

Chris De Barr, a friend of Peter Vazquez, is even larger and could indeed pass himself off as a bouncer or security except that his knees are weak and he has supposedly left New Orleans for better health coverage in Portland Oregon. He became famous as the evening half of The Green Goddess, an almost infinitesimal restaurant, on Exchange Alley, which combined the exotic with the local. I loved the extreme flavors of duck sausage, mashed sweet potato, garlic and Steens syrup (a Louisiana specialty that is almost impossible to find and use elsewhere because so sweet). There was an Indian vegetarian uttapam, a watermelon soup with crabmeat and avocado, and other dishes which made the no reservation policy and maneuvering for a table bearable. Chris liked to talk especially to single women, liked to offer samplings of his excellent wines, and foolishly left The Green Goddess for a larger restaurant doomed before it started. The location was unpromising and his partner was unlikely to wait for it to attract the sophisticated following it needed. The food at Serendipity was uneven in execution and I was not surprised when after differences of views with his partner Chris De Barr resigned to move to Portland. Like Peter Vazquez he had rapidly gone from having one of the ten best restaurants in New Orleans to having none.

There are now some great innovative restaurants such as Peche and Shaya, but a decade ago those chefs and restaurants before and that survived Katrina were less adventuresome. Susan Spicer was one of the first fusion cooks in the city and she taught many of the other significant chefs but the menu at Bayona is modern French influenced American and Louisiana food. Her fried oysters coated in cornmeal is a version of a dish common to New Orleans where many restaurants, some without pretensions, offer deep fried oysters, sometimes with pasta and garlic. The large, thick, juicy Louisiana oysters attract cooking and accompanying flavors unlike the smaller, more delicate, tasty oysters of France. The early twentieth century immigration of Croatians is said to have introduced the flaming of oysters with brandy, butter, and garlic, that is often imitated, but by others grilled with melted cheese or cheese and spinach.

I have not eaten at the famous old French restaurants in New Orleans for years. They may well have changed but I do not enjoy rich sauces, meats long cooked in wine, potatoes twice fried in fat. Some of the famous award winning New Orleans restaurants I find disappointing, even bad as there are too many different sauces on a plate and they all run together into a mess. There is a good inexpensive West African and an excellent chain of middle- eastern restaurants. I enjoy the food tucks around the city and small stalls that pop up at jazz festivals. I look forward each April to mango ice, fried soft shell crabs, quail gumbo, trout baquet, stuffed crawfish heads, and other specialties of Jazz Fest. Shrimp Monica is often imitated, but the recipe remains a secret. Some of the best food I have eaten in NOLA was home made. People enjoy and enjoy making food.

Much of the cuisine of New Orleans is a mix of the southern or Creole with immigrant cultures. The largest immigrant community is Italian which accounts for the New Orleans accent sounding more like New York than Atlanta and local food being a blend of Italian with African and Gulf. Eggplants, pastas, are topped with shrimp, oysters, crab, or other seafood and served with peppery hot tomato, garlic, or creamy pink sauces. Even remoulade sauce in New Orleans is African spicy hot and likely to be served with shrimp. New Orleans cuisine is African influenced Creole with okra, tomato, and peppers unlike the inland Cajun with its thick, rich roux and the scraping of the skillet to "dirty" starches.

There has always been a connection between food and jazz. It may be one of the few cities where food is offered for free at neighborhood jazz bars and where you could eat every night of the week by changing bars each night. Some even give away oysters. My favorite for a decade was the legendary Donna's Bar and Grill. It had great jazz every night and before the final set on Monday offered all you can eat of barbequed chicken along with red beans and rice, potato salad. Charlie Sims, Donna's husband and a former railway chef, enjoyed cooking although on most nights the menu was limited to fried chicken, French fried potatoes, and several pork dishes, but there were excellent purees.

Donna's was a small place fabled for its live jazz; the food was an afterthought.

Donna opened the bar on North Rampart Street in October 1993 as home to the new generation of brass bands that began emerging in recent decades; originally it had brass bands six nights a week. Then Donna's began to hire other musicians and became the place to hear jazz. Many of the audience were musicians or other entertainers who would sit in. When New Orleans jazz musicians became famous in the north and in Europe they would jam at Donna's or Vaughan's when they returned. It looked like the worst of dives and was in an area that could be dangerous at night, but jazz has often thrived in insalubrious surroundings, and at Donna's it thrived until a gunman demanded money from every one at the bar, the landlord refused to repair the increasingly dangerous building, and Charlie felt that he had had enough.

When you hear a performer in California or Paris say "now we go to New Orleans" they mean that the drummer will be playing that recognizable cross rhythm with its supposed Spanish tinge popularized by NOLA brass bands for the past half century. New Orleans is a great place for soul, blues, rhythm and blues, even zydeco, but that brass band drum beat has become characteristic. The modern brass bands are an off-shoot of the earlier New Orleans jazz revival. When the small group of intellectuals in the late 1930s and 1940s, including William Russell, began examining the early history of jazz and recording survivors they began a worldwide "trad" boom, which I and others I knew in my teens and college years participated. Besides recordings we admired New Orleans musicians then playing in New York and Chicago including Sidney Bechet and Albert Nicholas (before both moved to Europe), Pops Forster, Baby Dodds, and Danny Barker. Danny, a banjo and guitar player married to blues singer Blue Lu Barker, returned with his wife to New Orleans in 1965 and at the urging of a local pastor encouraged black youth to study the city's musical heritage and formed the Fairview Christian Marching Band which soon added a jazz ensemble.

Many jazz musicians who were to become significant to New Orleans in the near future played in these ensembles—Wynton and Bradford Marsalis, Michael White, Leroy Jones, Greg Stafford, Shannon Powell, Lucien Barbarian, Nicolas Payton, Kirk Joseph. The Fairview Marching Band lasted four years and was followed by musicians forming their own groups which kept sub-dividing to start further bands. Leroi Jones' Hurricane Brass Band gave birth to the Tornado which seeded the Dirty Dozen. The Dirty Dozen was the first to respond to the popular music to which young blacks were dancing; they played their own compositions influenced by soul, Latin, rhythm and blues. Younger musicians in the Rebirth Brass Band and then the Soul Brothers went further in this direction by including hip hop and singing. Michael White and Greg Strafford went in the opposite direction; Michael White's Liberty Brass Band attempts to continue older brass band traditions such as multi-themed compositions, keeping to the distinctions between trumpet lead, reed obligatos, and supporting trombone roles in ensemble, continuous ensemble playing and complicated harmonies. The Fairview bands and Barker, were important not only in renewing New Orleans brass bands, but in making young black musicians central to the local jazz scene as the traditional jazz revival had been keep going by white Europeans such as Sammy Rimington, Barry Martyn, Clive Wilson and Lars Edegran who moved to New Orleans to learn from and play with the older local musicians.

Adele and I were in New Orleans for a Modern Language Association conference over the Christmas holidays in 1988 and rented an apartment from Nina Buck. We had been at the MLA annual conference in San Francisco the previous year and visited Dick Hadlock who was still a clarinetist and had recently played with the singer Banu Gibson in New Orleans; he suggested that if we go there we should look up record producer and owner George Buck. He laughed when I telephoned to complain that George being blind kept his warehouse rooms dark as he knew where to find everything while I was the one who stumbled around blindly. Buck's British wife Nina would open the Palm Court Café the next year with black New Orleans traditional musicians, such as Danny Barker, who George was then recording, although his own taste was for the mostly white Chicago style jazz in which solo followed solo, common to northern jazz clubs of the 1940s and 50s. As the Café evolved into a restaurant it became predominantly white in clientele and musicians on stage. It is, however, a good place for us to dance between tables, which Nina, a veteran of the London trad jazz clubs, encourages as she likes to do so herself.

We shared the apartment with a French couple that were friends of Kirk Joseph, sousaphonist of the Dirty Dozen, and we followed them to The Glass House where the band played on Monday. I had heard them briefly in Paris and did not know what to think of their strange funky compositions and even the smaller than expected brass band size, about six musicians instead of the traditional eleven. Even their name was a dirty joke, the dozens being black obscene rhymes, insults traded within a group. The Glass House was wild with young blacks dancing madly to music that seemed designed for them. A few nights later The Rebirth Brass Band made the Dirty Dozen seem like the previous generation, their young second trumpet and singer Kermit Ruffins, was only fourteen years old and too young to be legally in the club. The Glass House was in a dangerous area, the doors were kept locked, and afterwards we were told to stay inside until the taxi arrived. I had no idea that the Glass House would become a New Orleans legend among musicians; it was where the new brass bands could play in a club instead of on the streets and where young blacks were creating a style of dancing more expressive than their parents' second lining. Despite the excellent swing dancing mostly by whites in New Orleans, black and white dancing styles remain different in body movements and kinds of display.

When we returned to New Orleans thirteen years later, in 2001, again for a Modern Language Association conference over the Christmas holidays, the Dirty Dozen Brass Band was now world famous and we danced to the Treme Brass Band at Donna's. The Treme was traditional in repertoire, and was co-founded by Benny Jones, who was one of the original Dirty Dozen, and Uncle Lionel Batiste. The lineup has varied over the years but until his death there was always Uncle Lionel on bass drum along with Benny, the manager, on snare and a powerful trumpet player to sing and lead the band. The way New Orleans band incubate and seed others can be seen from Kermit Ruffins once being the Treme trumpet and vocalist and Corey Henry their original trombonist. Corey joined Kermit's Barbeque Swingers, and then formed his own funket. Carolyn McKay and her husband, who we did not then know told someone else from the Ball State English Department about this older local couple who danced all night and the woman passed the tip jar for the band, and they were told that it sounds like Bruce and Adele, which it was.

We have danced over the years to the Treme Brass Band at Vaughan's, DBA, Candlelight Lounge, the French Quarter and Jazz Fest, most places they have played. The Wisdom Reception Center for a few years after Katrina was a favorite. On Sundays there was excellent food, a great ballroom, the evening would be shared by Treme and Kermit Ruffins. Afterwards we would go to Donna's for Shannon Powell's exciting jam sessions. I have a Treme Brass Band jacket embroidered with my name. Adele asked where outside New Orleans would I wear it, but ever since HBO's Treme was broadcast internationally I hear admiring comments in Paris and Croatia.

Uncle Lionel became famous less for his bass drumming than for his dandyish dress, courtly manner, obscene dancing, charm, and even for his singing that was featured on a CD. The Jazz and Heritage Festival devoted its 2010 Congo Square poster to him, and after his death a larger than life effigy was paraded around New Orleans for months like a strange ritual cult that had mysteriously emerged from the voodoo underworld. Lionel Batiste was indeed a legend, an icon, terms now often used for him, and his life says much about the role of music in the city. He began as a shoeshine boy who tap-danced, charmed and had children by several women, dressed with shiny fake jewels accenting his cuff links, shirt front, shirt collars, wore sharp but dignified stripped suits, carried an elaborate carved cane like an African king, wore spats over his shoes, and seemed to be continually walking the streets of the French Quarter like someone left over from pre-World War I Storyville when famous jazzmen might also be pimps, card sharks, wear noticeable clothing and jeweled rings. It is a black New Orleans tradition of distinguishing yourself by dressing, an individual trade mark that existed before Jelly Roll Morton, who had a diamond embedded in a tooth, and which Uncle Lionel taught Kermit Ruffin.

Uncle Lionel and myself after greeting each other would go into a contest of dance steps as a kind of friendly welcome. He had some clever, sophisticated shuffles and cross steps, but I was more agile and could shake, go lower, use more parts of my body including my fingers so I always won. Sometimes he would say that he wanted to go home with Adele which I pretended to take as a joke but I knew he had a reputation of staying late at parties, heavy drinking, and bedding women who often found him charming, even if, perhaps because, he became increasing obscene with his cane as the evening progressed. At the Candlelight Lounge, when he was old and often asleep, a black woman would mount him and appear to do her thing. The dance floor gave the appearance of an orgy as the bar became famous and changed from a Treme dive into a necessary part of a Scandinavian's tour of New Orleans. I would start the dancing with Adele, but after about thirty minutes we were lucky to have space on the tiny floor in front of the band, and after ninety minutes I would listen to women being propositioned and wonder whether they would rejoin the tour bus waiting outside. It was still difficult to get a taxi to take you home, as was true at many black clubs, but the owner has her friends although you might need to wait twenty or thirty minutes for your turn. It has become too popular and even has jazz on Sunday.

Before there was the Candlelight, one of the last black Treme bars with live music, there was Joe's Cozy Corner, which opened in 1992, a hangout also on North Robertson, for musicians in the Treme. It was a shack, a dive of a dive, with a bar, a few tables, and a level space for music and dancing. On Sunday's the Rebirth played there followed by Kermit Ruffins and his Barbeque Swingers. Taxis did not want to take you and you could forget about calling a taxi for the return. You earned respect by going there; several years later in Indianapolis I mentioned having been to Joe's to one of the Rebirth band and I was suddenly somebody. The taxi drivers were not wrong. The club was dangerous. Kermit's pianist would no longer play there after the night that someone walked in and shot another man in the back of the head. People knew who did it but would not tell either from hatred of the police or fear of themselves being killed as a witness to the murder. A few years later the bar was closed after the owner killed another man during an angry dispute. He claimed to have been defending himself and the victim did indeed have a history of violence. Joe died while awaiting trial it was said as a result of the police not allowing him the medicine he needed while in custody; others said that he would have been killed by the family of the man he shot, so it made no difference how he died.

Thursday night at Vaughan's unfortunately became a legend and soon hip college kids arrived from Ivy League universities and it did not take long before everyone from every educational institution arrived looking for a rave and we moved our dancing elsewhere. Some friends who usually hung out at Donna's continued on at Vaughan's but they were not dancers and seemed only slightly concerned that a New Orleans party had turned into a frat club booze up. Perhaps the best time at Vaughan's was the year after Hurricane Katrina when Cindy, the owner and manager, decided to have the Treme Brass Band on Sundays. There would be space for dancing and she would join.

Eventually Kermit Ruffins, now married and with children, trying to give structure to his life, decided that he wanted to begin playing much earlier in the evening and this meant that audience left early as did bar sales. So after over twenty years he gave up Thursday night to Corey Henry, now with his Treme Funktet, on alternate weeks with Travis "Trumpet Black" Hill, recently released from prison after nine years for armed robbery as a teenager. That lasted about two years until Trumpet Black died and himself became a legend.

The way New Orleans musicians belong to an extended family was brought home when I danced at the newly opened Treme club Ooh-Poo-Pah-Doo named after the hit song recorded by Jessie Hill. His daughter opened the bar and it at first featured James Andrews who used to play with the Treme Brass Band and who before taking up trumpet, like Uncle Lionel, shined shoes and tap danced. He is the older brother of Troy Andrews (Trombone Shorty), and the grandson of Jessie Hill. Travis Hill was a member of the family. The singer-trombone star Glen David Andrews is a cousin as is the great drummer Herlin Riley. Because so many are related New Orleans musicians are like a clan, there is always someone to instruct you how to play an instrument, let you sit in, teach you the business, hire you as a sideman until you can stand on your own. Someone else might have left prison a hardened criminal, Travis Hill could look forward to rejoining the music scene and making it his profession.

Thursday night, besides Kermit at Vaughan's, once also offered Tom McDermott and Evan Christopher at Donna's which was then presenting more straight forward and modern jazz. As Donna taught science in a high school in Florida and often was away from New Orleans, sometimes a waitress would become bar manager and chose the music program. Tom was originally a ragtime pianist from St Louis and one of a group of white musicians from around the USA that had come to New Orleans, learned local musical traditions and become part of them. He was especially attracted to the blues and boogie of James Booker, a homosexual drug addict whose strange harmonies suggested a local Thelonious Monk and whose piano fingerings were distinctive.

Evan, part Asian, part black, but raised in California by a white family had played clarinet in nationally known traditional jazz bands before deciding that he should move to New Orleans as it was the home of the style and had a history of great clarinetists. A serious dedicated musician he had become a great clarinetist himself, some say the best New Orleans clarinetist of all time, a title Evan was more like to concede to Sidney Bechet or Barney Bigard. We had first met Tom and Evan in Paris when they were living near us on fellowships that the French provided to some Crescent City musicians after hurricane Katrina. Evan, with Bechet as his model, was talking of remaining in France and would form his own European quartet with two guitars and a string bass modeled on the Hot Club of France although with a more inclusive repertoire including his own compositions. Tom decided that New Orleans was unlikely to lead to fame but it allowed him to perform with excellent musicians every night. Ironically he was to become famous through his many appearances in the HBO television series Treme. Tom and Evan, however, were inseparable to many jazz fans because of their CD performing Brazilian music and because of their excellence in playing the music of Jelly Roll Morton; they became the Fred and Ginger of the NOLA musical scene, expected to perform together despite differences. Evan became a wandering star often appearing around the world, we usually hear him once or twice each year in Paris.

I am recalling what already seems a golden period of New Orleans although it was only a few years ago and there are now newer places where we listen to jazz and dance such as the Tableau on Jackson Square. Monday night at Donna's was Bob French's. Besides being a drummer he had a jazz record program on the local radio station WWOZ and an outspoken personality. His band included Big Ellen Smith, a large tan sexy singer, and Fred Lonzo, a trombonist and vocalist with his own strong personality and who usually wears a suit and now plays with the Preservation Hall Jazz Band. It was good for dancing and there were always musicians sitting in. French had a wicked obscene tongue, fought with almost everyone, including his brother George French, who plays bass, and was often fired. After Donna's he played at Ray's Boon Boom Room (now the Maison), and elsewhere on Frenchmen Street until he took over Monday night at Irvin Mayfield's Jazz Playhouse where he continued to badmouth everyone including Mayfield. He claimed to lead the oldest band in New Orleans, the Original Tuxedo Jazz Band, started in 1910 by Oscar Papa Celestine after whose death it was led by Bob French's father who passed it on to his son. It is now led by Bob's nephew Gerald, George's son, another loud mouthed, at times obscene, drummer, who has changed the personnel but it still has the Monday night slot at Mayfield's Jazz Playhouse and we still dance to it when we are in New Orleans.

We have danced to Gerald's drumming at Palm Court, and to the singing of Topsy Chapman, the mother of Gerald's vocalist Yolanda Windsay. Topsy, Yolanda and another daughter perform together as Solid Harmony. Although Topsy and Yolanda often welcome our dancing we never really spoke with them until Adele praised Topsy's performance in the movie *12 Years a Slave* after which we were invited to a crawfish boil at her home.

Sunday at Donna's was the Shannon Powell jam session that usually included Jason Marsalis on Vibes and Evan Christopher or Charlie Gabriel (now of the Preservation Hall Jazz Band) on clarinet. Everyone was there: Jon Cleary and Greg Stafford, actors, musicians, drug dealers, tourists, squeezed around the bar and band stand. Out of town musicians would come for the final set after finishing at Snug Harbor. It could get crazy. Shannon is a great drummer, famous, often tours with his own band and with the Preservation Hall band, holds down Tuesday night at Preservation Hall, often has a night for his trio at Mayfield's Jazz Playhouse, but who is that sitting in, replacing him for a set, but Herlin Riley, New Orleans' other great drummer, more modern in approach than Shannon, as well-known, and equally capable of long drum solos that make anyone else on stage unnecessary.

Shannon belongs to a tradition of singing and performing drummers. I do not remember him ever having a vocalist, and most numbers feature himself, often singing, as a soloist, and chatting up the audience. He likes dancers. Watching us at Donna's he noticed that we followed the structure of tunes, anticipated breaks and chord changes, and kept time. Others have sometimes asked whether we were professionals, taught, or were retired professional dancers; Shannon immediately asked what instrument I used to play and when I replied drums he said he thought so. We became his dancers. Seeing us in the audience at a concert of The Preservation Hall Jazz Band at the Salle Pleyel he motioned us to come on stage to dance for the last number. I dislike the commercialization of Preservation Hall, but he will tell us to come to his shows on Tuesday, leave our name at the door, and expect us to dance on that tiny shaking stage while I worry about a board breaking and becoming a cripple. At Mayfield's we are also his dancers. When he has a party at the end of Jazz Fest we are invited. The food is good (his wife, an excellent cook, is a caterer), and there will be a jazz session. I will ask naively who is that and someone will say oh he plays bass with Dr John, he used to play for Ray Charles.

Dancing and music means as much to me as my career as a university professor and the writing and editing of books. We go to New Orleans for a month each year during the French and Jazz and Heritage Festivals to hear jazz and also to dance. As I age it is important to my health. During that month I lose weight, my waist loses at least 2 usually 4 inches, my muscles ache, I begin to look better, I make love more often. I enjoy the temporary celebratory status when people take photos of us at Jazz Fest and say they remember us from former years. Davell Crawford will interrupt his set at Snug Harbor to introduce us as lovers and ask us to dance, and I am even more amazed when listening to the Stanton Moore trio at Snug Harbor the pianist David Torkanowsky will suddenly turn around and say this is a number to which we should dance. We do not go to Snug Harbor often as it is one of the few clubs in New Orleans at which there is supposedly no dancing. At the beginning of the evening the owner makes an announcement telling the audience that they are there to listen to the musicians. Moore's trio includes James Singleton, an excellent string bass player, who Adele admires, and who often can be found in avant-garde performances. Moore's trio might be described as deconstructionist as many numbers kick around the older New Orleans jazz repertoire. I was surprised by enjoying the trio so much that we pay for both sets two weeks in a row.

    We started regularly going to New Orleans for a month each year after Adele retired from Ball State and we moved to Paris. The first year we purposely avoided Jazz Fest as too crowded, the next year we broke down and our routine included the two, then, three, now four days of the French Quarter Festival (free and supposedly limited to New Orleans musicians), Easter, and the two weekends of Jazz Fest (the second of which has expanded from 3 to 4 days). Besides Donna's, The Spotted Cat, and Vaughan's we used to dance to Marva Wright at a club on Bourbon Street club. It was a great blues band with Davell Crawford the pianist and leader until he left and Benny Turner, the bass player, became leader. We used to dance our tails off, shake, get down, posture, do tricky steps to the blues and boogie, and Marva, her musicians, audience and the club owners appreciated us. We created "atmosphere" and had an open invitation for free drinks. Adele was surprised when at the end of an evening the not so young female taxi driver told us that she was waiting for her baby who liked his greens warm when he got home. The baby was Davell, already famous for his gospel compositions and choirs.

    After Katrina Marva was literally homeless, unhappily rented a house in Virginia, and began to cry on stage when she saw us dancing to her at the Meriden Hotel Jazz Club in Paris. She told the audience we were friends from New Orleans and came down to greet us. Davell would do something similar a few weeks later at the same Paris jazz club and has told New Orleans audiences about it. Then Marva was singing once more in New Orleans, now at an open lounge at the Ritz hotel where it was difficult to hear her and Benny. We danced on the thick rugs but it was easier to dance to Jeremy Davenport on the dance floor of the restaurant. One night walking along Frenchmen Street we heard "psst, psst, Bruce", and it was Marva's manager who wanted us to come into Ray's Boom Boom where she had a large band, with horns, to celebrate the release of a new album. Embarrassed, feeling we were "raining on her parade", we tried to sit out a set until it was clear that Marva was waiting for us to dance. The music was great. With so many fine dancers in New Orleans I really do not know why anyone would want us, maybe it is our age. I still sometimes hear recordings of Marva and think what a privilege it was to hear and dance to such a great singer. We sometimes see Benny and his The Real Blues, but without Marva they play in clubs with ear shattering sound systems, bad acoustics, and have mostly dropped off the scene.

Hurricane Katrina changed New Orleans. Until then New Orleans was "chocolate city" with a black mayor and an overwhelming majority of black voters, many living on government assistance. To say that the city was poor and politics corrupt is to say no more than it is part of Louisiana, but there was a racial defensiveness against reform. The flooding of the city resulted in many blacks staying away and being replaced by Latinos and Asians employed for construction work who found opportunities to open small shops and gave New Orleans other cultural directions. There were now Latin bands, tango nights, South American sea food restaurants. The closing of the Projects, subsidized housing mostly for poor blacks, was fiercely contested, and remains politically sensitive, but it made areas that were mostly no-go become welcoming to new businesses, especially to small hip cultural places like restaurants, jazz clubs, and organic food shops crowded out of the Faubourg Marigny as Frenchmen Street replaced Bourbon Street for tourists. The city elected a white liberal reforming mayor (opposed by reactionary white racists even more than by African-Americans who hoped for the return of a "chocolate city"—now the flavor of an ice cream), and the former black mayor went to prison because of corruption.

Immediately after Hurricane Katrina New Orleans was desolate. Friends in Paris said do not go, there will be nothing left, but they were wrong. Jazz fans and musicians from around the world wanted to save New Orleans more than the American congress or President Bush. Almost every jazz cub around the globe found ways of hiring New Orleans musicians and sending money, usually cash handed over personally to avoid the expected delays and ineptness of government. Jazz fans came and volunteered to build houses for musicians. While politicians bickered and some white Senators announced that the city was not worth rebuilding and black politicians saw everything as an attack on Black Power, the seeds of renewal were planted. New Orleans was saved by jazz. Later the city would be rediscovered, the French Quarter and Marigny, old areas built above sea level, which had been barely touched by the flooding, became increasing valuable as would adjacent areas like the formerly black impoverished Treme. Louisiana passed a tax break to attract films and soon New Orleans was producing more films than Hollywood. Hollywood stars bought houses and were photographed at such places as The Candlelight Lounge and Vaughan's, new restaurants opened, and magazine articles claimed that for young professionals New Orleans was the place to be, the new Portland. As the crime rate remains high and everyday there is a murder, mostly due to drugs or young black gangs, and violence has moved into the French Quarter and Frenchmen Street the reports of a new paradise are exaggerated but as locals say, "We are Back".

The rapid rise of a new New Orleans did not seem likely when we returned the April after Katrina when our friends drove us to see the damage. Corlita was from a New York family involved in union and Black Power politics. Although growing tired of the interests and skills she had learned she claimed that plans for redeveloping New Orleans were intended to disempower the black families that had not yet returned and who should be receiving immense sums to come home before there should be local elections. Her husband David showed me the incredible desolation outside New Orleans where he was raised, a poor white racially prejudiced area, that remained flooded, and had suffered more than anything I saw in the city. Nothing was left. The impoverished, white and black, were victims of the flooding and just as they had both suffered in the past so they were likely to be ignored in plans for the future. Another friend who showed us the damage worked for the generally despised FEMA and explained to us the practical and bureaucratic problems she faced. She knew that the organization was thought incompetent and avoided telling people about her employment.

Everyone had a Katrina story. Irving Mayfield's father did not want to leave and had died. Karen's family in Algiers, across the Mississippi river, armed themselves with guns for protection against what they assumed would be looting crowds fleeing the city. (The racial polarities and assumptions will be obvious.) Rollan Garcia, owner of Bullet's Sports Bar in Treme, said that he told his family to evacuate but he had stayed on to protect the bar as he assumed Katrina would be no worse than previous dangerous hurricanes. Then the flooding began and he retreated to the second floor, then the attic and when the water reached his shoulders he broke a hole in the roof, climbed out and was saved by someone with a rowboat trying to find people like himself. I read that afterwards he had hired a man to help protect the bar against looting, but marauding hungry dogs rather than armed human gangs were the problem.

Bullet's gets its name from Rollan supposedly having a bullet embedded in his head. It is one of the places that before being made know through HBO's Treme series was thought too dangerous for tourists or for anyone not from the 'hood'. I and others used to go there because Kermit Ruffins played on Tuesday when it is packed with people drinking, dancing, partying, having a wild time. If you do not reserve a table, or are not lucky enough to be invited to squeeze in, you need to stand, but as people are always getting up to dance, join friends, leave for food, eventually there are seats. It is not classy, but certainly not a dive. Some people dress up and there are often entertainers, professionals, a cross section of black New Orleans in the audience. There are very good food trucks outside. There is music most nights, but

Tuesday is the night just as The Candlelight is where you go on Wednesday and Vaughan's on Thursday. One Tuesday there was a party of Saints footballers suited at the tables next to us with their beautiful and beautifully attired women.

    The trouble is that Bullet's can be difficult to reach without a car and even more difficult to leave. Several taxi drivers have been robbed and killed in the neighborhood. The second time we went the Arab taxi driver insisted we made a mistake and would not stop once he saw a group of black men standing outside Bullet's. We had to get off a block away and walk. Most taxis will not come late night and if they do you might need to run as they idle their engine some distance away. But that's New Orleans.

    I keep alluding to the French Quarter Festival and the Jazz and Heritage Festival; they are why we spend a month in New Orleans each year, indeed we go to listen to jazz and dance. But where and how changes as we age. During our first years of visiting New Orleans we would dance to the Treme, other brass bands, and Kermit Ruffins at the stages along the river front during the French Quarter Festival. After losing sight of each other in a Trombone Shorty crowd we settled for Jackson Square and felt we were opening the festival dancing to Pete Fountain, Banu Gibson and Don Vappie before dancing to Shannon Powell's trad band at the French Market and other traditional bands on Royal or Bourbon; we would go on another day to the Old Mint to dance to Latin music and listen to the Tin Men—an unusual witty trio of washboard, tuba and guitar. Evenings would be spent at Irving Mayfield's Jazz Playhouse, We spend more time now at the Playhouse dancing on the tiny floor to Tom McDermott and Evan early evening and usually Shannon's modern trio at night. We have become "regulars", often welcomed to sit at the few reserved tables in the front of the bandstand, even on the Wednesdays when Mayfield has an all-star group.

    Jazz Fest mostly means Economy Hall as the sound is good, there is a fresh breeze and most important there is a large dance floor where we see other dancers we have come to know over the years. We are photographed and people will send us the pictures of ourselves. Out of the blue we will receive on-line several years of photographs taken by a person we do not know and who will not reply to our messages saying thank you and asking who are you. I look forward to hearing and dancing to the Treme Brass Band, Michael White's Liberty Brass Band, Greg Stafford's Jazz Hounds, Don Vappie's Creole Serenaders, and Lars Edegdran Ragtime Band, but there are always surprises. We might move to the Blues, Gospel and Jazz Tents but the sound systems are not as good and there is no dancing. Occasionally we will stand in the sun at one of the outdoor stages because I want to hear someone like Jonathan Batiste, but afterwards I swear never again, how do the large crowds survive? The food is a gourmet's paradise, but if it rains, and it usually does at least one weekend, even the next day the ground is like a swamp. Getting home after the shows is a nightmare best avoided by snaring a law breaking taxi in a prohibited area instead of waiting in lines for hours for either a bus or a lawful taxi.

In the week and a half between the two festivals we dance at Palm Court, Mayfield's Jazz Playhouse, D. B. A., and perhaps see a film or two at the Prytania, Canal Place or Zeitgeist theatres. I look forward to Evan Christopher's annual International Jam Session the first Monday after the French Quarter Festival, although he has moved it from the Chickie-wah-wah, a favored neighborhood music bar with excellent food where we danced non-stop, to the Little Gem, a restored tavern supposedly going back to the earliest years of jazz, which, although now a regular jazz club, I find emotionally cold. The Prytania allows us to have a meal and shop at the Cheese Factory, which has Canadian, Vermont and British cheddars I cannot find in Paris. Adele finds short skirts and dresses at Frock Candy where I also found my Chinese all paper hat (for women) that survived several years' wear and rainstorms, often drew raves and I was once told looked like a pimp's hat, before I replaced it with a similar hat. Much of New Orleans fashion judged by Parisian standards is openly erotic and I love the loose low cut tops, sexy cuts, exposed thighs, bright colors.

New Orleans has some of the best dancing in the world, although Sweden where the swing dance revival is said to have begun, and where many foreigners attend summer swing dance camps has its own champions. Swing dancers come to New Orleans, sometimes first as part of national contests, and then they stay. I first saw Nicole Rochelle in Paris performing as Josephine Baker in a musical biography that toured internationally; she stayed on in New Orleans where I saw her dancing at Maison. I was surprised by Sylvia Howard in Paris, jiving around on stage whereas previously her style was to sit in a dignified manner on a chair and fan herself. She recently had been performing with Nicole Rochelle who returned to Paris. In other countries I mention writers I know, in New Orleans musicians and dancers.

I am awed by the young swing dancers at the Spotted Cat and on the floor at Economy Hall; their steps are complicated and require much practice, energy, trust, and physical ability. Even when much younger I could not dance like that. I recently saw a young black dancer, who moved from Seattle to New Orleans, who had the grace of black dancing and the skills of the best swing dancers. I thought him the best dancer since Fred Astaire. I continue to be amazed when we are regarded as good dancers and the young ask if they can be our partners. I am not in the same class as those I admire. But for some reason, perhaps age, perhaps my weight, perhaps Adele's grace, perhaps because I have learned how to fake with my body what I can no longer perform with my feet, people take photographs of us, and treat us as famous. Every year a German trumpet player who leads a brass band introduces himself to us; it was only when he sent us photos of our dancing that I gave his name attention—"Doggie" Hund.

We try not to miss Seva Venet's Storyville Stringband, a revival of a New Orleans tradition that goes back to 1890s. Seva Venet, an excellent guitar player, from Los Angeles, along with James Singleton, Tom McDermott, Christopher Evans, the trumpeter Duke Hertiger, is one of the extremely talented players who have moved to New Orleans and revitalized its music. The level is extremely high, perhaps the best in the country outside of New York where the chance of fame, fortune, and the future attracts some of the better local musicians such as Henry Butler, Jonathan Batiste and Wynton Marsalis. Evan Christopher and Davell Crawford try to have an international career while New Orleans remains home.

We used to dance at the Jazz National Historical Park on North Peters Streets before they moved most of their concerts to their better appointed, more modern, performance space at the Old Mint where events now seem educational. I regret that Bruce Sugarpie Barnes, a marvelous harmonica and accordionist, a master of the blues and zydeco, uses his great talent for short lecture performances on Canjun, Caribbean, African and other influences on jazz. Barnes, one of many children of a black Arkansas share cropper who knew formerly famous blues musicians, made his way to being a nationally famous football player and for a time one of the NFL's Kansas City Chiefs to a marine biologist, and as a naturalist a National Park Ranger, eventually coming to New Orleans where the group he leads plays one or two nights a week in clubs and at festivals. Handsome, domineering, he also has appeared in films and on TV commercials. He once told a friend that he knows when the Festival season has arrived when he sees us dancing,

Jazz and dancing have led us to many of our friends. Corlita is the niece of Rasul an American musician we know in Paris. Lynda, an often married and divorced world traveler and now a California food caterer, formerly lived in New Orleans, often saw us dancing, and when she returns accompanies us to Treme jazz bars and the city's fancy restaurants. Jean-Marie and Geneviève Hurel, observed us dancing in New Orleans and then in Paris to their friend Don Vappie ; they were our first personal introduction to Vappie and they became friends who we now see on both sides of the Atlantic and share information about who is playing where. Jean-Marie plays trumpet and leads the Fidgety Feet, a French New Orleans style band. Geneviève teaches music. Their daughter, Juliette Hurel, is an internationally known flautist.

Although NOLA has a reputation as a city of artists we do not know any with the exception of Oscar, a local legend who makes inexpensive jewelry from colored plastic. Each pin or earring is dated by year and his pieces are collected and compared by fans who discuss how and when to find the best selection of new work. It would seem logical to go to the French Mart but he saves new pieces for Jazz Fest and also sells on-line. Besides each year having new themes the older designs change, so last year's hummingbirds, angel fish and parrots may be different from those he is selling this year. There are always designs themed with sports stars, the new Jazz Fest posters, and other local events. Many pieces show musical instruments sometimes played by alligators, owls or other creatures. They can be seasonal, such as penguins on ice skates or snowmen, and often that is the last of them. If you ask for him to make one he is too busy, no longer has that color, or his assistant does not know how—although he claims to make each piece himself. At Jazz Fest many people wear hats covered with Oscar's pins. We give his pins as presents to Parisians who often buy more for their friends on line. Oscar is part of a lineage of local folk artists who make jewelry; Adele has a pink pin with a front view of a pig made from dyed old vinyl recordings and the maker, who has disappeared, claimed to have taught Oscar. Now there is someone selling bad imitations of Oscar pins in the French Market.

The *New York Times* carried an obituary of Cosimo Metassa, whose primitive recording studio, a converted grocery store, in New Orleans was one of the origins of Rock and Roll. As a recording engineer he invented what was known as the New Orleans sound, with strong bass, guitar, drums, and weaker piano and horns. Many of the hits issued by Atlantic, Chess, Savoy and others companies were recorded in his studios He recorded Little Richard, Big Joe Turner, Professor Longhair, early Fats Domino. Hit records included "Shake, Rattle, and Roll", "Tutti Frutti". "Good Golly Miss Molly", "Land of a Thousand Dances" and "I hear you knockin' ". Many were covered by other artists, but Metassa's simple equipment which did not allow splices meant that records had to be a complete take with no stops, nothing added later. Aiming for enthusiasm, he would tell musicians to imagine that they were performing live in front of an audience.

I danced in my teens and twenties to songs he first recorded and even now in Paris they get the French to their feet. After he retired I would see him, an unassuming man unconscious of his increasing fame as Rock and Roll became cultural history, passing time at the tightly packed family owned Metassa's Market, a grocery at the corner of Dauphine and St Philip in the French Quarter where I would shop for luncheon meats, asked for large cartons of orange juice or packages of ice cream to be delivered to where we were staying or pick up a bottle of beer or wine on my way to Eat, a restaurant on the next block.

# Chapter 20: Summer Home

*Hvar, Croatia 2005–15*

At first after we stopped spending summers in Deia there was no problem as I needed to do research in the Caribbean, explored Indianapolis in summer, and was glad to return to Paris. I had often heard that the best time in Paris was the summer after the French went on their holidays. No one had warned me that the metro smelled, the shops and restaurants were closed, you needed patience to wait in line for tickets to evening open air modern dance recitals that soon after they began and were no longer refundable would inevitably be cancelled by rain. The summer jazz concerts at the Parc Floral meant standing in line for hours before the gate opened and fighting the crowd for seats out of the hot sun. It was necessary to do as the French and have a long summer holiday.

Croatia seemed the place to go, if only because it was not favored yet by the French, but where in Croatia? After a year of searching websites we flew directly Paris-Split and were met at the airport by the taxi driving husband who was to take us to his wife's apartment house on Ciovo, near the medieval island Trogir. He kept saying "hurry up" in broken English and I hurried so much that I three times in a row mispunched my number and my credit card disappeared. It would require several hundred dollars in telephone calls and taxi rides and much pleading before it was returned. Villa Edita however seemed at first paradise, just across an unfortunately busy road, from clear blue water in which we swam for hours twice a day. We had lunch at the apartment from what was available at a local grocery, and at night we carried a torch while walking up to and back from an open air restaurant near a camping ground. Villa Edita was, however, without means of transportation except by Edita's busy erratic husband, although we did manage to get to Trogir where all the famous monuments were closed for repair or just maliciously closed, although that did not stop

British tourists from reading in their guide books what they came hoping to see. There was a food market just outside the old city with good bakeries and I wished we had daily access to them. We bought made in the third world imitations of fashionably labelled sun hats. After a week we moved on to Split which had good shops, museums, many restaurants, the famous Roman ruins of Diocletian's Palace, the incredible Temple of Jupiter, and supposedly good beaches, but they and the museums required a car or better transportation than reliance on busses.

We would return to Villa Edita the next summer, but the traffic seemed heavier at night, we were walking longer distances for variety in restaurants and food shops, and Adele disliked the sharp incline where we skidded down to and struggled up from the small beach. We tried Brac Island with its lovely famous long pebbly beach at Bol. There was little to do at nights, the restaurants were good but not exciting, and the beach required deck chairs; the current shifted radically at times so you had to change sides of the beach. Brac was not for long stays.

We went to Komiza on the island of Viz for two weeks to avoid the crowds of tourists in Hvar during early August. We were told by Croatian friends that Komiza was undiscovered and like Hvar in the past as until 1989 the island was a naval base and forbidden to tourists. Yes it was not yet like Hvar but smaller and on its way to becoming another international holiday resort. The tourists were Croatians, Slovenians, Poles, British and Australians by way of England instead of Hvar's Italians, Germans and Nordics. The three best restaurants were always full, difficult to book, and bunched together at the other end of a spread out village, near the only hotel. As in Hvar there were still fishermen; smoked fish, especially tuna, was a local specialty. Another specialty is a tomato, onion, green pepper, olive, closed pasty, versions of which can be found around the Mediterranean. There were already all night discos, newsstands that carried foreign newspapers, but only one bakery which when it ran out meant it was impossible to purchase bread, and it was not very good. There was swimming from small rocky beaches, the town was attractive, there were some very strong winds, and a hill that required strong legs to ascend to the apartment that we rented, and eventually we gave in and started taking taxis when we could find them. One night there was a festival with live music and our landlady was proud that we had been shown dancing on television; her friend in Chicago had forwarded the picture.

I liked Komiza, but not as much as Hvar. Although ferries go between Split and Viz daily, they only stop at Hvar one morning each week. We returned before the cafes were open and decided to go swimming at the cloister beach until our room would be available. There were four French young women barely dressed, dancing strangely, obviously stoned, still singing the previous night's house music anthems. A local policeman was trying to get them to leave the area quietly. French tourists had discovered Hvar.

It seems incredible that we would spend many summers in Hvar town when the first one began so badly by having booked with a pension recommended by the French because the owner's wife was French. He had a farm, donkeys, and thought the ideal holiday for foreigners was to smell the donkeys, brush away the flies they attracted, eat small portions of bland food, and go early to bed after many free drinks so as to go on donkey tours after a very early simple breakfast. It was not my notion of a holiday and we almost came to blows before he agreed that we could get up later and have a small tasteless lunch before going swimming in the afternoon and to restaurants at night. We and many locals danced to a duo of keyboard and bass and watched a woman said to be over a hundred years old who arthritically tangoed with a younger escort. She claimed that swimming every day kept her alive and physically mobile but someone suggested it was the younger man. Afterwards we would need unlock the pension gate, and try silently to navigate the building to our room by moonlight. A friend of the musicians, Rina said we should rent an apartment from her family next summer, which we would for many years.

On that first trip to Croatia we had booked a room in Dubrovnik for a week but everyone said we must stop on the way at Korcula, another beautiful medieval and Renaissance fortress town When the boat got there all rooms were taken and someone at the tourist office telephoned a friend who agreed to take us in for a night or two. We woke up the next morning to a beautiful sight, we were on small inlet and the entrance to the water for swimming and the owner's boat was by way of the stairs that lead down from the villa. The apartment seemed to have been reserved for friends and visiting family. There was no transportation to the old town, but it was manageable with a thirty minute walk down or longer on return up hill. We returned to Korcula for much longer visits two other summers and even for a time were tempted to buy an apartment in old Korcula. The restaurants were mostly not as good as in Hvar but there was an excellent fish restaurant we went to regularly. It was owned by a family; the mother cooked and her two sons waited on tables: one would fish underwater with a harpoon in the morning, which as Croatia has a law against underwater fishing with oxygen masks meant holding his breath.

Locals kept saying Korcula was best in September and October after the tourists left and fish were plentiful, but a very cold windy two weeks when we stayed under covers to keep warm and the water was too cold for swimming made it another paradise lost. There were sandy beaches further along the island less affected by the cold, and we were one day taken to one, but you needed a car. I liked the owner of the villa, his wife often cooked extra which she gave to us for lunch, and we still send photographs at Christmas, but Hvar offered more to do, including a regular summer concert season and better restaurants. I thought the fresh food market outside Korcula's old city disappointing and was unwilling to spear fish underwater even if I could have held my breath.

Dubrovnik was beautiful although pock marked where shelled by the Serbs and their allies in Montenegro. The large walled old town was now a fortress for tourism with its workers mostly living outside close to the coast. Split's ugly high rise housing was a smaller version of the horrors outside Paris, but Dubrovnik sprawled attractively like a series of pretty suburbs with small houses, the areas connected by bus and available taxis. There were many gardens and it was one of the cities that claimed Lord Byron said it was had most beautiful view in the world. It would be a good place to live. No doubt others also thought so as I was told that the cost of property in or near the old town was Paris and London expensive. The main problem, surprising, was the lack of beaches. The one we went to on our first visit required a walk to a bus, a hotel elevator to a beach and a small payment. When next year the hotel no longer allowed non-guests to use the elevator a long long walk in the sun down and later up a long winding road was necessary, and that beach was finished for us. The city beach just outside the old town was crowded, dirty, packed with rental chairs and umbrellas, and half that space had been taken over by a club. Many people took a bus to a village outside Dubrovnik but even there it was a longish walk to swim and it was really a day trip on weekends for families.

Our Paris doctor visited Dubrovnik and claimed he never had fresh fish there and while I imagine he only ate where the French gathered it is true that the food was often disappointing. There was an inexpensive restaurant at the tip of the fort, just outside the walls, on the harbor that served good small local fish, but too many restaurants offered the same fish I could have bought in Indiana. I even had to send back a sadly cooked dish at an expensive restaurant I was told was one of the best. There must have been fresh food markets, but as I tourist I never saw one except through the window of a bus when travelling through what I imagined to be a working class area where ships were docked far from the fortress.

During the day there was a drill of tourists moving through the old town and on the walls. The shops were packed with international brands. In the evening one went to the old town for entertainment unless dining above in a hotel restaurant. The concerts were expensive but included known musicians. Unfortunately the palace used for concerts could become very hot and one night the pianist, dressed formally in a stiff white shirt with white bow tie and tails, succumbed to the heat during an evening of Beethoven and segued after the second movement of one sonata into the third movement of the next on the program. As he was well applauded at the end of the concert I wondered whether it was possible that no one else had noticed. Surely the audience must have realized he only performed two, not three works. Perhaps they also were overcome by the heat and the applause was so that they could leave?

There was an open air jazz club owned by a string bass player that claimed never to have closed during the war for independence. When the city was shelled, the patrons and musicians moved inside. The jazz was mostly jazzy pop except when visiting musicians sat in and the bass player would join and it became more serious. Several nights there was a good American trumpet player who was then based in Italy and whose sexy attractive American wife kept naively telling us that she wanted children and a family. We danced, were applauded, and then one night a blonde erotic Swedish dancer, a girlfriend of one of the musicians, followed us when we sat down to rest. I joked that we were the warm up act and Adele replied "no, she is the warm up act".

On one visit to Dubrovnik we took a day trip by coach to Montenegro as we were wondering about summering there in future. Antagonism between the two former parts of Yugoslavia continued as locals remembered being shelled, during their war of independence. Shelling served no purpose and was senseless destruction. Our landlady claimed that the recent summer fires on the hills above Dubrovnik were caused by Montenegro. At the boarder our bus seemed to be held up for an interminable period while papers were checked and re-examined. The Montenegro coast was as beautiful, even more dramatic, than that of Croatia, but the villages with housing and hotels were far from main roads and would require a car to reach. A famous village for the rich not only needed a car but was guarded against trespassers. Most of the seacoast had soviet style bunker-like hotels, fast food stands, and beaches with identical sun umbrellas set close to each other at identical distances. This was soviet style tourism and the Russians were said to be investing heavily in Montenegro, especially the sandy beaches in the south where Albanian was mostly spoken. I thought the country dirty and toilets non-functioning after the Nordic-like cleanliness of Croatia. There were few road signs and there were police everywhere, traffic appeared needless jammed and I was glad to return to Croatia but felt I really should have taken longer and found the perfect place instead of making a hasty negative judgment. Perhaps I was too influenced by the Croatian view of Montenegro? Montenegro had immense stadiums for international jazz and pop music concerts, visiting theatre groups; in Korcula I sometimes watched Montenegro television for its musical shows although I could not understand a word.

## II   Hvar

Hvar, like Deia, is often said to be one of the ten most beautiful places in the world. That means I have lived for long periods in a fifth of the world's most beautiful places. Both Deia and Hvar town are on islands, are surrounded by contrasting hills, and have beaches for swimming, but whereas Deia is a small village with dynamic views, Hvar is a larger town notable for its thirteenth century walls dominated by a fortress, its Gothic, Renaissance and Baroque buildings around an open car-free plaza and narrow winding side streets. There is a Baroque cathedral, a seventeenth-century theatre, a Renaissance cloister with an impressive bell tower and a resident monk. The Arsenal, now used for occasional concerts, jewelry displays, and other activities was built in the early seventeenth century to repair ships and was then thought the most beautiful building in Dalmatia. What would be tourist sights elsewhere are functioning; the cathedral is used for mass, the cloister has a research library, is used for concerts during the summer and is near a small rocky beach where we usually swim.

We used to swim further out of town in the opposite direction at the now restored 1927 "bonj" beach club across from the immense Hotel Amfora. During the 1930s it was where the rich, fashionable, and aristocratic of central Europe and the Balkans bathed during the summers, itself an historical monument to a style of luxury in the past. We first went there, before it was restored. There were still platforms from which you could dive into the clean blue water, rusty somewhat dangerous steps and ladders for climbing in and out of the water, and the rooms for changing clothes were in a poor state but usable. The white walls with their inset changing rooms and the long concrete "beach" was still impressive, an impressive ruin from some earlier era. The second year there was a change. Some thuggish looking guys—I understand one was a national wrestling champion—and their appropriate looking women and an attack dog seemed at home. They even knew how to turn on the water and we could use a hose to wash off the salt. Unfortunately the next year the Amfora Hotel had taken over Bonj and began to transform it into a luxury beach club. There were now clean modern rooms for changing, a series of showers, a toilet room for men and one for women, deck chairs with umbrellas, and a restaurant that became increasingly expensive each year without improving. The first Amfora year it was possible to sit around a table in a wet bathing suit, the next year there were cushions on the chairs, the tables were laid out as if dining were expected, and you had to be in dry clothing.

We had to adapt as the Bonj and rules changed. The walk to and from Bonj was pretty but long. We would go before the sun became too hot, swim, rest, swim again, and return home as the sun was going down. But once it became necessary to sit around in dry clothing we limited ourselves to one swim and reading while having something inexpensive to eat and drink, only occasionally treating ourselves to a meal. We also had to make friends with the waiters and leave decent tips. Still it was worth it as the swimming was good and Bonj was a pleasant place to sit and read, although increasingly there was loud "house" music late afternoon from across the water. If we stayed on late we would sometimes have an early meal at the "Two Fishermen", on the walk back to the town along the harbor.

Then one afternoon Adele finished swimming before me, climbed the concrete steps near the restaurant, looked back for me and I saw her fall down the equivalent of two flights of steps landing on her head. She was unconscious and I was afraid that she was dead. I watched as an American and his girlfriend who seemed to know what they were doing kept speaking to her, asking her questions until she regained consciousness and responded. She had an immense bump on her head. The doctor in town put some bandages on Adele and telephoned our very strong landlady to help her climb the steps to our apartment. After a couple of painful days in bed she took a taxi to the small hospital outside of town, near the Amfora and Bonj where she was X-rayed and found to have five broken ribs on the side of her remaining lung and a broken bone or two in her hand. The doctor in charge of the X-ray panicked, said she might die immediately if a broken rib punctured the lung, and telephoned the car ferry leaving for Split to wait until our ambulance arrived. If the wild ride through the hills to the other end of the island did not kill Adele nothing should; that the ferry officer put us in a room with a chapel was more depressing. Of course there was no waiting ambulance in Split, indeed no one had any idea of what was going on until we located a policeman who telephoned the Hvar hospital and took Adele to the public hospital in Split where she stayed for several nights while being further X-rayed until the doctors decided that there was nothing they could do for her and we returned to Hvar.

Which is why we changed from swimming at Bonj to the pebble beach in front of the cloister. Adele used to be a strong swimmer and wanted to try Bonj again, but found she was no longer powerful enough and I had trouble getting her from the water back to the steps. For a couple of years she regained her strength and courage and was swimming for up to two hours, then after mini-strokes and discovering that she had a heart condition she became weaker and swims very little. But then I no longer can swim as long, an hour of paddling is my limit. We are not going to be like that 100 year old woman who danced the tango and supposedly swam every day.

Although I miss the walk to Bonj and observing the changing color of the flowers on the cactus plants, the women walking in their bikinis, and the small boats that dock on that side of the harbor, there are other great views in Hvar including from the water at the cloister beach. There are palm trees, the monastery, the bell tower, and the seagulls that like to congregate on a hotel roof where I count them, the starlings that startle me as they whiz by my ears, even the dogs that seem to have their personalities as they look for other dogs. If we swim early enough there is a woman with two dogs that like to swim, indeed one is always edging her on to throw a stick or ball into the water so the dog can make a running leap and retrieve it. Then there is the monk on a motor boat lowering stale bread into a fish trap and pulling up yesterday's trap to see how many fish he has caught. The miracle about turning loafs into fish explained! After swimming we usually have a cappuccino at the beach café, joke with the young waitresses, and start back to our apartment for lunch, perhaps stopping on the way at the Nonica cake shop, and the open air market and perhaps the "cheese" shop. I need be careful about carrying too much up flights of stairs, as I too have started to get dizzy climbing. I am still too proud to ask the local old ladies to help me, although they seem able to go up and down the stairs with packages as if they were goats.

The "cake" shop, Nonica (grandmother), is a classy patisserie, better than most in Paris, started by the wife of an Hvar restaurant owner who at first only baked cake and bread for her husband's restaurant. Bored with the lack of social life on the island during the winter she decided to take cordon bleu baking courses in London and has continued to do so every winter as her shop has become successful, an attraction, with full time staff, and mentioned in guide books as not to be missed although it only has two tables inside and two more outside. Customers sit and socialize on the steps of the Baroque church next door. Thin, well dressed often with an interesting necklace, nervous, bird-like, serious, she cannot keep up with demand. I tell her she needs another bakery with Orientals in the basement, and she inevitably replies that all her staff, ingredients and recipes are local and everything made that morning by hand.

Ivor from whom we buy aged local cheeses, anchovies and olives has a small delicatessen on the side of the open air market. It looks like a stereotypical Mediterranean bodega with antique decorations, smoked hams, many bottles and kinds of wine, and no doubt is meant to appear that way to attract the tourists whom Ivor charms with outrageous stories, to whom he offers glasses of wine, tastes of ham and cheese, and who he charges more than locals. "I need to make a living". He especially hopes to supply the visiting yachts that stock up in Hvar. Although Ivor's shop is regarded by Hvar natives as a place to avoid, among tourists he is a legend and is photographed, even shown on foreign television programs. He is overweight, always tasting and drinking his own supplies and often on a diet. His lips glisten from the smoked ham that he says is better than any from Italy or Spain and which even Italians include among the purchases they take home from their holiday. His claims are unending ranging from being of noble stock to a self-made man, from having ran restaurants in Vienna and other European cities, to having left Croatia to avoid military service. Although he supposedly has retired and handed the business over to one of his sons (the other has a late night wine bar) he is usually at the shop from opening to closing. At one time he even talked of opening branches in Vienna, London and Paris, but a winter in Vienna ended that fantasy. Similar to many from Hvar he speaks several languages that his generation were more likely to learn while working abroad than in school.

He really does have excellent cheese and knows his best cheeses. Every other week I will ask for his goat milk feta; he will tell me to wait for better olives next week, and always tells me that he made a bargain price especially for me even if I suspect I have over paid. When I am leaving Hvar I will get a winter's supply of anchovies and the local wine vinegar cured capers. If I have spare time I like to drop in for a glass of wine, and listen to his views about the softness of the young, although I am tired of being introduced to others as "Gershwin", the dancing American in Paris, which he laughs at as if he never said it before and as if it made much sense.

Our first years in Hvar we regularly ate at the terrace restaurant at Konoba Luviji near where we stay. The chef was then Adam, and although we have known him for a decade we still refer to his wife as "Eve". Adam was and still is one of the better cooks in Hvar, perhaps the only one who improvises on traditional Dalmatian methods of cooking. He will in large pots cook chopped onions and garlic in lots of olive oil, stew with white wine sliced local mealy potatoes, and put the mixture in the oven with mussels, mixed fish, and several kinds of shrimp in various combinations from day to day. It will be finished with finely chopped strong local fresh parsley. He does many different variations on this basic dish as in the past local cuisine was normally cooked in ovens; there were no burners, grills, steamers or frying pans. His garlicky tuna fish pate remains the best I have tasted. I still remember a rapid throw together lunch of large pasta, mussels, small shrimp in that basic onion-garlic-olive oil-white wine sauce. One of the reasons we decided for Hvar over Korcula was his cooking.

Unfortunately Adam enjoys drinking too much for his own good and can be self-destructive. He is said to once have had the first or best gourmet restaurant in Hvar town which he lost through gambling. When one of the owner's daughters wanted to take over the terrace restaurant Adam left—she claims because of his drinking, he says because she wanted to him work longer hours without any increase in salary. They could both be correct. For the next summer he did not work in a restaurant although we once ate at his apartment. He owns a stone house on the narrow street behind the Cathedral between Luviji's and the Apartmani Haracic where we stay. His family lives on the second floor, the ground floor is rented at an extremely low rent to Croatian tourists. Then he became the chef at a somewhat luxurious hotel just outside of town that was too far to walk. I would see him go there on the back of a motorbike and sometimes the wife of the owner, a licensed nurse at the local hospital, would come to town for us. We are usually alone or perhaps there would be a largish party around a long table alongside the open air swimming pool. It always seemed to me a strange place, almost empty of tourists at the height of the season. I was told that it was packed in winter with contractors and others who had to work in the off-season; and it was used for celebrations such as weddings and anniversaries. There were also rumors, as there often are in Hvar, about being built with illegal money. Adam remained an excellent cook when partly sober, but at least one meal suggested that that part had given in. After three courses each piled high with fried fish we asked for our bill and to be driven home before more oily fried fish arrived.

Next I heard that a drunken Adam had badly cut his arm and was now a carpenter at the hotel which no longer served meals. He had been unlucky with his children who were said to be drug addicts. His daughter died of an overdose and the druggie father sent the grandchildren to Hvar to live with Adam and Eve who herself broke down, quit working in a bakery, and appeared glassy eyed when we saw her. After the fall life continues. The wife has recovered and we see her with the grandchildren sitting outside the cathedral; Adam looks sad as he wanders around Hvar town and sometimes has a drink at a bar. It is said that he will open his own restaurant in the ground floor of his house next year.

Men in Hvar are expected to be competent at building, laying pipes, fixing the electricity as well as being a farmer or fisherman. The Luviji tower was built by the father, a strong muscular man, I would at any cost avoid picking a fight with. The upper restaurant is managed by a son, Rocco, and is locally famous for dishes cooked in the wood burning oven and the great view over the village and the well-lighted fortress and its walls. It is tiny and for years reservations were required in advance, usually through the daughter who runs the terrace restaurant.

Then there was a quarrel between brother and sister and now the terrace restaurant is packed until long after midnight with drinkers socializing and partying and eating local dishes from an unchanging menu but the roof restaurant is sometimes empty as before you go up the stairs the sister will remind you that you can eat below with her and she serves the same excellent wine from the family vineyard. Although we have known her for many years and we sometimes bring her an Oscar pin from New Orleans I had thought her unhappy, unattractive, and angry, then suddenly she seemed relaxed, smiling, and happy; she had a boyfriend.

Alviz run by Katarina an energetic woman I often see at 8 AM in the market or shops, who prepares lunch for her family, bakes desserts and swims in the afternoon, and is in charge of the restaurant, taking orders, from six in the evening until midnight. Afterwards she likes to sit and chat. She has a good sense of humor, appreciates oddities, likes to hear amusing gossip, and seems to know everything about the lives of those in or who visit Hvar. She needs to be energetic, tolerant and amused as Vjekoslav, her husband, notoriously seldom helps with the restaurant except perhaps a night or two at the height of the tourist season if one of the waiters is too ill to work. He can usually be found laughing at some story while drinking at an outside table at other restaurants; usually the owner is a relative of his. He seems related to most of the restaurant owners in Hvar and even to the owner of the best known restaurant in Komiza who's faded Jastozera tee shirt he wears. He jokes that his job is Public Relations for Alviz. I was surprised when he told me that he had served in the army during the war for independence from Serbia and receives a pension. He claims to have been so frightened of shooting an enemy that he planned to flee his post.

Katarina is his opposite. She talks of their having been so poor as not to expect wedding presents, of them both having to work as waiters in England to survive and learn English. There is a family farm that occupies her after the tourist season where she grows grapes for the restaurant's wine and vinegar and has olive trees. Similar to others who have worked their way up from village farmers to restaurant and property owners in Hvar she is training an heir, and building an apartment for retirement. In her case the heir is her daughter who helps wait on tables during the tourist season. The son can go to university, become a professional and move away. Those who have come to Hvar to open restaurants during the summer and spend most of the year elsewhere enjoying the profits she and other locals regard as little better than invaders who deserve trouble with regulations, taxes, disputes about ownership.

Another Katrina runs the Two Fishermen (Dva ibara), named after her husband and brother who I can still remember bringing freshly caught fish into the restaurant at night. She is a source of unexpected piscatorial information. Mackerels do not like moonlight, you need to know the tide to catch this fish, and so on. Now the husband has retired as a fisherman and spends his evenings supervising the kitchen, the son has become responsible for the restaurant although Katrina is still outside trying to bring in customers and manages the family farm where besides the wine, olives, fruit and vegetables for the restaurant she is in charge of their bee hives and the honey they bottle and sell. She also loves classical music and plays flute in an orchestra during the winter, but during the tourist season is never free for rehearsals or to hear concerts. The two daughters are the ones here who can escape the family inheritance although the older one who has married and has a child helps in the restaurant whenever she returns from Split and the younger daughter before leaving for university developed a surprising interest in baking cakes. She uses the local carob beans for a cake that looks similar to chocolate but tastes better.

The restaurant is small, packed, on a terrace and looks on to the harbor and sometime attracts the yachting crowd including a Russian gangster who reserves tables for ten friends and six body guards, Katrina says he is like a child when touring her garden smelling herbs and plants that remind him of his youth. I like to sit watching the lights from the yachts and the moving moon and the unending parade of people walking (often with small dogs) into or out of the center of the town. Her son seems to regard us as locals and our excellent fish meals are often about half the normal price. He buys fresh from the fishing boats every day and apologizes if he suspects some scallops among our cooked shell fish might have been frozen, or warns us if the mussels are wild and possibly from polluted water and need more cooking to be safe. After the whole breams (of every possible stripe, pun intended), St Peter's fish, and even a surprisingly tasty local horse mackerel, sea bass and even monk fish can seem dull. We almost always have a whole fish or a plate of shell fish perhaps with spaghetti; when I return to Paris I cannot afford a whole fish and fish fillets seem pathetic. Some of the local fish I am told are only good the day that they are caught but I see them on sale in French markets. Adele likes the black squid ink risotto here better than in any of the other restaurants, and when we were watching our money we often tried to lunch on the packed with fish soup and rice. Now they have a new fashionable chef and the fish soup is twice the price.

Another favorite restaurant with a good view is George's Garden (Dordota Vartal) on a terrace above the small beach near the cloister. It is family run and all the waiters and waitresses are either members of the family or friends who have gone to school with them. The father cooks and the two daughters supervise, take orders, and wait on tables. Some of the waitresses work during the day at the café restaurant downstairs near the beach. It is about an hour's walk from our apartment and requires walking up several flights of stairs so we come here nights when there are concerts at the monastery, especially in mid-summer when the music starts latter. The restaurant is good enough that the other cooks in town speak highly of it. Similar to Alviz and several other restaurants it makes good pizzas, and this is another place we often eat shellfish with spaghetti. I sometimes will have for a first course or for lunch their warm sautéed zucchini and feta salad, or the Croatian specialty of deep fried breaded cheese. It is difficult not to follow the progress of the waitresses from school to university on the mainland, return to Hvar to work for summer, and their talk of going with boyfriends to jobs in America.

We probably would not have known Ivona if she had not written a Master's Degree thesis at Split University on the British novelist Kazuo Ishiguro for a joint degree in English and Italian. The English part was on contemporary fiction, the Italia, on Dante, she disliked. I was sufficiently intrigued that I read her thesis and was impressed that it was the standard of, say, a British Honours Degree essay, the equivalent of a third year course at a good university. She was then supervising and waiting on tables at her father's restaurant Luna where we liked to eat downstairs to avoid the waits and crowd on the roof terrace. We became friends and went to her graduation party late one night after the restaurant closed. I liked her liveliness, cynical attitudes, and that she is obsessed with clothes—I never see her twice in the same dress and she tells me where she bought it and her shoes. She is flirtatious, likes to act sexy, knowing, and also for a time had an affair with a musician who was a mutual friend, until she understood that he preferred to be unattached and not interested in taking over her father's restaurants; there are now two as he bought the glamorous more upscale Giaxa with its 15th century setting. She, however, seems increasingly conservative, reads less, and has become obsessed with running Giaxa. One year Luna was closed due to a dispute between two members of the family who claim to be the owner. Such endless legal battles over who is heir have become more common since property in Hvar has increased in value and those who live outside the country realize that the current worth of what they might have left behind.

Ivona's father is likeable, sympathetic but in devoting himself to his restaurants is said to have lost his school teacher wife and is perhaps increasingly out of touch with his other daughter, Ivona's twin, who earned a doctorate in marine biology and has a scientific job in Split. He wants Giaxa to become a world class restaurant and takes Ivona abroad to courses in wine tasting, to famous restaurants, but it is a labor of Sisyphus. The fish, vegetables, and fruit in Hvar are too good to require elaborate preparations, Croatian wine is among the best in the world and each farm on Hvar produces distinctive wines, often complex tastes, that the French with their obsession with territory would die for. Even the wait for dishes to be finished in the kitchen of Giaxa means that service can be disappointing. Ivona's father seems always worried and harassed and says Adele and I are on a continuing honeymoon. I told him that as Ivona rapidly takes over and enjoys running Giaxa he needs to find himself a hobby or something to do to avoid drinking too much; retirement can be fatal to men. He replied that he already drinks too much, and was on his way to drink with a friend.

There is the temptation to become accustomed to a life of six months hard work during the summer followed by six months leisure and travel. Many of the waiters and waitresses in Hvar have no choice but to work in summer and then return to university courses or even construction work the remainder of the year. Unlike the coastal towns with their European standard of living, those on the mainland remain impoverished. For restaurant owners in Hvar there is a choice of what to do. Many have farms to keep them busy, but on another island I have been told "we watch television for six months". A younger friend travels with her parents during the off season and will take over her father's restaurant instead of pursuing a career as a violinist. Her artistic side for the present is fulfilled by a talented, emotional, concert hall pianist, but as his reputation takes off and they become older, music may become her romantic past.

Another young woman avoided such temptation by going to art school and returning to Hvar to design clothing and open a successful boutique. She now has three boutiques. Her designs and what she buys from others to sell are excellent and up to date fashion. Adele often buys at her shops and she once designed a summer handbag for Adele that everyone in Paris admires. We buy some of her witty dolls and jewelry as presents. Her father also had a restaurant where she would eat but not work and when he retired he sold the business to his waiters who have made Mediterraneo one of the better seafood restaurants in Hvar. A wood burning oven to cook fresh fish, a grill in the kitchen for fresh vegetables, excellent recently pressed olive oil from a farmer in a local village, Hvar island wines, and an open court yard for tables is all you need. The nest for their young that migratory birds build on the underside of a roof every year is a further, but unnecessary, attraction.

Not all are so self-protective. A woman who had her own small museum and who hand painted for Adele two marvelous tee shirts (much admired when she wears them) gave up painting and sculpture to start a small restaurant with her husband. The stone stairway that goes up the hill leads by the restaurant exhibits some of her paintings, but those days are over.

Yacht watching is part of life in Hvar, especially along the docking area where they can get water and oil. Some carry landing areas for helicopters. I imagine some even have hidden weapons to use against pirates. Near the Arsenal there is an ice cream shop with outdoor tables that is a good place to watch. One time there was a very sleek world champion racing yacht from New Zealand and I asked the owners how they got it to Hvar; they shipped it to Europe for international competitions. I expect a few times during the summer to see two similar yachts, one for the rich owner, the other for friends, but once there were three very distinctive yachts, unusual in design, similar but each larger than the other. They were said to belong to the richest Chinese man in the world—the largest for himself, the middle size for his guests, and the smallest for the crew. Some chefs rent apartments where they cook private meals for the yachting crowd. A restaurant that imported a chef from Thailand lost her to a yacht five days later. Since then Asian food in Hvar is cooked by Croatian chefs.

During six months of the year there are concerts each week in the court yard of the monastery, which I imagine as a place to stage *Hamlet* with the prince challenging his father's ghost on the walls that could also double for soliloquies, short dialogues, and back stage scenes. The main action would take place on the narrow stage facing the audience and my attention sometimes wanders to the hanging wild caperberries. There are usually two "classical" concerts, an hour or a bit more of $19^{th}$ and early $20^{th}$ century chamber music, and a folk or klapa concert. Croatians have an extraordinary interest in and ability to sing together in harmony. Klapa suddenly start up in Alviz, Luviji or other restaurants around Hvar, and both local and nationally known groups from other cities perform at the monastery. The man who everyone admires for keeping the concert series going continuously for over a decade and who can be seen putting up posters, distributing hand bills, and complaining to friends about the difficulties of booking on his budget, and the lack of audience, especially by the young, is also said to hate his job which was originally given to him as make work when his career on the local council ended. I do not know if that is true, but I have noticed that he seldom listens to the music.

Most concerts have a Croatian or Serbian connection, such as a famous violinist who teaches in Chicago but who when young was a Yugoslavian child prodigy. I was stunned when our Parisian-American drummer friend John Betsch and an American bass player from Berlin accompanied a well-known Croatian former Eurovision and television singer who now lives in Berlin where she belongs to the avant-garde. I had difficulty knowing the songs she was singing as she deconstructs them. I asked her how and why she had changed from a popular to an avant-garde singer and she said that the American saxophonist Steve Lacy had changed her life, the same Steve Lacy who had started as a Dixieland clarinetist in New York, was converted to Free Jazz by and recorded with Cecil Taylor, and who lived in Paris for decades until the French wanted to tax his MacArthur Fellowship and he returned to the USA.

For several years there was a jazz trio six nights each week during the summer on the terrace of the Adrianna Hotel where we could dance. Joe Pandur is an excellent guitarist who lived for years in the USA and Canada. He recorded a CD with Maja, a tall attractive woman in designer clothes whom he had taught to sing, that was awarded the Croatian equivalent of a Grammy; we were given the CD by listeners who appreciated our dancing. Joe and Maja were a couple until she wanted to go to London to improve her jazz singing and left him after she returned. The next year we danced to him, an upright bassist, and a variety of singers, ranging from excellent to boring. He felt he was wasting his abilities, moved to Vis where he had rich friends, and played occasionally. I heard he was living with a university student in Dubrovnik where he performed during the summers when not touring China, Hong Kong and Thailand.

Joe was followed at the Adrianna by a trio from Split that softly played for the diners swinging versions of Miles Davis's electric jazz, Herbie Hancock, Theolonius Monk, standards and blues. The leader said it was the first time he was paid for playing jazz every night instead of the usual covers of hits, rock, and blues, but being a musician he still was unsatisfied and thought he might try Vienna next year. Before the season ended there was a loud jam session with his son, a drummer from Split, and an American saxophone player who lived and taught in England. That was the last year of our dancing on the terrace of the Adrianna. It has not had live music since, instead it often has empty tables for expensive steak diners and expensive wines for foolish tourists who appreciate the view of the bay.

At first the action shifted to the new Jazz Club started by the Park Hotel. The trouble was that the self-taught architect from Split who designed the club had in his youth tried to be a musician and kept telling the young blues band he brought from Zagreb how play—faster, slower, more loudly, more softly, cover hits, do not cover hits—and gave little attention to where they would stay or transportation. He was moody, insecure and ranged from reserving two chairs for us as if we were visiting dignitaries to refusing to speak with Adele whom he accused of telling him what to do. Still it was an interesting time. Regis the leader who played electric piano was a black Frenchman studying medicine at the University of Zagreb, had played in some good Paris jazz clubs, and was a friend of Joe. There were visiting musicians sitting in. After several years of the daughter of the Park Hotel hiring a band that loudly covered hits from the 1980s and 90s (the repertoire seldom changed and I could sing along to the arrangements), there are some jazz groups again to which we dance.

Occasionally there are jazz concerts at the walled fortress overlooking the town. That drummer from Split who led the jam session one night on the Hotel Adrianna terrace reappeared, once on tour with the same American-British saxophonist, the next time with a remarkable group of Croatian and Slovenian musicians who had played professionally in the USA and learned the irregular rhythms that are now in fashion and which I love as you can dance to the propulsion while the accents keep changing. We were photographed from wristwatch cameras and the next day everyone in town knew we were dancing the previous night at the fort.

We met Anne-lise, a friend of Regis and his other musicians, at the Park Club and she became a good friend. She was also studying medicine in Zagreb in English. Whereas Regis will probably drift into being a full time musician, Anne-lise kept returning to France for short courses, and practice examinations, and passed her Croatian and French medical examinations and interns at a Paris hospital. If she had not done well enough in the competition to come to Paris she would have chosen the Caribbean where her brother already lives and claims to do nothing beyond his job except swim. Her passing her French and Croatian examinations was celebrated by about fifteen members of her family, mother, father, brother, aunts, coming to Hvar, and among other entertainments dancing the twist at the Park Club. I find Anne-lise amusing. She is small and fears having twins which run in the family; how will they fit into her? She has no hesitation in partying all night in Hvar and sleeping through the day, but when she found the first man who made her heart pound she decided to limit their relationship to kissing as that was romantic. After he returned to London they exchanged loving messages until he predictably found someone else. Some of her stories are more French than America; after they undress and go to bed the man wants to stop as he does not know her well enough

The art scene in Hvar has almost disappeared. The Slovenian who exhibited large minimalist colored photographs was evicted by his local woman who turned the gallery into a clothing shop. Lavender grows wild on Hvar and at the shop I buy lavender scented cloth cats as presents for friends in Paris. The surrealist whose amusing, primitive, painting and large sculptures of sailors with long twirled mustaches on which seagulls sit, no longer has a gallery. The absurd long blue sausage nosed man with prominent red veins on his face was not a product of surreal imagination but an older local singer. The larger than life nun brandishing a stick is similar to the nuns I watch from our terrace chasing their chickens in the garden behind the Cathedral. His Hvar is comic.

Marina sometimes comes to Hvar with Ducan her well know Slovenian artist boyfriend. As Ducan like many painters prefers the hill villages to the town Marina's knowledge of the island is different than our own and includes several communities of foreign artists, as well as gossip about the drug scene in Hvar town where she was once a waitress. Her Hvar resembles the Deia we knew, but then in the past Hvar was a hippie stop between Mallorca and the Greek islands. Friends from Deia came during the Communist period for the nude bathing on the small islands, and that terrible Donkey Pension we first stayed in was recommended to us by a French friend we knew from Deia. It was by way of the French wife of the owner we met perhaps the last of the older Deia-Hvar connection. She had invested in clothes from Bali and Goa that did not sell well in her boutique and after a few years she quietly disappeared. I cannot even remember her name but recall her faraway look and the obvious dope heads who were her friends. Many celebrities visit the island but she had no clue who was John Malkovich until her assistant bragged about him coming to the boutique twice in one day.

The apartment where we stay is in a small street behind the Cathedral. When young Lepsa came on holiday to Hvar and fell in love with a fisherman. His family had land in the town and on a small island off Hvar. He stopped fishing and they lived from the rentals to tourists of the two apartments he constructed attached to the house he built. The war between Serbia and Croatia left their daughter Rina stranded in Belgrade where she had been sent to Lepsa's relatives while attending university.

Then the father and son built a larger third apartment, a still larger fourth apartment and were planning on building a fifth when the father had a heart attack on Christmas Eve. He died in the helicopter before it reached Split. The family was in debt and by the time they found a bank willing to lend them money for the last apartment their neighbors had a change of mind, objected to more building as blocking their view, and further expansion was abandoned. Rina quit her job with the tourist bureau in Split and stays with her mother and two children by a Latin American sailor who works on yachts during the tourist season but who she and the children live with half of the year. During the summer she helps with the apartments. Igor rents boats to tourists in summer and fishes during the winter as well as taking care of the electricity, repairs and other problems concerning the apartments. He is very tall, almost seven feet, and married an attractive, thin, short, emotionally complicated woman with whom he has two children. They moved into the largest of the apartments. We prefer the second largest as it has a terrace where we can eat lunch, sit under an umbrella to read, and Adele hangs laundry to dry. We stay seven to twelve weeks every summer. The other two apartments are often booked by returning guests such as 18 Norwegians who need to lodged in several houses along the street and who take advantage of the warm nights and inexpensive drink for noisy parties.

After a decade we swim only occasionally, otherwise we read, shop and I write at the computer during the days; at night we eat out, listen to classical music at the cloister, dance, and four nights each week there are films in a no longer used nightclub, supposedly forced to close by Carpe Diem, a famous nightclub that also has all night parties on an island and is said to have political influence. Most people in the apartments are treated as part of the family and if Lepsa cooks we might be given fish, cooked ricotta cheese cakes, or salad made from a small octopus Igor caught in the harbor while at work. This autobiography was mostly written during summers in Hvar.

It is difficult to watch so many people walking their pocket sized dogs without feeling that the most significant and lasting relationship in contemporary Europe is canine companionship. I noticed three very large black highly groomed black French poodles in the company of two large white French poodles and two smaller black poodles. What is this about? I had to ask the owners. The three large black poodles were European champions, one bought in Russia, the other in Germany, and the third was their off-spring, and yes they were kept in Hvar when not travelling to shows or being rented for breeding, The white pair belonged to neighbors, the smaller black poodles were brought by visiting friends. This was a different perspective on Hvar than I had during the past decade, a championship poodle owner's view of the island's social life and place in Europe.

# Chapter 21: Reprise and an Unconclusive Conclusion

*Paris, France 2015–?*

*An Interesting Life*, the title of my autobiography, echoes a remark my wife Adele made decades ago when she had lung cancer and was told she would likely die soon. My unconcluding conclusion alludes to the final chapter of Samuel Johnson's novel *The History of Rasselas-Prince of Abissinia*—in which nothing is concluded, (unconcluding is a non-word, should I use instead the less accurate, misleading, inconclusive?); indeed no narrative that concerns reality should conclude as life is motion and change and what is said to be finished always has an afterlife. That I cite Johnson and allude to Thomas Hobbes, will remind some readers that I was a scholar of traditional Seventeenth- and Eighteenth-Century British literature before I became a promoter of what is variously termed Commonwealth, New National, Postcolonial, Global or World English Literature and that like many influenced by T. S. Eliot and the Modernists I believed that interest in contemporary writing goes hand in hand with appreciating earlier literature; it is important to see that the past co-exists with the present and to see the present as part of an evolving tradition. So my reading and teaching of contemporary literature co-existed with studying older English literature. This would have been standard for an intelligent person in most art forms mid-Twentieth Century.

Many of us assumed that art builds upon art, that revolutions in culture draw upon preexisting tendencies, that there always is a prequel. Perhaps unique was my involvement in jazz, drumming, and Black culture, and that my professional career led me to become involved with contemporary literature (and its social, cultural and political contexts) outside England, Europe and the United States. I had become a homeless, wandering Jewish intellectual, perhaps I had always been one since my childhood in Gloucester, New Jersey, and one of those talented birds of passage who contribute to rapid development and modernization wherever they land. As I looked around the new literatures I noticed there were many Jews such as Ezekiel in India, Brasch in New Zealand, Cohen and Richler in Canada and I wondered if helping to create a culture was a way of being assimilated and accepted, similar to how Jews had contributed to American culture through comics, Hollywood, popular music, literature and cultural criticism.

No, I am not telling you how to read my life and autobiography, but this is how I have come to see it. I am still a product of my early loneliness, feelings of being smarter and better read than others among whom I lived, set apart because my parents were known as Jews, influenced by my later reading of Thomas Hobbes, John Dryden, John Wilmot (the Earl of Rochester), Jonathan Swift, Samuel Johnson, and T. S. Eliot. Reading John Dryden, Ben Jonson, Restoration Comedy and Brecht led to feelings of familiarity with Wole Soyinka's plays; also we had both studied at Leeds University with G. Wilson Knight. I had gone to Leeds because of Knight, and that would eventually lead me into Commonwealth Literature and teaching at Commonwealth universities in Nigeria, Canada, and New Zealand, along with jobs in Bristol and Sterling and research in India. The metaphysical poets, James Joyce, and modern poetry contributed to my interest in Derek Walcott. W. C. Williams and New Directions annual anthologies made me familiar with Arun Kolatkar's poetry; the surreal-lyrics of Robert Bly, and James Wright foreshadowed Jayanta Mahapatra's explorations of the moods that produce poetry. When I see or read Soyinka's plays I feel he has learned the structures and conventions of the theatre from Ben Jonson, Wycherley, Ibsen, and other dramatists I have studied, seen performed and read.

My reading of Classical, European, American and British literature meant that the new world of postcolonial world literature was an exciting extension of the familiar. Even now in writing to a friend I mention Hobbes and Epicurus without pausing. Will younger literary critics and writers have a similar literary and cultural range? A West Indian friend who lives in the USA complained to me that the only American history her daughter was taught in school was re slavery, which was studied in not so different guises year after year.

There is a relationship between education and the arts. Modernism was a product of an era of when schools taught Latin and Greek, people travelled to Ruskin's Italy to study churches and paintings. The mid-Twentieth Century popularization of Modernism was made possible by the near universal teaching of Great Books and the Humanities in universities. The narrowness of postcolonial literary degrees—there are British universities that do not offer courses in Shakespeare—might suggest a new world of ignorance ruled by politics and Theory, but somehow some people learn the culture of the past as well as the present. Among my younger friends, the Indian poet and dance producer Karthika Nair often leaves me surprised with her range of allusion, and among French scholars and literary critics Laetitia Zecchini is well educated and grounded in literature. Both command a larger vocabulary that I have lost in aiming for simple clarity. Cultural traditions continue but in ways that keep changing. Life is never-ending change as long as there is life. And no matter how ideologues try to limit information to their views of politics, literature and history, somehow the past escapes and is found by the curious and intelligent.

Samuel Johnson still and probably will always interest me as he understood the dangers of sexual and other desires, including how need for motion to combat the depression created by stagnation fuels the imagination, but the lack of limits to the imagination is also dangerous. There is no end, no conclusion to the human fall into desire. Although some people will fantasize or be persuaded of final satisfaction in an after-life, life is always a paradise being lost, there is always something else that attracts. Freud thought the engine is sexual desire.

Thomas Hobbes and Samuel Johnson knew it was larger, greater than a primitive impulse, it was the continual feeling that there is more to be gained or satisfied that is inherent to being human, the essential human dissatisfaction. I have no idea whether other species also feel it, but the fact that I wonder is itself an example of how human knowledge is bound to unlimited curiosity, desire and imaginings,

The previous chapters central to the factual side of my autobiography were written in Paris, France and Hvar, Croatia, mostly over two summers, and finished late August 2014. Then September 2015 I had a stroke and spent several months in hospital before returning to our apartment overlooking the Canal St Martin in Paris. I have always hated hospitals but it did not help sharing a room with someone who woke early to recite Islamic prayers, who sometimes could be heard masturbating, and who continually watched television give-away programmes; that was the other side of globalization, boring TV programmes that were similar regardless of the language and nation. The first night home I slipped and fell down, as the elevator door was too heavy for me to hold. I shouted up the stairs for Adele to bring neighbors to help me. Two came and while lifting me up discussed who else was around who might have helped. All I knew was that I did not want to return to the hospital, but I would need be careful in future.

My life had changed; I was lucky to have survived a stroke and had survived without being turned into one of those monstrous cripples I had seen in hospital. I tried encouraging one towards whom I felt friendly but only received animal grunts in reply. I feared such a future confined to a wheel chair. Within weeks I returned to jazz clubs dancing with two canes for support, and after that with one or no canes, but I will not pretend that I had the same energy as in the past. After five or ten minutes I rested until my body and ears responding to lively grooving music made me get up to dance again. I was aware of people applauding me but that did not matter, what mattered was that I was afraid to try complicated steps, especially if I had to cross my feet back and forth, or bring one foot behind the other and turn; and I did not have the confidence to give Adele the strong lead that she requires. I used to be proud that I could get down, squat, and lower than others, now I wondered if I could get up afterwards. My physiotherapists were amazed and asked for dancing demonstrations, but I knew that my new normal was limited and likely to remain so for years. My body would need to relearn, my muscles would need improve and remember. I was not myself, I was suddenly aging.

One must adjust. I tried doing without a cane for short distances, but I do not have the confidence to do so for long periods. Friends said I needed a "walker"—"ambulator" in French—but I wanted independence from support, mechanical or human. Keep away, I do not need your help. I remembered that my mother was like this as she became older and once more I recognized my mother in myself. Childhood and early youth do implant themselves in you, your memories, and habits. Phillip Larkin's poem about the lasting effects of parents has a universal truth.

Sexual intercourse became a problem. Why had no one mentioned this to me before, but after strokes erections become difficult, even impossible, and the pornography of the imagination disappears. The porn will eventually come back, especially in dreams, but not as strongly. What do you do if you have a beautiful sexy wife and you are impotent? Luckily there are other satisfactions and pleasures besides penetration, and while Adele enjoys sex and you are wild with desire for her, you think of all the times you passed up or failed until too late to understand invitations to bed by other women. Will this ever change, has my sexual life become this memory of what it might have been? I am told that those with diabetes have similar problems. Is this true? Can X have risked his fame, position and savings by harassing women he could not penetrate, if so was he driven by the need to display his power? And could Y settle for a husband who offered her security, but dissatisfaction in bed? Is that what she meant when she mailed me to remember the glories of the past? Did I just now give in to contemporary usage with "mail" for email? It is like writing "Hi" instead of "Dear" at the opening of a correspondence. My heart beats when female French friends address me as "my dear Bruce", although

I know "my dear" is simply translated from a common French expression.

An unattractive student once told me she thought absurd Pater's notion of burning with a hard gem-like flame. I have never understood those who accept a quiet unassuming life of survival. Thomas Hobbes writes of glorification and I think the need for fame and honors drives us as much as sexual desire. In America and probably most of the world honor is mostly a matter of money; if you are rich you can buy your way into society, culture, politics and bodies. Enough money and new money rapidly becomes old money, money will buy romance, estates, titles, a place in the aristocracy. My world is cultural. I am a writer, editor, literary and music critic, scholar, former university teacher and professor. For years I thought of myself as not quite making it. I was not a tenured professor with a named chair at a famous ancient university. I joked that I was a sometimes distinguished, now extinguished professor. I felt rejected. Recently I became aware that I often had been told I was famous, and people made a point of mentioning that they knew or studied with me. Instead of graduate students I keep getting emailed requests from strangers for information, sources, guidance, meetings. It was not as good as "I slept with him" but it would need do instead.

Three books published during 2016–17 confirmed my new status. When Janet Wilson started a new series of books on World Literature she asked me to be one of its first authors and it at first seemed a chance to republish miscellaneous works from the past. I consulted with several friends about what to include and was surprised that essays I had forgotten about were now influential, prized, kept, and had to be in the book. I puzzled over inclusions. Should there be chapters from my books, essays already anthologized, what about my writings on Seventeenth-Century literature (my Seventeenth-Century life), my essays on music. I had even published later anthologized essays on Thoreau and Mark Twain. A personal favorite was about W. B. Yeats' prose that I was told at the time was the first critical commentary on his prose as creative literature, I thought of one, then two 300-page books. Eventually a 632-page narrative, a purpose beyond mere republication, became clear. *From New National To World Literature* (***ibidem*** 2016) traced the evolution of English literature from the time when African literature was the most prominent sign of cultural and political decolonization after the Second World War and a model for other collective assertions of difference (—Black Power, feminists, gays). Their relationship to African liberation is discussed in my Introduction, "New Centres of Consciousness", to the book I edited, *New National and Post-Colonial Literature* for Oxford University Press, 1996, through the assertion of Canadian, Australian, Caribbean, and other Anglophone former colonies and Dominions, the "Internationalism" of British English literature itself as recent immigrants and their children became a normal part of the British cultural scene (unlike the earlier immigrants) to the challenges presented to the foundations of the secular liberal West by the expansion of Islamic extremism and of Islamic immigrants. In the long "Introduction" to *From New National to World Literature*, itself a major new essay, I tried to show how my own publications, often based on where I was teaching at the time, had tracked the development of these new literatures and by implication the changing world politics that had begun with decolonization fifty years previously.

I see the word "Modern" keeps appearing in my titles, but I have never considered myself an avant-gardist, a real trailblazer. I respect traditions, feel sympathy for the past, and am in many ways conservative. I have an Edmund Burkean sense of the value of stability and community and the difficulty of putting the inner workings of societies back together once they are taken apart. I recognized and appreciated a similar wariness in the early writings of V. S. Naipaul. Decolonization often brought horrors and social injustices to those outside the new elites and majorities who now ruled. I suppose the correct term for me is reformist, rather than revolutionary or avant-gardist, but reformist seems too meek, too unchallenging, and I do not think that the past must change, that life consists of transformations, of constant renewal. I like older New Orleans jazz styles but I know that you cannot revive them and when I hear young musicians copying recordings from the 1920s I regard that as more a new movement in taste than a real revival of past styles. Fashion often includes the retro, the retrospective imitation and look backwards. To return to the same restaurants and have the same meals soon becomes boring no matter how good they are. Great chefs knew that they must keep inventing new dishes, new tastes, and have a relationship to new fashions in foods, taste, and eating, whether less fat, small plates, organic, environmental, microbiological, or renewals of dishes from earlier times.

I became aware of the necessity of change through the poetry of Wallace Stevens, and at first it puzzled me although it was obvious. How could such an abstract, philosophical poet concerned with the eternal verities of perception and our relation to the natural world give such weight to fashion and change? This was not the mutability that older writers recognized and warned against. Stevens was himself an ultramodernist as a poet, he was also in his themes concerned with the metaphysics of reality. How did the actual, the fashionable, the opinionated come into it? Why must art be influenced by politics, economics, mass culture?

I have since come to recognize that art is mostly about the human, indeed its themes are humanity and itself. Greek tragedy, the medieval morality play *Everyman*, Shakespeare's *Hamlet*, *King Lear*, *Measure for Measure*, *Coriolanus*, *Tempest*, Jane Austen's novels, the novels of Saul Bellow and Philip Roth are about various aspects of living, of humanity, whether it is pride, governing, illusions, uncertainty, creating a place in society, marriage, or the needs, humiliations and conditions of the human body. Why else should we read, go to plays, or films? And to be about art and its conventions is also about us as we seem to be born with instincts to dance, make music, illustrate, tell stories, make myths as much as the potential to speak or bond. Those who claim we are in a period of post-humanism are as wrong as those who claimed all arts of the past were naturally part of the spiritual and religious. Intellectuals pass along such nonsensical assumptions from generation to generation, citing and footnoting earlier sources, without themselves thinking about or testing the idea. One would have thought that the millions killed, tortured, and imprisoned by Fascism, Communism and now Islamists would be sufficient warning against absolute, intolerant and totalizing claims.

I have praised V. S. Naipaul for the clarity, liveliness, harmony and economy of his prose, his interest in his characters and their social position, his originality in opening new areas to literature, such as the lives of the formerly colonized and for seeing that irony, caricature, humor, amusement, and satire were more interesting and honest than lying, sentimentalization, propaganda, or racial and national assertion. He helped revitalize travel writing, and journalism to formerly remote places. He has Graham Greene's ability to suggest the unprepossessing by noting a few details, whether it is a landscape or the blemishes on a woman's skin. He can write of repressed desires, shameful behavior, and he enlivens his writing with characters and anecdotes. Naipaul has the journalist's instinct to investigate, to travel to interview, see, compare and learn. Who else made two long journeys through the Islamic world to understand why it was in revolution, violent and once more expanding. He noted how oil money was fueling competition between two major Islamic sects and that they were increasing their influence through Arabization, and internationally financing Islamic schools in which modern knowledge was ignored, except as illustrating heresies that believers must devote their life to destroying. Naipaul will undoubtedly be seen in future as one of the great realists, someone who wrote of the social relationships, political and sexual relationships between people. The problems of meeting and courting the opposite sex, the tensions, fantasies, and pains as well as the pleasures of their relationships, the conflicts of marriage, the striving and competition for social advantage and power whether in education, dress or politics. That he can be irritable, vain, outrageous does not matter, I do not live with him.

Man is a maker, an inventor, a creator. We wish to be gods, we create ourselves as well as our gods. By the time I was 21 I had decided to be a literary critic and not a creative writer or a jazz musician. Such lives were risky and I knew I needed the security I did not have as a child. Earning a living as a university teacher was a way to support my literary criticism. I was attracted towards and studied with the then famous Lionel Trilling and G. Wilson Knight, but I was not one of the greats such as Samuel Johnson, I was one of those intermediaries between writers and readers, who introduce and persuade. If I had a model in mind it would be the undervalued, even derided, Victorian Edmund Gosse who usefully explained to the public what they should see in the texts of great writers that they were supposed to read. I admired George Saintsbury's comprehensive anthologies of minor Seventeenth-Century poets. He was a discoverer and tastemaker. Samuel Johnson was also useful, but he was more original and more in the way of theory, biography, description, morals, summarizing. If I somehow could be useful I would be satisfied: that I helped introduce a change in literary taste was a bye product, almost an accident. I hoped but never struggled or aimed to be famous. Recently a young woman we know wrote to me that I am a superstar and I wondered what it means; I am not well-known, was she also taking into account my dancing, my wife, my enjoyment of life. We are often told by the young that we are role models, which is upsetting as it is a reminder of aging.

I have always been a writer, whether the writing was for my high school newspaper, jazz criticism, about food or dance, or literary criticism and scholarship. I have been told by many good writers, even by Nobel Laureates in literature, that I write well. I hope, dear reader, that you think my autobiography interesting and well written. No it is not great imaginative fiction, not inspiring deeply felt poetry, not a display of literary fire-works, not great art, but it is in a smaller way my attempt at writing literature, something to be enjoyed and perhaps mentioned to others. Is that a bit presumptuous? I can do many voices. I have over the decades written and edited many books, mostly about literature or theatre. I seem to remember (should I mention names?) Frank Moorhouse telling me that he was uncomfortable with Thomas Keaneally (or was it Peter Carey?) writing and publishing so much.

But I always assumed a writer wrote and wrote, what else did it mean to be a writer. I had a vague figure in my head like a hundred books I should write or edit and I have come close to it if I include the three series of books I edited. If I felt myself qualified to discuss the writing of others, including the greats of the present, it is because I thought I had a sense of style and rhythm, could write clearly, had an ability to precise and summarize, could judge, avoided clichés and readymade opinions, I had curiosity, and that I had other writerly virtues—such as an interest in the orchestration of words—that made me intrinsically more than a book reviewer and university professor. I also had a love of words, but I see now what I lost in my aiming for clarity, economy, and the natural. I am aware of how much I have sacrificed when young friends use an accurate but learned word or expression. I have impoverished my vocabulary for what? If I had no great original talent that would create a new literary age, I felt I was building upon sensibility and some ability as I read and tried to describe writings by Donne, Marvell, Lovelace, Carew, Dryden, Johnson, Thoreau, Twain, Narayan, Hope, Walcott, Jussawalla, Daruwalla, Mehrotra and Naipaul. At least that is what I hoped.

In my teens and twenties I had wanted to be a famous influential literary critic like some of my teachers and those I read. Without knowing it I had become a contemporary version of such a leader, less by being cited, than by being one of the pioneers opening up new literary and cultural areas that were also part of the main political and cultural changes of my time. I thought of myself as someone seeking a better job and recognition, but people who had recognition were telling me that I was the best scholar of my time, that I knew the great writers of the world, that I had become famous. I remembered the compliments but took more seriously the cloak room attendants at jazz clubs who told me that Adele and I had become legends from our dancing, and I am still puzzling the meaning of someone having said that instead of people being honored by meeting famous chefs, chefs were honored to meet me. Even in repeating such praise I am aware of showing my insecurities and trying to convince you my life is worth reading. This chapter is a reprise, a bringing up to date, a demonstration, argument, and offers a poetic. Wow!

New *Soundings in Postcolonial Writing* showed my new unexpected status: subtitled *Critical and Creative Contours* and co-edited by Janet Wilson and Chris Ringrose (Brill 2016) "in Honour of Bruce King", the book was a secret, otherwise I would have questioned why there were no essays about African writing or creative contributions by Africans, but it included New Zealand and Australia and a large contribution of poems by Indians. It does seem that from *Modern Indian Poetry in English* onwards my role in the opening of Indian literature to serious critical and scholarly inspection would be regarded as central to my life, although as the reader of my autobiography will know Wole Soyinka and Derek Walcott, were among those I knew and I thought myself as engaged in Africa and the Caribbean as India. Among the creative contributors I appreciated rediscovering John Haynes who I knew in Zaria. I think since then he lost his Nigerian wife, acquired a new Nigerian wife, and won the Costa Award (2006) and was shortlisted (2010) for the T S Eliot poetry prize. While I was pleased to see new pieces by, among others, David Dabydeen (part of a novel in progress based on his years as Guyana's ambassador to China), Jeet Thayil (the amusing surreal parody "The Book of Bruce"), what I assume was Adil Jussawalla's metaphoric praise in "Memoirs for Bruce", Karthika Nair's lovely blending of Western and Indian poetic conventions in "Habits: Remnants" (which should be widely anthologized), I was especially pleased by Fred D'Aguiar's "For Bruce (& Adele)":

> Dance floors never had it so good.
> Lit up on both sides of the Atlantic,
> Libraries too, as well as our lives,
> Thanks to you.

Seventeenth-Century poets could address and praise others without appearing artificial, D'Aguiar has a similar colloquial immediacy that too often has been lost.

No doubt we, Adele and myself, have aged since the periods in the chapters about Hvar and New Orleans. We did not return to either during 2016 and we keep having messages from friends in both cities claiming that life is not the same without our presence. We did travel to Nice for two weeks during the summer without even once going to the beach. Adele was afraid of the rocks and the waves; we were applauded for our dancing at a festival in a village; but Adele hurt her back getting in and out of cars and developed a colon infection that required hospitalization when we returned to Paris. For months she could barely hobble.

Still I feel lucky, I have a beautiful, lively, extremely well dressed, intelligent, talented wife with whom I have lived for over sixty years, have a beautiful apartment in a fashionable attractive lively part of Paris, we have somehow enough money to be comfortable, we have attractive, lively, culturally active, younger friends, many visitors, we often go out, to friends, to modern dance recitals, music concerts, jazz clubs, and restaurants. We know the now famous and up and coming chefs. I used to look on with envy at those who knew Hemingway and Stein and the modern painters during the 1920s and 30s and I see that for the young and the generation after us we have had fabulous, envious lives. Success. But Paris now does not have the same radiance it had during the early, and mid-twentieth centuries. This is not Paris of the early Modernists, the Lost Generation, the Existentialists. New York, London and Berlin are said to be more exciting and influential. I cannot imagine writing a book or making a moving picture about us.

And if they did after we died, how would it matter? I think of Derek trying to place himself in the pantheon of poets, but how will he know after he is dead if his reputation survived?

I feel like an aging character in a Philip Roth novel. How is it possible that Roth was never made a Nobel Literature Laureate? Whether writing of sexual desire or the humiliations of aging, he is the author of books about Everyman; taken together his works are the most complete study of the human condition by any contemporary writer. But he is also while writing of the human condition writing of the process by which social outsiders become part of society, how minorities, and those who begin as excluded assimilate and eventually are included. His portraits of Jews have their historical context, a context affirmed by the massive eruption of writing by Jews during mid-20$^{th}$ century America, of which my autobiography is perhaps a late off-shoot, although like many of the writers I did not set out to be a sociological illustration but to record my life and its contexts; but also because I am a writer with the need to write about something that I know and also because like most humans I have a need for recognition and glorification.

I see my autobiography as completing a trilogy that began with the republished essays, lectures, and reviews in *From New National to World Literature* (**ibidem** 2016) and continued by *New Soundings in Postcolonial Writing: Critical and Creative Contours* co-edited by Janet Wilson and Chris Ringrose (Brill/Rodophi 2016). Taken together the three books are a memorial of my life, marriage, writings, reputation, and life. It is a start.

# Index

## A

Achebe, Chinua  132, 137, 156, 283, 349, 381, 386, 420
Adam, Ian  324, 478, 479
Ahmadu Bello University (Nigeria)  135, 198, 203, 211, 213, 290, 349
Ahuja, Sarayu  414
Alberta (Canada)  125, 126, 323
Alexander, Charlene  303
Alexander, Meena  285, 334
Ali, Agha Shahid  266, 317, 334
Alviz, Vjekeslav  480, 482, 486
Andrew, John  382
Andrews, Troy ('Trombone Shorty')  452
Archibald, Douglas  378
*ARIEL*  324
Arnabaldi, Paul  168, 183
Arnold, Nick  173, 178, 183, 189, 403
Association of Commonwealth Literature and Language Studies (ACLAS)  202, 347
Atlantic City (New Jersey)  15, 26, 31, 33
Auden, W.H.  194, 274, 376
Aurobindo, Asbram  408
Awoonor-Williams, Kofi  242

## B

Baker, Florence  22
Baldwin, James  219
Bali  257, 259, 281, 489
Ball State University  248, 255, 293, 302, 311, 321, 323
Bangalore (India)  280, 289, 403–405, 429
Barbados  353, 357, 370, 374, 376, 379
Barbarin, Paul  72
Barnes, Bruce 'Sugarpie'  123, 463
Batiste, Jonathan  461, 463
Batiste, Lionel ('Uncle Lionel')  449, 450
Baugh, Edward ('Eddie')  359
Bearden, Romare  428, 432
Beauvoir, Simone de  67, 222
Bechet, Sydney  48–51, 447, 453
Beier, Ulli  136, 141, 157, 291
Bellagio (Italy)  291, 413, 419
Bellow, Saul  13, 500
Ben Gurion University  248, 323, 326, 329, 331, 345
Bennington College  81
Bentley, Eric  74, 231
Betsch, John  423, 428, 486
Bhandari, Jane  399
Bhatt, Shatki  403
Bhatt, Sheela  406
Biemel, Walter  180, 181
Birmingham (Alabama)  251, 402
Birmingham (UK)  86
*Black Orpheus*  136, 157
Blackburn, Julia  171
Blackburn, Thomas  107
Bly, Robert  157, 259, 494
Bombay (India)  258, 259, 264–266, 268–271, 273, 274, 276–278, 282, 285, 286, 290, 332, 334, 343–345, 397–401, 405, 407, 410, 411, 413, 414
Bonj Beach (Hvar)  474–476
Boston University  298, 351, 352, 381, 382, 389, 390
Bowe, Walter  49, 122, 151, 242
Bown, Lalage  201

Brac (Croatia) 468
Brasch, Charles 494
Bristol 100, 144, 145, 147–150, 152, 159, 228, 253, 298, 494
Brodsky, Joseph 388, 391
Brydon, Diana 297
Buck, George 448
Buck, Nina 70, 448
Burke, Kenneth 19, 79, 81
Bushrui, Suheil 104, 133, 207, 213
Butler, Henry 424, 463

# C

Cage, John 323
Caillard, Claire 428
Calcutta (India) 10, 11, 229, 265, 267, 269, 286, 329, 343
Calgary, University of 247, 323
Camden (New Jersey) 21, 22, 28, 29, 33, 34, 37, 38, 45, 48, 49, 51–57, 59, 69, 71, 151
Camus, Albert 40, 44, 67, 110, 111, 144, 204, 222, 294, 298, 393, 418
Canal St Martin (Paris) 38, 187, 415, 496
Caveau de la Hachette (Paris) 425
Cawson, Frank 158
Césaire, Aimé 204, 275
*Chandrabhaga* 258, 260
Chapman, Topsy 454
Chase, Leah 420
Chaudhuri, Amit 269, 410, 411, 430
Child, Juju 423
Chitre, Dilip 343
Choudhury, Malay Roy 278
Christie, William 180, 426
Christopher, Evan 420, 424, 452, 454, 461, 463
CIA 52, 94, 136, 138, 206, 386, 387, 389
Ciovo (Croatia) 467
Clark, Ebun 159

Clark, John Pepper 136, 138, 155–157, 163, 386
Cleary, Jon 454
Cockcroft, George ('Luke Rinehart') 172
Coe, Ada 227
Coe, Richard 101, 106, 108, 111, 116, 123, 205
Cohen, Jan 206
Cohen, Neal 206
Collier, Gordon 394
Collier, John Lincoln 97
Columbia College 8, 24, 37, 65, 76, 81, 333, 428
Columbia University 120, 174, 198, 380, 428
Connick, Jr, Harry 72, 428
Cook, Harry 29
Cook, Herbert ('Buddy') 29, 31, 32, 60
Cook, Rose 30
Cowasjee, Saros 63, 106, 107
Crawford, Davell 224, 455, 456, 463
Crotto, Karen 179, 180, 326, 459
Crotto, Paul 178, 183, 189
Crowder, Michael 205

# D

Dabit, Eugene 416
Dabydeen, David 298, 321, 348, 409, 504
Daruwalla, K. N. (Keki) 260, 261, 266, 271, 407, 411, 419, 503
Das, Kamala 278, 284, 290
Dasgupta, Rana 410
Dash, Cheryl 210, 359
Dash, J. Michael 215, 351
Dathorne, O. R. (Ron) 133, 136, 155, 157, 321, 370
Davidar, David 97, 273, 332
De Barr, Chris 444
de Paris, Sidney 48
de Souza, Eunice 272, 276, 277, 282, 318

Deb, Siddharta 410
Deia (Mallorca) 9, 77, 144, 165–169, 171–189, 224, 225, 239, 259, 279, 281, 347, 403, 467, 473, 489
Delacourcelle, Philippe 437
Desai, Anita 263
Dhurssen, Alfred 172
Dickman, Mike 432
Dickman, Vera 367, 431
Direen, Bill 235, 236
Dobree, Bonamy 87, 89–91, 99, 104
Docker, John 321
Dodds, Warren ('Baby') 48, 52, 53, 73, 112, 236, 447
Dollar (Scotland) 217, 228, 229
Doniger, Wendy 334
Doran, Greg 382, 391
Douglas, Noble 354
Drozdowski, Ray 55
Dryden, John 57, 88, 90, 91, 104, 116, 144, 147, 215, 230, 241, 246, 320, 494, 503
Dubrovnik (Croatia) 470–472, 487
Dunne, Tommy 229
Dupree, Jack 52, 61

# E

Ebejer, Francis 241
Edegran, Lars 447
Ellington, Mercer 53, 98, 304, 305
Ellrodt, Robert 217
Epicurus 495
Esslin, Martin 245
Etherton, Michael 205
Etrange Festival 427
Evans, Ken 165, 201, 211, 462
Evaristo, Bernadine 297
Ezekiel, Nissim 266, 270, 335, 337, 410, 494

# F

Fabre, Geneviève 360, 421
Fabre, Michel 214, 218, 219, 298, 350, 353, 366
Farah, Nuruddin 321
Feldman, Irving 77
Feldman, Matthew 430
Fell, John 97
Fernando 105, 281
Fields, Boney 424
Figes, Eva 169, 230, 240, 245
Figuerola, John 347
Fine, Jack 62, 121
Fine, Janet 399
Fisch, Harold 89
Florence (Alabama) 27–29, 31, 33, 34, 52, 116, 233, 248–251, 253–255
Fountain, Pete 305, 460
Frame, Janet 245
Frances, Esteban 44, 187
Fraser, Ann 106, 108, 149, 184
French, Bob 453
French, George 454
Freund, Gerald 387–389, 391
Friel, 'Jazz' 52, 53
*From Africa* 421
*From New National to World Literature* 499, 506
Frost, William 104, 215
Furtado, Joseph 282

# G

Gabriel, Charlie 454
Gaddy, Bob 60
Garcia, Rollan 459
Gates, Jr., Henry Louis 313, 382, 388
Gavin, Robert 198, 203, 209
Gérard, Albert 218
*Geste* 89, 109
Gibson, Banu 448, 460
Gibson, Richard 242

Gilbert, Henry 21, 38
Gilkes, Michael 357
Ginsberg, Allen 10, 11, 78, 113, 114, 329, 334
Giron, Peter 423, 428
Givner, Duffy 62
Givner, Joan 62
Gloucester (New Jersey) 15, 20, 33, 37, 40, 41, 45, 46, 48, 54–57, 71, 494
Goa 173, 244, 259, 270, 279–282, 286, 288, 289, 290, 344, 401, 402, 413, 489
Gordimer, Nadine 10, 193, 314, 365
Gosse, Edmund 502
Gowda, Anniah 282, 283
Graves, Robert 9, 77, 106, 166, 168, 173, 177, 185, 211, 230, 261, 339, 419
Greene, Graham 197, 501
Guedj, Dominique 433
Guha, Ramachandra 405, 406

## H

Hacker, Marilyn 422
Hamner, Robert 350
Haracic, Ivor 477
Haracic, Lepsa 489, 490
Haracic, Rina 469, 489, 490
Harlow, Michael 238
Harris, Bob 60
Harris, Rodney 204
Harris, Wilson 365, 367
Harrison, J. P. 388
Harrison, Tony 89, 104, 107, 135, 230
Hasan, Anjum 405
Hashmi, Alamgir 242, 247
Haverford College 50
Haynes, John 210, 347, 504
Hayward, Berry 428
Hayward, Timothy 428
Heaney, Seamus 391

Hearne, John 359
Heath-Stubbs, John 89, 104, 107, 184
Henry, Corey 449, 452
Hertiger, Duke 462
Hess, Linda 272
Hill, Geoffrey 89, 107, 259, 339
Hill, George 224
Himes, Chester 219
Hirson, Denis 366, 432
Hobbes, Thomas 493–495, 498
Honnelgere, Gopal 289
Hope, A. D. 241, 258, 370
Horn, Andy 205
Hoskote, Ranjit 318, 414
*Hotel du Nord* 416
Howard, Charlotte 313, 320
Howard, Sylvia 304, 423, 462
Humphries, Barry 245
Hurel, Geneviève 463
Hurel, Jean-Marie 463
Hurel, Juliette 463
Hvar (Croatia) 12, 187, 419, 421, 431, 437, 467–470, 473, 475–486, 488–491, 496, 505
Hyman, Stanley Edgar 19, 81

## I

Ibadan (Nigeria) 61, 100, 126, 128, 131, 132, 134, 137–139, 141–144, 146, 147, 152, 155, 156, 158, 160, 162, 165, 166, 198, 201, 207, 236, 291, 321, 324, 339, 347, 359, 370, 387
India 10, 63, 91, 97, 107, 127, 146, 169, 188, 197, 248, 255, 257–260, 262–268, 270, 271, 273–277, 281, 288–290, 311, 317, 321, 323, 326, 327, 331–334, 336, 338, 339, 341, 343, 344, 346–348, 358, 359, 361, 371, 397–402, 404–414, 418, 419, 421, 428–431, 494, 504
Indianapolis 8, 300, 303–310, 321, 396, 423, 451, 467

*Internationalization of English Literature, The* 105, 320, 421
Ireland, Kevin 317
Israel 7, 9, 16, 17, 69, 89, 133, 142, 168, 169, 184, 221, 243, 248, 252, 286, 311, 321, 323, 326–330, 332, 346
Izevbaye, Dan 155

## J

Jackson, Shirley 81
Jamia Milla Islamic University 407
Jaroussky, Phillippe 427
Jazz Kitchen (Indianapolis) 304–306, 309, 310
Jeffares, A. Norman ('Derry') 99
Jeffares, Jeanne 110, 112
Jhabvala, Ruth Prawer 229
Johnson, Bunk 53, 73, 115
Johnson, Samuel 146, 213, 493–495, 502
Johnson, Stephanie 237
Jones, Benny 449
Jones, Edward Horatio 325
Jones, Eldred 155
Jones, Errol 355
Jones, Leroi 447
Jonson, Ben 494
*Journal of Postcolonial Writing* (JPW) 316, 409, 437
Joyce, James 54, 77, 89, 91, 192, 193, 195, 196, 323, 324, 376, 494
Judt, Tony 298
July, Robert 387
Jussawalla, Adil 263, 266, 271, 274, 290, 503, 504

## K

Kakar, Sudhir 413
Kashmir 339, 340, 344
Katmandu 257
Katrak, Kersey 276

Kay, Jackie 297
Kaye, Sam 182
Kemp, Janet 236
Kerala (India) 245, 281, 283–286, 333, 429
Kermode, Frank 100, 106, 145, 146, 148, 149, 152, 298
Kettle, Arnold 88, 116
Key Palace Theatre (Indiana) 301
Khair, Tabish 318, 410, 411
Khajuraho (India) 412
Killam, Douglas (D. W.) 155, 158, 247, 316, 323
King, Adele 7, 22, 24, 25, 45, 52, 57, 63, 72, 75, 89, 92–96, 108–110, 115–117, 122, 123, 125, 127, 130, 132, 135, 142–144, 148, 150, 152, 153, 157–159, 161, 165, 170, 172, 176, 180, 183, 184, 186, 189, 191, 192, 195, 196, 198, 199, 206–208, 210, 212–214, 221–227, 230, 234, 236, 239, 240, 244, 248, 251, 255, 257, 258, 264, 265, 270, 272, 277, 279, 286, 287, 289, 290, 293, 294, 296, 297, 298, 300, 311, 312, 314–317, 321, 323, 324, 327–329, 332, 333, 338, 339, 341, 345, 346, 348, 350, 353, 355, 356, 359, 362–364, 369–371, 393, 395, 398, 399, 401, 403, 412, 413, 415, 417–419, 421, 424, 427, 428, 430, 433, 437, 440, 442, 447, 449, 450, 454, 456, 461, 462, 464, 468, 472, 475, 482, 484, 485, 487, 490, 493, 496, 497, 503–505
King, Lillian 22, 24, 27, 33, 34
King, Miriam ('Mimi') 18–20, 54, 240
King, Nicole 9, 20, 145, 148, 152–154, 157, 165, 168, 171, 175, 186, 187, 192, 195, 199, 200, 208, 210, 212, 214, 224, 229, 235, 240, 243, 244, 248, 250, 255, 257, 286, 287, 289, 290, 299, 312, 339, 346, 350, 358–365, 369, 417, 462
Knight, G. Wilson 81, 86, 88, 90, 141, 388, 494, 502
Knight, Steve 62, 63

Koestler, Bob 95, 383
Kohli, Devindra 339
Kolatkar, Arun 263, 264, 269, 276, 277, 343, 401, 410, 411, 430, 494
Komiza (Croatia) 468, 469, 480
Korcula (Croatia) 470, 473, 478
Kuizenga, Donna 213
Kumar, Shiv 267

# L

La Chappelle, Carol 354
Lacy, Steve 49, 122, 486
Lagos (Nigeria) 101, 127, 139–141, 145, 152, 153, 155–161, 163, 166, 203, 208, 211, 214, 229, 247, 249
Lagos University 141, 152
Lal, P. 229, 265, 266, 271
Laveau, Albert 355, 380, 381
Laye, Camara 213, 214, 222, 223, 294
Lea, Marge 440
Leeds (UK) 11, 24, 63, 70, 81, 83, 85–87, 89, 90–92, 96, 99–101, 103–113, 115–117, 123, 140, 149, 150, 165–167, 194, 201, 227, 228, 259, 324, 333, 339, 347, 348, 494
Leeds University 11, 24, 63, 70, 81, 96, 99, 259, 339, 347, 348, 494
Levenson, Christopher 397
Lewis, Arthur 375
Lewis, George 72, 114
Lind, Jakov 168
Lowell, Robert 381, 385, 386, 389
Lutz, Jacques 179
Luviji, Rocco 478, 479, 486
Lynton, Jan 166, 171, 179

# M

Maarlin, Caroline 374
Madame Walker Ballroom (Indianapolis) 304
Maes-Jelinek, Hena 247
Mahabalipuram (India) 408

Mahaffy, Sarah 230
Mahapatra, Jayanta 258, 263, 266, 347, 494
Mahood, Molly 100, 131
Malherbe, Didier 175, 176
Malroux, Claire 423
Manceau, Simone 431
Markham, E. Archie 366, 422
Marsalis, Bradford 447
Marsalis, Jason 454
Marsalis, Wynton 463
Martinique 358
Marvell, Andrew 163, 167, 197, 241, 246, 290, 503
Matassa, Howard 465
Maxwell, Desmond E. S. 133, 184
Mayer, Peter 93, 97, 333
Mayfield, Irving 459, 460
Mazrui, Ali 356
McBurnie, Beryl 354, 378, 379
McCarthy, Mary 13, 35, 103, 150
McDermot, Galt 382
McDermot, Tom 420, 452, 460, 462
McGhee, Brownie 8, 59, 62
McKay, Carolyn 449
McKinley, Jim 172, 173
Mehrotra, Arvind Krishna 260, 262, 276, 430
Mehrotra, Palash Krishna 262
Menon, Ritu 431
Meric, Rosalie de (Blackburn) 171
Metivier, Norline (Walcott) 356, 380
Mexico 57, 66, 70–72, 92, 125, 126, 168
Michaux, Henri 431, 432
Millgate, Jane 194
Millgate, Michael 106
Mishra, Pankraj 411
*Modern Indian Poetry in English* (MIPE) 266, 268, 317, 335, 337, 354, 393, 397, 401, 410, 419, 428, 504
Modi, Narendra 334

Montenegro 470, 472
Moore, Stanton 424, 455
Moraes, Dom 271, 273, 318, 335, 411
Moraes, Leila 414
Morris, Cathy 304, 307
Morris, Mervyn 359
Morrison, Wanda 45
Moss, Howard 386, 389
Muncie (Indiana) 11, 248, 255, 293, 294, 297, 300–303, 310, 312, 316, 321, 323, 350, 362, 363, 396, 404, 436
Murthy, U.R. Anantha 283
Muscle Shoals (Alabama) 188, 249, 250, 252, 254

# N

Naipaul, V. S. 90, 246, 325, 331, 349, 361, 375, 392, 394, 395, 410, 421, 499, 501, 503
Nair, Karthika 428, 495, 504
Nambisan, Kavery 405
Nambisan, Vijay 318, 405
Nandy, Pritish 258, 265, 278
Narayanan, Vivek 401, 407, 408, 410
*Narcopolis* 413
National Endowment for the Humanities 320, 352, 371
Nepal 206, 257, 290
*New National and Post-Colonial Literature* 319, 499
New Orleans (Louisiana) 8, 25, 49, 51, 53, 62, 65, 69, 72, 73, 111, 114, 119, 235, 295, 302, 304, 309, 310, 418, 420, 421, 424, 428, 439–464, 480, 499, 505
*New Soundings in Postcolonial Writing* 504, 506
New York (New York) 8, 9, 15, 17, 18, 24, 26, 34, 37, 45–47, 49–52, 53, 55–57, 59, 63, 65, 71, 72, 78, 82, 85, 89, 93–95, 97, 108, 114, 119–121, 123, 127, 134, 149, 151, 160, 163, 168, 174, 176, 178, 182, 186, 187, 213, 229, 236, 244, 262, 263, 329, 331, 333, 336, 350, 354, 378–380, 382, 385, 387–389, 393, 398, 400, 401, 404, 419, 424–426, 428, 439, 443, 445, 447, 458, 462, 464, 486, 505
New Zealand 7, 8, 20, 56, 117, 188, 233–240, 243–245, 247, 315–317, 326, 348, 366, 437, 485, 494, 504
Nice (France) 359, 433, 435, 436, 505
Nicholas, Albert 114, 447
Niven, Alastair 101, 227, 229, 349
Nizan, Henriette 221
Nizan, Paul 180, 192, 221, 294
Noble, Charlie 301, 302, 311, 354
Nonica 476
Norden, Stanley 94
Novak, Max 214

# O

Oates, Joyce Carol 192, 195
Okigbo, Christopher 136, 137, 152, 155–157, 381
Okri, Ben 349, 365, 367
Omra, Mra 423–425
Oxford University Press 203, 258, 259, 263, 317–319, 335, 371, 391, 392, 406, 499

# P

Padhi, Bibhu 265, 318
Palma de Mallorca 92, 165, 166
Paname Tropical 434
Pandur, Joe 486
Paniker, Ayyappa 245, 282, 287
Papp, Joseph 379
Paris (France) 8, 10–12, 16, 24, 27, 28, 38, 49, 53, 55, 62, 66, 68, 69, 74, 80, 83, 92–94, 97, 101, 103, 113, 114, 116, 117, 122, 127, 140, 144, 165, 173, 175, 176, 178–180, 183, 184, 186–188, 191, 213–215, 217–225, 227, 236, 240, 245, 248, 250, 255, 274, 281, 294, 298–300, 302–304,

307, 309–311, 314, 317, 321, 329, 332, 336, 339, 346, 347, 349, 353, 355, 358, 360–363, 365–370, 396–398, 406–409, 412, 415, 417–424, 428–431, 433–437, 439–441, 443, 446, 448, 449, 453, 456, 458, 461–463, 465, 467, 471, 476, 477, 478, 482, 484, 486–489, 493, 496, 505

Parthasarthy, R. 259, 263, 290
Patel, Gieve 263, 271, 276, 401
Payton, Nicholas 304, 447
Peden, Margaret-Sayers (and William) 212
Peeradina, Saleem 272, 297
Perkins, Pinetop 303
Petersen, Kirsten 202, 247
Poland 17, 30, 115
Pole, Jack 315
Pondicherry (India) 407–409
Port of Spain 353, 355, 356, 358, 359, 378, 379, 383, 385
Potts, Steve 428, 432
Poutre, Paul 427
Povey, John 214
Powell, Shannon 115, 447, 449, 454, 460
Prasad, Mathusudan 263, 318
Pune (Poona) 339, 341, 342

# R

Rachell, Yank 303
Ramanujan, A.K. 259, 271, 272, 283, 297, 318, 334–338, 406
Ramraj, Victor 325
Rayaprol, Srinivas 267
Redkey (Indiana) 295, 300–303, 321
Rhone, Trevor 359
Richards, Ian 237
Ricks, Christopher 152, 298
Riley, Herlin 452, 454
Ringrose, Chris 504, 506
Roberts, Katy 423, 424
Rochelle, Nicole 462

Rockefeller Foundation 160, 214, 378, 387, 388, 391, 420
Rodrigues, Santan 277
Rosen, Marika 433
Ross, Alan 385, 392
Roth, Philip 13, 500, 505
Rousseau, Madeleine 67, 221
Ruffins, Kermit 448, 449, 451, 452, 459, 460
Rushdie, Salman 243, 271, 276, 333, 348, 365
Russell, William 49, 72, 73, 206, 214, 446
Rutherford, Anna 202, 247, 360
Ryder, Joy 188, 189, 249

# S

Saffire: the Uppity Blues Women 301, 303
Saintsbury, George 147, 502
Sander, Reinhard 290, 350
Sargeson, Frank 245
Saro-Wiwa, Ken 156
Schroeder, Betty 46
Schuh, Russell 206, 214
Sehene, Ben 421
Selvon, Sam 325, 349
Senghor, Léopold Sédar 162, 204, 386
Senior, Olive 366, 367
Shahmoon, Sassoon 66
Shakespeare, William 57, 75, 81, 88, 91, 132, 147, 217, 235, 253, 314, 330, 331, 380, 382, 398, 495, 500
Shankar, Ravi 410
Shetty, Manohar 273, 276, 277, 344, 402
Shivdasani, Menka 345
Shortland, Gaye 201
Silkin, Jon 107, 112, 238, 241, 254, 297
Simmons, Harold 107, 135, 140, 375, 377
Simon, Paul 382, 393

Sims, Charlie 446
Singh, Kushwant 333
Singh, Ravi 334
Singleton, James 53, 424, 456, 462
Slippery Noodle (Indianapolis) 303
Smith, Ellen 453
Smith, Norleen 184
Smith, Ray 192
Solomon, David 167
Sontag, Susan 391
Soyinka, Wole 82, 107, 112, 132, 136, 137, 139, 141, 156, 157, 162, 242, 272, 311, 314, 349, 350, 356, 381, 386, 387, 494, 504
Sparrow, 'Mighty' 370
Spicer, Susan 444
Split (Croatia) 467, 469, 470, 475, 481, 483, 487, 488, 490
Sprung, Roger 61
Srivatsa, Sarayu 411
St Lucia 187, 215, 314, 350, 352, 354, 357–359, 370, 373–375, 377, 379, 382, 389, 391
Stafford, Greg 447, 454, 461
Stead, C.K.. 238, 245
Stevens, Wallace 73, 76, 106, 108, 238, 500
Stirling (Scotland) 51, 101, 227, 228, 229, 230, 258, 313
Stokle, Norman 204
Stone, Judy 352, 355
Strauss, Robert 221, 385, 389
Sturgess, Charlotte 367
Subramiaman, Arundathy 398
Sugar Blue 303
Suleiman, Susan 221
Surendran, C. P. 318
Sweden 179, 433, 462
Sweet, Stephanie 166, 180, 187

# T

Tanker, André 95, 355, 382–384
Terry, Sonny 8, 60

Thailand 168, 183, 257, 443, 485, 487
Thakore, Anand 400
Theroux, Paul 155
Thieuille, Anne 367
Thompson, Bob 63, 122, 359
Thoreau, Henry David 335, 498, 503
Three Indian Poets: Ezekiel, Moraes, Ramanujan 318, 332, 335, 397, 410, 428
Tindall, William York 76–78, 80
Todd, Olivier 221, 222
Torkanowsky, David 455
Trilling, Lionel 76, 78–80, 90, 242, 502
Trinidad 95, 187, 215, 314, 325, 350, 352–354, 357, 359, 361, 370–373, 375, 378–382, 384, 385, 387–391, 394
Trinidad Theatre Workshop 350, 352–354, 357, 359, 371, 372, 379–382, 384, 387, 388
Tucker, Susie 152
Twain, Mark 96, 131, 136, 335, 498, 503

# U

University of Paris III (New Sorbonne) 214
University of Paris VII (Charles Cinq) 353

# V

Van Den Hoven, Adrian 191
Van Herk, Aritha 325
Vappie, Don 460, 461, 463
Venet, Seva 462
Vis 286, 407, 411, 487
Visvanathan, Susan 286, 407, 411
Vuskovich, Joe 310

# W

Walcott, Alix 374, 377, 379

Walcott, Derek  9, 220, 246, 248, 272, 300, 311, 313, 314, 319, 349, 353, 371, 379, 395, 428, 494, 504
Walcott, Margaret  353, 357, 379
Walcott, Roderick  352, 391
Walcott, Warwick  374
Waldren, Bill  176, 189
Waldren, Jackie  177
Warren, Robert Penn  18, 96, 388
Watts, Charlie  8
Wedde, Ian  238
White, Cliff  306
White, Michael  447, 461
White, Patrick  370
Whiteman, Paul  50
Whitman, George  66, 113, 235
Wilbur, Bob  48, 51
Wilding, Michael  241
Williams, William Carlos  40, 268
Wilmot, John (Earl of Rochester)  146, 494
Wilson, Clive  447

Wilson, Janet  315, 316, 409, 419, 430, 498, 504, 506
Wilson, William  409, 421
Windsay, Yolanda  454
Windsor (Canada)  8, 191–197
Windsor University  8, 191
*WLWE*  316
*Women in French Studies*  297
Wright, Marva  456
Wright, Richard  219

# Y

Yats  309, 310
Yeats, W. B.  101–104, 184, 336, 498

# Z

Zaria (Nigeria)  135, 140, 158, 191, 197–199, 202, 204, 206, 207, 211, 504
Zecchini, Laetitia  278, 430, 495

# Bruce King: List of Publications

*From New National to World Literature* (***ibidem*** Press, Stuttgart, 2016.)

*ReWriting India: Eight Writers* (Oxford University Press, New Delhi, 2014.)

*Robert Graves* (Haus Publishers, London, 2008)

*Three Indian Poets: Ezekiel, Ramanujan and Moraes* (Oxford University Press, New Delhi, 1991, New edition enlarged, revised, 2005)

*The Internationalization of English Literature 1948–2000*, (Oxford University Press, Oxford, 2004. Volume 13 of The Oxford English Literary History) (Foreign Language Teaching and Research Press, People's Republic of China, 2007.)

*V. S. Naipaul* (Palgrave, Basingstoke, 1993; revised, enlarged edition 2003.)

*Modern Indian Poetry in English/ Revised Edition* (Oxford University Press, New Delhi, 1987, revised 1989. New edition 2001.)

*Derek Walcott: A Caribbean Life* (Oxford University Press, Oxford and New York, 2000.)

*Derek Walcott and West Indian Drama/ 'Not Only a Playwright But a Company'/ The Trinidad Theatre Workshop, 1959–1993* (Oxford University Press, Oxford and New York, 1995.)

*Coriolanus* (Macmillan, London; Humanities Press, Atlantic Highlands, New Jersey, 1989.)

*History of Seventeenth-Century English Literature* (Macmillan, London; Schocken Books, New York, 1982.)

*The New English Literatures: cultural nationalism in a changing world* (Macmillan, London, and St. Martin's Press, New York, 1980.)

*Marvell's Allegorical Poetry* (Oleander Press, Cambridge, 1977.)

*Dryden's Major Plays* (Oliver & Boyd, Edinburgh; Barnes & Noble, New York, 1966.)

## Series editor

Literature, Culture, and Identity. Continuum, London & New York

English Dramatists. Macmillan, London

Modern Dramatists. Macmillan, London; Grove Press, and St. Martin's Press, New York

## Books edited

*New National and Post-Colonial Literatures: An Introduction* (Clarendon Press, Oxford and New York, 1996.)

*West Indian Literature* (Caribbean Macmillan, Basingstoke 1979, enlarged second edition 1995.)

*The Later Fiction of Nadine Gordimer* (Macmillan, London and St. Martin's Press, New York, 1993.)

*Post-Colonial English Drama: Commonwealth Drama since 1960* (Macmillan, London and St. Martin's Press, New York, 1992.)

*The Commonwealth Novel since 1960* (Macmillan, London, 1991)

*Contemporary American Theatre* (Macmillan, London and St. Martin's Press, New York, 1991.)

*A Celebration of Black and African Writing* (Ahmadu Bello University Press and Oxford University Press, 1976.)

*Literatures of the World in English* (Routledge and Kegan Paul, London and Boston, 1974, reprint 1985.)

*Introduction to Nigerian Literature* (University of Lagos Press, Nigeria and Evans Brothers, London; Africana Publishing Corporation, USA, 1971.)

*Dryden's Mind and Art* (Oliver & Boyd, Edinburgh, 1969.)

*Twentieth Century Interpretations of All for Love: a collection of critical essays* (Prentice Hall, Englewood Cliffs, N.J., 1968.)

*ibidem*.eu